Marine Interstitial Poecilostomatoida and Cyclopoida (Copepoda) of Australia

CRUSTACEANA MONOGRAPHS constitutes a series of books on carcinology in its widest sense. Contributions are handled by the Editor-in-Chief and may be submitted through the office of KONINKLIJKE BRILL Academic Publishers N.V., P.O. Box 9000, NL-2300 PA Leiden, The Netherlands.

Recently published in this series:

CRM 007 - *Ivana Karanovic*	Candoninae (Ostracoda) from the Pilbara region in Western Australia
CRM 008 - *Frank D. Ferrari & Hans-Uwe Dahms*	Post-embryonic development of the Copepoda
CRM 009 - *Tomislav Karanovic*	Marine interstitial Poecilostomatoida and Cyclopoida (Copepoda) of Australia

For a full list of titles published, please see the final pages of this volume.

In preparation (provisional titles):

CRM 00x - *Chang-tai Shih*	Marine Calanoida of the China seas
CRM 00y - *Katsushi Sakai*	Axiidae of the world (Decapoda, Thalassinidea)
CRM 00z - *Darren C. Yeo & Peter K.L. Ng*	Southeast Asian freshwater crabs

Author's address:

Dr. T. Karanovic, School of Zoology, University of Tasmania, Private Bag 5, Hobart, Tasmania 7001, Australia; e-mail: tomislav.karanovic@utas.edu.au

Manuscript first received 1 August 2007; final version accepted 27 September 2007.

Cover: A collage of fig. 82, the map of Australia, and of fig. 65H, *Neocyclops tropicus* n. sp., female, see pp. 293 and 241, respectively.

Marine Interstitial Poecilostomatoida and Cyclopoida (Copepoda) of Australia

By

Tomislav Karanovic

CRUSTACEANA MONOGRAPHS, 9

BRILL

LEIDEN • BOSTON
2008

This book is printed on acid-free paper.

Library of Congress Cataloging-in-Publication data

The Library of Congress Cataloging-in-Publication Data are available from the Publisher.

ISSN: 1570-7024
ISBN-13: 978 90 04 16459 8

PRINTED IN THE NETHERLANDS

CONTENTS

PREFACE AND SUMMARY

The Australian marine interstitial was sampled in 2002 and 2003 at 103 different beaches, all around mainland Australia, and the results for poecilostomatoid and cyclopoid copepods are presented here. A total of 26 species was recorded and classified into 14 genera and seven families.

A new cyclopoid family is established (Abrsiidae n. fam.) to accommodate a single new genus and species from New South Wales: *Abrsia misophrioides* n. g. et sp. This new free-living family is characterized by a misophridiform habitus (with enormously inflated cephalothorax and first pedigerous somite, and the rest of the body cylindrical and relatively slender), conical caudal rami, disproportionately large mouth appendages, absence of aesthetascs on the antennula, two-segmented endopod of the maxillula, two chitinous beak-like structures on each side of the copulatory pore, first and second swimming legs with both rami one-segmented, third and fourth legs positioned laterally and lacking intercoxal sclerites, and third and fourth legs without endopod and with one-segmented exopod, the latter with fused coxa and basis. Gorging as a feeding strategy was postulated for this strange cyclopoid which, due to its primitive mouth appendages, shows many similarities with the primitive families Archinotodelphyidae Lang, 1949, Cyclopinidae G. O. Sars, 1913, Mantridae Leigh & Sharpe, 1934, and with the most primitive genera of the Notodelphyidae Dana, 1853.

A second species of the family Polyankyliidae Ho & Kim, 1997 was found in South Australia and is herein described as a new species. This family was previously known only from a tidal pool on mud-flats in Korea. One new poecilostomatoid genus (*Cemihyclops* n. g.) was established in the family Clausidiidae Embleton, 1901, for a newly described South Australian species. It seems closely related to the genera *Hemicyclops* Boeck, 1873 and *Giardella* Canu, 1888, but could be distinguished by a combination of apomorphic and plesiomorphic character states.

No cyclopoid or poecilostomatoid copepods have previously been reported from the Australian marine interstitial, so all species are new records for this environment. Two poecilostomatoid and one cyclopoid species found are well-known planktonic marine animals, apparently buried in the sand by the wave action, and two other cyclopoid species were known previously from Australian continental waters, but 21 species are new to science and ascribed to the following 12 genera: *Hemicyclops* (one new species), *Cemihyclops* n. g. (one species), *Polyankylis* Ho & Kim, 1997 (one), *Cyclopinella* G. O. Sars, 1913 (one), *Hemicyclopina* Herbst, 1953 (one), *Allocyclopina* Kiefer, 1954 (one), *Cyclopina* Claus, 1862 (six), *Heterocyclopina* Pleşa, 1968 (one), *Cryptocyclopina* Monchenko, 1979 (one), *Abrsia* n. g. (one), *Neocyclops* Gurney, 1927 (six), and *Halicyclops* Norman, 1903 (one new species).

As a result of taxonomic analyses of various genera and families, some new taxa are established and others synonymized. The following three species are described as new: *Neocyclops plesai* n. sp. [syn. *Neocyclops* sp. of Pleşa (1981)], *Neocyclops monchenkoi* n. sp. [syn. *Eurycyclops remanei vicinus* (Herbst, 1955), cf. Monchenko (1974)], and *Neocyclops dussarti* n. sp. [syn. *Neocyclops affinis* nov. sp. of Dussart (1974)]. The genus *Heterocyclopina* Pleşa, 1968, which was regarded as a synonym of the genus *Procyclopina* Herbst, 1955, is reinstated. In addition, five genera are synonymized: *Muceddina* Jaume & Boxshall, 1996 and *Barathricola* Humes, 1999 are synonymized with the genus *Cyclopinella*; *Glareolina* Huys & Boxshall, 1990 with *Hemicyclopina*; and *Troglocyclopina* Jaume & Boxshall, 1996 and *Heptnerina* Ivanenko & Defaye, 2004 with *Cyclopina*. The subgeneric division of the genus *Neocyclops* is abandoned and the subgenus *Protoneocyclops* Petkovski, 1986 established as its junior synonym. *Oithona attenuata* Farran, 1913 is synonymized with *O. nana* Giesbrecht, 1893, while the following five species are here established as synonyms of *Halicyclops* (*H.*) *thermophilus* Kiefer, 1929: *H. spinifer* Kiefer, 1935; *H. japonicus* Itô, 1956; *H. latus* Shen & Tai, 1964; *H. dedeckeri* Brownell, 1983; and *H. antiguaensis* Herbst, 1983. *Cyclopina heterospina* Shen & Bai, 1956 is formally excluded from the genus *Cyclopina* and placed as incertae sedis in the family Cyclopinidae, while two former members of the genus *Procyclopina* Herbst, 1955 are transferred to the genus *Hemicyclopina*: *P. feiticeira* Lotufo, 1995 and *P. uguaipuku* Lotufo, 1995. An overview of 17 cyclopoid families is given when justifying the new family and, also, copepod orders are discussed and listed with their corrected authors. The Poecilostomatoida Thorell, 1859 and Cyclopoida Rafinesque, 1815 are treated as separate orders, and the Thaumatopsylloida G. O. Sars, 1913 are also accepted as a valid order.

A key to species worldwide is provided for the genus *Neocyclops*, while further keys are given for the Australian representatives of *Cyclopina* and *Halicyclops*. The genera *Polyankylis*, *Cyclopinella*, *Hemicyclopina*, and *Cryptocyclopina* are reported for the first time from the Southern Hemisphere, while the current study provides the first Australian records of *Allocyclo-*

pina, *Cyclopina*, and *Heterocyclopina*. *Neocyclops petkovskii* De Laurentiis, Pesce & Halse, 1997 is reported for the first time from South Australia.

While vicariance was recognized as the most acceptable hypothesis for explaining zoogeographic connections of the Australian freshwater subterranean copepods with disjunct distribution patterns, dispersal seems to be a much more important mechanism in the marine interstitial. For example, dispersal, facilitated by the West Wind Drift, seems the most plausible explanation for the current distribution of the species of *Allocyclopina*. However, most genera reported here are either cosmopolitan or very widely distributed in marine habitats, and the zoogeographic connections of the Australian marine interstitial with other regions of the world are not very obvious. The importance of looking at small-scale patterns when inferring Gondwanan biogeography is emphasized, and a number of zoogeographic regions are becoming apparent in Australia, even when only copepods are analysed. This contrasts to the recognition of no more than two Australian regions for marine interstitial and subterranean inland waters, as proposed in "Stygofauna Mundi", a widely used reference work (Botosaneanu, 1986). The division of Australia into nine distinct biogeographic regions, based on mangroves, is the most suitable model for marine interstitial copepods, although, using copepods, Shark Bay in Western Australia as well as eastern South Australia are additionally recognized as separate regions. The very distinct freshwater copepod fauna of two neighbouring Western Australian regions, the Pilbara and the Murchison, is reflected in distinct marine interstitial faunas on their shores. The genus *Halicyclops*, the most speciose cyclopoid genus in the world's marine interstitial, is barely represented in this habitat in Australia, although many species are known from Australian continental waters. The diversity and abundance of the genus *Neocyclops*, which is probably competing for the same resources and may be a relatively recent arrival, is suspected as a possible explanation for the near absence of *Halicyclops* in the marine interstitial of Australia.

INTRODUCTION

Microscopic life in the interstices of sandy sediments is a discovery of the 20th century (Remane, 1933; Nicholls, 1935; Wilson, 1935a). Twenty-three higher metazoan taxa have some representatives in the interstitial (Vincx, 1986), but nematodes and copepods are the most abundant and diverse groups (Hulings & Gray, 1971) in this habitat. Interstitial copepods tend to become more abundant as the particle size of the sediment increases, and in coarse sands they often outnumber nematodes (Hicks & Coull, 1983). Today, several hundred scientists are working on interstitial fauna (www.meiofauna.org), which is sometimes also called meiofauna (a term that defines groups of organisms by their size and has very little relationship with taxonomic or ecological groupings (Preker, 2005)), but wide areas in Africa, South America, Asia, and Australia remain terra incognita in terms of interstitial fauna (Giere, 1993).

The Australian freshwater interstitial fauna was poorly known until recently, since arid zones in general have been considered to have poor prospects for supporting subterranean faunas, owing to the lack of water and the low nutrient input from xeric plant communities (Moore, 1964; Howarth, 1980). However, studies of the freshwater interstitial fauna of arid Australia over the past decade show that it comprises one of the world's most diverse and notable subterranean faunas (Humphreys, 2001). A number of stygal copepods has been described from ancient freshwater habitats (Pesce et al., 1996a, b; Pesce & De Laurentiis, 1996; De Laurentiis et al., 1999, 2001; Karanovic, 2003, 2004a, b, 2005a, 2006) and some also from anchialine waters (Karanovic et al., 2001; Karanovic & Pesce, 2002), although the latter habitat is still in an early phase of exploration in Australia (Bruce & Davie, 2006; Humphreys, 2006; Humphreys & Danielopol, 2006). In Karanovic (2004a) alone, 24 new copepod species, and five new genera were described from the Murchison region (Western Australia), while a monograph about the Pilbara region (Karanovic, 2006) included descriptions of five new genera, two new subgenera, 22 new species, and one new subspecies. Many more publications regarding freshwater copepods from Queensland, Northern Territory, and Western Australia are in preparation.

The marine component of the Australian interstitial fauna, on the other hand, remains poorly known, despite research on marine interstitial copepods beginning relatively early in Australia through Nicholls (1945a, b). He was one of the international pioneers in studies of interstitial copepods (Nicholls, 1935, 1939) and was responsible for the introduction of the term "interstitial fauna", after the German "Sandlückenfauna" (Giere, 1993). He described four harpacticoid copepods from Western Australian beaches: *Ectinosomoides longipes* Nicholls, 1945; *Psammopsyllus operculatus* Nicholls, 1945; *Neoleptastacus longipes* Nicholls, 1945; and *Paramesochra longicaudata* Nicholls, 1945. *Ectinosomoides longipes* was described as a new genus and species (Nicholls, 1945a) from Dongara (about 350 km north of Perth, Western Australia) and is still a valid species and genus in the family Ectinosomatidae G. O. Sars, 1903 (see Bodin, 1997). *Psammopsyllus operculatus* was also described as a new genus and species (Nicholls, 1945a) from both Perth and Dongara, and also still is valid in the family Leptopontiidae Lang, 1948. *Neoleptastacus longipes* was likewise described from Perth and Dongara, and also as a new genus and species (Nicholls, 1945a). Today, *Neoleptastacus* is accepted as a subgenus of *Arenopontia* Kunz, 1937 in the family Leptopontiidae Lang, 1948 and the species is still valid (Bodin, 1997; Karanovic, 2000). *Paramesochra longicaudata* was described from Cottesloe Beach in Perth (Nicholls, 1945b), as a new species in the genus *Paramesochra* T. Scott, 1892. In the same paper, Nicholls (1945b) also described a new family, the Remaneidae, to accommodate *Paramesochra* and three other genera, but this family is today considered a synonym of Paramesochridae Lang, 1944.

By 1986, no other marine interstitial copepod had been described from Australia, whereas 392 species of copepods were recorded worldwide in this ecosystem (54 cyclopoids and 338 harpacticoids) (see Herbst, 1986; Wells, 1986). However, recent investigations of other groups in the Australian marine interstitial have yielded most interesting results for nematodes (Stewart & Nicholas, 1994; Nicholas & Stewart, 1995), kinorhynchs (Sorensen et al., 2000), tardigrades (Boesgaard & Kristensen, 2001), and gastrotrichs (Nicholas & Todaro, 2005, 2006).

Realizing that there potentially also would be a great diversity of Copepoda, I sampled 103 different localities around mainland Australia in 2002 and 2003, 88 of which yielded copepods. Calculating the length of the Australian coastline and the time available for a proposed study, it was considered possible to collect a sample approximately every 100 km of coastline. Unfortunately, some parts of the Australian coast are not easily accessible by road, and some had only a limited number of sandy beaches (cliffs or clay bottoms), so the original plan was modified according to the field conditions encountered. Only 55% of the Australian beaches are accessible by road, and only half of those have a sealed or a good-condition gravel road, while the rest are publicly accessible only by rough four-wheel-drive tracks (Short, 2006). The material collected in 2002 and 2003 formed the basis for the grant I received in 2004 from the Australian Biological Resource Study (ABRS). This monograph, containing poecilostomatoids and cyclopoids, reports results of the first part of the investigations. The second part is in preparation and will also be presented as a single monograph, containing 70 species of Harpacticoida from Western Australia (including one new family). Harpacticoids from other Australian states will be identified and published at a later stage, but here it is sufficient to mention that Queensland and South Australia are also major centres of harpacticoid diversity, comparable with tropical Western Australia. As none of the Australian islands have been investigated, possibilities for student theses will be enormous once the whole project will be finished.

The Australian mainland coast, including Tasmania, is 29,900 km in length, with 49% of it consisting of 10,685 predominately sandy beach systems (Short, 2006). The typical Australian beach is relatively short (average length of 1.37 km; maximum 213 km at Coorong Beach), experiences lowered waves at the shore, has a variable beach morphology, and tends to have features such as rocks, reefs, and islets at one or both ends. Australian beaches are predominately composed of fine to medium sand (0.1-1 mm), with a mean grain size of 0.4 mm, while only 4% consists of coarse sand. The sand ranges from pure quartz to pure carbonate, with a mean carbonate content of 39% (Short, 2006). The northeast and east coast are dominated by quartz-rich beaches (>80%), while the south, west, and northwest coasts are rich in carbonate and some even are entirely composed of carbonate sands. Wave-dominated beaches predominate around the higher wave energy at the southern coast, while tide-modified and tide-dominated types are most prevalent amongst the higher tides and lower waves of northern Australia. The Australian coast does not possess tide-modified beaches exposed to higher ocean swell and storm seas, or any ice-affected beaches. It also has relatively few gravel and cobble beaches (less than 1%). In a sandy beach section, the individual copepod species occupy a specific area of colonization, whose horizontal and vertical limits may be constantly affected by environmental influences (Mielke, 1976), but genuine interstitial species occur in only small numbers. Only extensive ecological studies (e.g., Harris, 1972a, b, c; Mielke,

1976; Coull & Dudley, 1985) can give data on the true spatial and temporal distribution of a species of copepod; the occasional sampling undertaken for this study obviously cannot provide such information.

For the purposes of this monograph, I have considered cyclopoids and poecilostomatoids as two separate, valid copepod orders, contrary to some recent views (Boxshall & Halsey, 2004; Huys et al., 2006). The basis for this decision is provided in the discussion section for the new family, described below.

MATERIAL AND METHODS

All samples were collected using the Karaman-Chappuis method (Chappuis, 1942), which consists of digging a funnel-shaped excavation into the sediment down to the groundwater level, allowing water to fill the bottom of the pit, and then collecting that water with a dish and filtering it through a plankton net. The distance between the pit and the water's edge depended on the wave action and the slope of the specific beach. The Karaman-Chappuis method is very simple, easy to use on a long field trip in remote areas, and it eliminates many of the littoral species (many benthic copepods also explore the first few centimetres of the sediment and especially their juveniles can be frequent there). Samples were preserved by adding a few drops of 30% formaldehyde (until it makes approximately a 5% solution with the sea water) and the copepods were later separated using a dissecting microscope and transferred into 70% ethanol. At most localities a GPS position was recorded and standard environmental data (including temperature, salinity, pH, O_2 concentration, etc.) were taken, and every beach was photographed. All the field work was done during several small trips along the southwestern coast of Western Australia (6 days and 3,800 km altogether in 2002) and two long trips: first from Perth to Sydney anticlockwise along the coast and back across the Nullarbour (13 days and 10,601 km, from 21 December 2002 to 2 January 2003) and the second one from Perth to Sydney clockwise along the coast and back across the Nullarbour (20 days and 17,691 km, from 1 to 20 August 2003).

Below is the list of localities where copepods were collected (beginning from Perth, clockwise), with those containing cyclopoids and/or poecilostomatoids marked with an asterisk:

(1) Western Australia, Jurien Bay, 25 October 2003, 30°18.156′S 115°02.248′E
(2) Western Australia, Geraldton, Drummond Cove, 1 August 2003, 28°40.055′S 114°36.787′E
(3) Western Australia, Kalbarri, Chinamans Beach, 1 August 2003, 27°42.620′S 114°09.664′E
*(4) Western Australia, Shark Bay, Nanga Bay Resort, 2 August 2003, 26°15.277′S 113°48.333′E (fig. 83)
(5) Western Australia, Shark Bay, Denham, 2 August 2003, 25°55.841S 113°32.373′E
(6) Western Australia, Shark Bay, Monkey Mia, 2 August 2003, 25°47.622′S 113°43.224′E
*(7) Western Australia, Shark Bay, Shell Beach, 2 August 2003, 26°13.032′S 113°46.632′E (figs. 84-85)

(8) Western Australia, Gladstone (near Wooramel Roadhouse), 2 August 2003, 25°57.227′S 114°14.895′E
*(9) Western Australia, Carnarvon, town beach, 2 August 2003, 24°52.959′S 113°39.302′E
*(10) Western Australia, Coral Bay, 3 August 2003, 23°08.577′S 113°46.163′E (fig. 86)
*(11) Western Australia, Exmouth, 3 August 2003, 21°56.726′S 114°08.383′E (fig. 87)
*(12) Western Australia, Exmouth Gulf, Pebble Beach, 3 August 2003, 22°02.479′S 114°06.889′E (figs. 88-89)
(13) Western Australia, Onslow, 3 August 2003, 21°38.156′S 115°06.784′E
*(14) Western Australia, Dampier, 4 August 2003, 20°40.051′S 116°42.244′E (fig. 90)
*(15) Western Australia, Karratha, jetty, 4 August 2003, 20°43.298′S 116°53.458′E (fig. 91)
*(16) Western Australia, Point Samson, 4 August 2003, 20°37.516′S 117°11.795′E (figs. 92-93)
*(17) Western Australia, Port Hedland, town beach, 4 August 2003, 20°18.553′S 118°35.359′E (fig. 94)
(18) Western Australia, Bidyadanga, 5 August 2003, 18°40.281′S 121°45.527′E
(19) Western Australia, Broome, town beach, 5 August 2003, 17°58.207′S 122°14.164′E
*(20) Western Australia, Derby, jetty, 5 August 2003, 19°17.459′S 123°36.549′E
*(21) Northern Territory, Dundee Beach, 7 August 2003, 12°46.555′S 130°21.377′E (figs. 96-97)
(22) Northern Territory, Mandorah, 7 August 2003, 12°26.611′S 130°46.005′E
*(23) Northern Territory, Darwin, Mindil Beach, 7 August 2003, 12°26.754′S 130°49.856′E (fig. 98)
(24) Northern Territory, Bing Bong (near Borroloola), 8 August 2003, 15°38′S 136°22′E
(25) Queensland, Cooktown, rescue jetty, 11 August 2003, 15°27.766′S 145°14.990′E
(26) Queensland, Wonga (near Mossman), 12 August 2003, 16°23.251′S 145°25.130′E
(27) Queensland, Cairns, town beach, 12 August 2003, 16°54.583′S 145°45.979′E
(28) Queensland, Kurrimine Beach (near Innisfail), 12 August 2003, 17°46.581′S 146°06.493′E
(29) Queensland, Forrest Beach (near Ingham), 12 August 2003, 18°42.685′S 146°17.895′E
(30) Queensland, Townsville, Strand, 12 August 2003, 19°15.053′S 146°48.961′E
*(31) Queensland, Alva (near Ayr), 12 August 2003, 19°27.232′S 147°28.924′E
(32) Queensland, Bowen, Queens Beach, 13 August 2003, 19°58.739′S 148°13.932′E
*(33) Queensland, Midge Point (S. from Proserpine), 13 August 2003, 20°38.983′S 148°43.338′E (figs. 99-100)
(34) Queensland, Sarina Beach (near Mackay), 13 August 2003, 21°23.418′S 149°18.777′E
(35) Queensland, Clairview, 13 August 2003, 22°07.196′S 149°32.158′E
*(36) Queensland, Yeppoon, 13 August 2003, 23°08.582′S 150°45.659′E (fig. 101)
*(37) Queensland, Gladstone, Barney Point Beach, 14 August 2003, 23°50.850′S 151°16.378′E (fig. 102)
*(38) Queensland, Hervey Bay, 14 August 2003, 25°16.728′S 152°50.400′E (figs. 103-104)
*(39) Queensland, Tin Can Bay, 14 August 2003, 25°54.466′S 153°00.555′E (fig. 105)
(40) Queensland, Caloundra, Golden Beach, 15 August 2003, 26°48.685′S 153°07.345′E
(41) Queensland, Brisbane, Victoria Point, 15 August 2003, 27°34.812′S 153°18.937′E
(42) New South Wales, Ocean Shores (N. from Ballina), 15 August 2003, 28°29.847′S 153°33.037′E
(43) New South Wales, Iluka (S. from Ballina), 15 August 2003, 29°24.772′S 153°21.289′E
*(44) New South Wales, Moonee Beach (near Coffs Harbour), 15 August 2003, 30°12.523′S 153°09.387′E (figs. 106-107)

(45) New South Wales, Crescent Head (near Kempsey), 16 August 2003, 31°11.253′S 152°58.734′E
*(46) New South Wales, Harrington, 16 August 2003, 31°52.324′S 152°41.453′E (fig. 108)
(47) New South Wales, Hawks Nest (near Nelson Bay), 16 August 2003, 32°40.874′S 152°11.081′E
*(48) New South Wales, Swansea, 16 August 2003, 33°04.820′S 151°39.994′E (figs. 109-110)
(49) New South Wales, Sydney, Botany Bay, 16 August 2003, 33°58.171′S 151°09.167′E
*(50) New South Wales, Wollongong, Boat Harbor, 17 August 2003, 34°25.326′S 150°54.296′E (fig. 111)
(51) New South Wales, Kiama, 29 December 2002, 34°41.384′S 150°51.024′E
(52) Victoria, Mallacoota, 29 December 2002, 37°35.158′S 149°44.238′E
(53) Victoria, Lakes Entrance, 28 December 2002, 37°52.950′S 147°58.285′E
(54) Victoria, Port Albert, 28 December 2002, 38°40.321′S 146°41.643′E
(55) Victoria, Shelley Beach (near Melbourne), 28 December 2002, 37°58.237′S 144°42.172′E
(56) Victoria, Lorne, 28 December 2002, 38°32.985′S 143°59.146′E
(57) Victoria, Peterborough, Boat Bay, 27 December 2002, 38°34.799′S 142°49.706′E
*(58) South Australia, Robe, 27 December 2002, 37°09.725′S 139°45.153′E (fig. 112)
(59) South Australia, Camp Coorong (near Meningie), 27 December 2002, 35°46.934′S 139°17.234′E
(60) South Australia, Aldinga, Aldinga Beach, 26 December 2002, 35°16.130′S 138°27.080′E
*(61) South Australia, Yorke Peninsula, Stansbury, 26 December 2002, 34°54.524′S 137°47.873′E
(62) South Australia, Yorke Peninsula, Marion Bay, 26 December 2002, 35°14.766′S 136°58.718′E
(63) South Australia, Yorke Peninsula, Hardwicke Bay, 26 December 2002, 34°53.493′S 137°27.295′E
(64) South Australia, Whyalla, jetty, 25 December 2002, 33°02.414′S 137°35.414′E
*(65) South Australia, Eyre Peninsula, Cowell, 25 December 2002, 33°41.180′S 136°55.786′E
*(66) South Australia, Eyre Peninsula, Tumby Bay, 25 December 2002, 34°22.417′S 136°06.262′E
(67) South Australia, Eyre Peninsula, Port Lincoln, 25 December 2002, 34°43.170′S 135°51.459′E
(68) South Australia, Eyre Peninsula, Hall Bay, 25 December 2002, 34°07.662′S 135°16.032′E
*(69) South Australia, Eyre Peninsula, Venus Bay, 25 December 2002, 33°13.896′S 134°40.347′E (fig. 113)
*(70) South Australia, Streaky Bay, camp, 25 December 2002, 32°47′S 134°13′E
*(71) South Australia, Smoky Bay (S.E. from Ceduna), 24 December 2002, 32°22.735′S 133°56.006′E (fig. 114)
*(72) South Australia, Ceduna, 24 December 2002, 32°06.800′S 133°40.319′E
(73) South Australia, Great Australian Bight, 19 August 2003, 31°38.955′S 129°15.500′E
(74) Western Australia, Great Australian Bight, Eucla, old jetty, 24 December 2002, 31°42.765′S 128°53.197′E
(75) Western Australia, Great Australian Bight, Eyre Bird Observatory, 23 December 2002, 32°15′S 126°18′E
(76) Western Australia, Condingup (E. of Esperance), 22 December 2002, 33°54.574′S 122°35.600′E

(77) Western Australia, Stokes Inlet (W. of Esperance), 22 December 2002, 33°49.080′S 121°09.016′E
(78) Western Australia, Two Peoples Bay (E. of Albany), 21 December 2002, 34°52.893′S 118°24.407′E
(79) Western Australia, Frenchman Bay (S. of Albany), Misery Beach, 16 November 2002, 35°05′S 117°54′E
(80) Western Australia, Denmark, Ocean Beach, 30 November 2002, 35°02′S 117°18′E
(81) Western Australia, Walpole, Sandy Beach, 30 November 2002, 35°01′S 116°46′E
(82) Western Australia, Windy Harbor (near Northcliffe), 1 December 2002, 34°50′S 116°02′E
*(83) Western Australia, Cape Leeuwin (near Augusta), 1 December 2002, 34°22′S 115°10′E
(84) Western Australia, Hamelin Bay (S. from Margaret River), 1 December 2002, 34°12′S 115°03′E
*(85) Western Australia, Witchcliffe (near Margaret River), beach near Conto Spring, 19 February 2002, 34°02′S 115°00′E
*(86) Western Australia, Busselton, Busselton Jetty, 8 December 2002, 33°39′S 115°20′E
(87) Western Australia, Mandurah, Port Bouvard, 8 December 2002, 32°35′S 115°40′E
*(88) Western Australia, Perth, Munster, Jervoise Bay, 8 December 2002, 32°08′S 115°46′E

Specimens were dissected and mounted on microscope slides in Faure's medium, which was prepared following the procedure discussed by Stock & Von Vaupel Klein (1996), and dissected appendages were then covered by a coverslip. For the urosome or the entire animal two human hairs of the appropriate thickness were mounted between the slide and coverslip, so the parts would not be compressed. By manipulating the coverslip carefully by hand, the whole animal or a particular appendage could be positioned in different aspects, allowing the observation of morphological details. During the examination water slowly evaporated and appendages eventually remained in completely dry Faure's medium, ready for long term storage. However, this dry medium can be redissolved with a drop of distilled water after more than 50 years (G. S. Karaman, pers. comm.) and probably even much longer. All drawings were prepared using a drawing tube attached to a Leica-DMLS brightfield compound microscope, with C-PLAN achromatic objectives. Specimens that were not drawn were examined in a mixture of equal parts of distilled water and glycerol and, after examination, were again preserved in 70% ethanol.

The morphological terminology follows Huys & Boxshall (1991), except for small differences in the spelling of some appendages (antennula, mandibula, maxillula, instead of: antennule, mandible, maxillule), as an attempt to standardize the terminology for homologous appendages in different crustacean groups. Also, a much simplified version of the armature formula of the swimming legs is used. The reason for this is that there are a number of transitional forms between spine and seta in copepods. To avoid possible confusion in some descriptions I sometimes use the term "armature element" (or just "element") instead of spine or seta. When a particular appendage or structure is not mentioned in the description (usually paragnaths) it means it was not observed, but when a drawing is missing and the appendage is described in the text it means that it was observed but not drawn, because either it was very badly mounted or it is almost identical to the same appendage of another species from the same genus in the monograph. Biospeleological terminology follows Humphreys (2000b).

Most of the material is deposited in the Western Australian Museum (WAM), while some paratypes are deposited in the Tasmanian Museum and Art Gallery (TMAG).

TAXONOMIC ACCOUNT:
CLASS COPEPODA H. MILNE EDWARDS, 1840
ORDER POECILOSTOMATOIDA THORELL, 1859

Family CLAUSIDIIDAE Embleton, 1901
Genus **Hemicyclops** Boeck, 1873

1. **Hemicyclops nichollsi** n. sp.
(figs. 1-4)

Material examined. — Types: Holotype, ovigerous female dissected on 2 slides (WAM C34238); allotype male dissected on 2 slides (WAM C34239); paratypes, 2 males + 3 females (2 ovigerous) + 4 copepodids (1 male + 2 ovigerous females + 4 copepodids in alcohol (WAM C37202); 1 male + 1 female on 1 slide in toto (TMAG G5771)): Australia, Queensland, Tin Can Bay, 14 August 2003, leg. T. Karanovic, 25°16.728′S 152°50.400′E.

Other material: Australia, New South Wales, Moonee Beach near Coffs Harbour, 15 August 2003, leg. T. Karanovic, 30°12.523′S 153°09.387′E: 1 male + 1 female in alcohol (WAM C37203).

Australia, New South Wales, Harrington, 16 August 2003, leg. T. Karanovic, 31°52.324′S 152°41.453′E: 5 males + 10 females (8 ovigerous) (4 males + 9 females in alcohol (WAM C37204); 1 male + 1 female on 1 slide in toto (TMAG G5772)).

Description. — Ovigerous female (holotype). Body length, excluding caudal setae, 1.712 mm. Habitus (fig. 1A) very robust, with prosome/urosome ratio 1.7 and greatest width at second pedigerous (first free) prosomite. Body length/width ratio about 2.4; prosome elliptical in dorsal view, about 1.4 times as long as wide and 2.3 times as wide as genital double somite. Posterior corners of fourth pedigerous somite expanded laterally, other prosomites without pronounced lateral expansions. Fifth pedigerous somite about as wide as genital double somite. Colour of preserved specimen yellowish; integument strongly chitinized. Nauplius eye not visible. Rostrum broadly rounded and furnished with two large sensilla.

Cephalothorax (fig. 1A) almost triangular, 0.74 times as long as its greatest width, representing 31% of total body length. Surface of cephalic shield with numerous small sensilla and cuticular pores, same as on other prosomites (fig. 2A). Hyaline fringe of prosomites relatively wide and frilled, but not serrated. First pedigerous somite completely fused to cephalosome. Arthrodial membrane between fourth and fifth pedigerous somites well de-

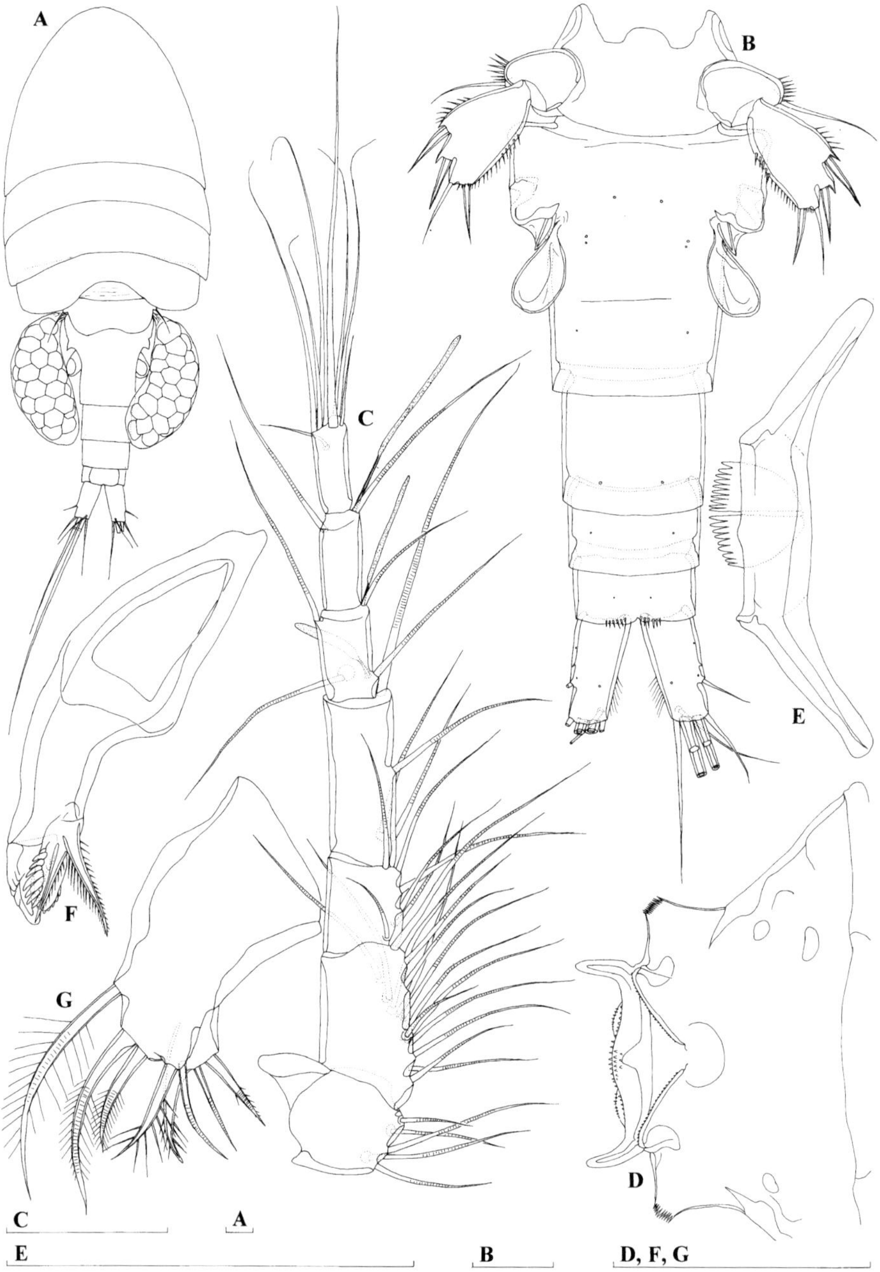

Fig. 1. *Hemicyclops nichollsi* n. sp., holotype female. A, habitus, dorsal view; B, urosome, ventral view; C, antennula; D, labrum; E, postoral plate with paragnaths; F, mandibula; G, maxillula. Scales = 0.1 mm.

veloped; no sclerotized joint between fifth pedigerous somite and genital double somite visible (fig. 1A, B).

Genital double somite (figs. 1B, 2B) with laterally expanded anterior part, about as long as wide, ornamented with characteristic pattern of eight pores ventrally and six pores dorsally. Hyaline fringe of genital double somite and of two succeeding somites narrow, frilled, and completely smooth (not serrated). No cuticular recesses, windows, or sensilla visible. Copulatory pores and ovipores lie together within lateral genital apertures, each with ovoid spermatophore attached somewhat laterally (fig. 2B). Copulatory ducts and seminal receptacles not clearly visible. Ovipores situated laterally, covered by reduced sixth legs, without any armature visible. Two egg sacs attached to anterior part of genital double somite, each kidney-shaped, with 23 eggs and reaching slightly beyond posterior margin of third urosomite. Third urosomite ornamented with two ventral, two lateral, and two dorsal pores in posterior half, while fourth urosomite with only two ventral and two dorsal pores. Anal somite (fig. 2C) with smooth, broad, very short, and almost straight anal operculum, which represents 61% of the somite's width and reaches to only 1/5 of the somite's length; ornamented with two large sensilla and two pores dorsally, three pores laterally on each side, four pores ventrally, and a transverse, short row of spinules on posterior margin ventrally; its posterior margin finely serrated where not covered with spinules. Anal sinus widely open dorsally, ornamented with several diagonal rows of minute spinules.

Caudal rami (figs. 1A, B, 2C) cylindrical, twice as long as wide (ventral view), divergent, with very small space between them; they represent 7% of total body length. Distal margin ventrally smooth and convex, with medial cuticular pore not visible. Ornamentation consists of several long hair-like spinules along posterior half of inner margin and three ventral, two lateral, and three dorsal pores; no lateral sensillum visible. Dorsal seta about 1.2 times as long as ramus, inserted in small recess close to posterior border and inner dorsal corner, uniarticulate at base and unipinnate distally. Lateral seta smallest, arising at 2/3, smooth, and somewhat longer than ramus' width. Outermost apical seta also smooth, 0.85 times as long as ramus and 0.7 times as long as innermost apical seta, which is strong and bipinnate. Principal apical setae with breaking planes and bipinnate; inner seta about 1.8 times as long as outer one and 0.4 times as long as body.

Antennula (fig. 1C) slender, not visible from dorsal view, seven-segmented, unornamented, with one short aesthetasc on fifth segment laterally, one slender aesthetasc on fifth segment subapically, one long as slender aesthetasc on

Fig. 2. *Hemicyclops nichollsi* n. sp., holotype female. A, pleural area of third pedigerous somite; B, right sixth leg and spermatophore attached, ventral view; C, anal somite and caudal rami, ventral view; D, antenna; E, apical part of mandibula; F, maxilla; G, maxilliped. Scales = 0.1 mm.

sixth segment subapically, and setal formula: 5.15.6.4.4.2.8. No setae with breaking planes or articulating on basal part and all setae smooth (but majority with internal ribs). Length ratio of antennular segments, from proximal end and along caudal margins, 1 : 1.5 : 1.4 : 2.1 : 1.2 : 1.3 : 1.3.

Antenna (fig. 2D) four-segmented, comprising coxobasis and three-segmented endopod, strongly geniculate with first and last segment pointing in opposite directions; length ratio of antennal segments, 1 : 0.7 : 0.5 : 0.2. Coxobasis about 2.8 times as long as wide, unarmed, with two short rows of long spinules at proximal outer corner and four short rows of spinules (two of which minute) along inner margin; seta representing exopod absent. First endopodal segment 2.3 times as long as wide, armed with one pinnate seta at 2/3, ornamented with longitudinal row of long spinules along inner margin; seta 0.7 times as long as segment. Second endopodal segment almost triangular, with very short outer margin, about as long as wide, ornamented with spinules along both margins (those along inner margin much stronger), armed with four setae subapically (three setae very strong, geniculate, two of them strongly pinnate; one seta slender and smooth). Third endopodal segment quadrangular, 0.8 times as long as wide, with produced spiniform outer corner, ornamented with arched row of minute spinules at middle, armed with seven apical setae (five of which strong and geniculate, one of them pinnate as well).

Labrum (fig. 1D) trapezoidal, but very complex three-dimensional cuticular plate, with relatively broad and somewhat convex cutting edge, which is trilobate, armed with many short and sharp spinules; lateral margins of cutting edge produced as blunt and elongate spiniform processes; distolateral corners sharply serrated; another two diagonal ridges with serrated (or spiniform) distal margins, starting between the distolateral corners and the produced margins of the cutting edge.

Postoral plate (fig. 1E) also trapezoidal, but unornamented, with small, foot-like and sharply serrated paragnaths almost completely fused to it.

Mandibula (figs. 1F, 2E) very simple, unornamented, slightly geniculate, about three times as long as wide and without any trace of a palp. Coxal gnathobase cutting edge armed with four clearly distinct armature elements: two foliaceous spines and two bipinnate setae. Outer foliaceous spine with both margins serrated, middle spine with very strong spinules along convex outer margin and minute spinules along inner. Innermost seta 1.3 times as long as middle seta and 1.8 times as long as middle spine; outermost spine 1.4 times as long as middle spine.

Maxillula (fig. 1G) also very simple, unornamented, unsegmented but bilobate. Inner lobe much smaller, armed apically with two bipinnate spines and one smooth seta in middle, all of about same length; outermost spine strongest, geniculate, and fused to lobe. Outer lobe (reduced palp) larger but not longer than inner one, armed with five elements: three very strong and bipinnate setae along outer margin and apically with one strongly bipinnate spine and one smooth seta; outermost seta longest.

Maxilla (fig. 2F) two-segmented, comprising large and ovoid syncoxa and much smaller and quadriform basis. Syncoxa ornamented with short row of spinules proximally, armed with two subequal, strong and bipinnate setae on presumably proximal endite; distal endite reduced and unarmed. Basis about 1.5 times as long as wide, ornamented with two cuticular pores and armed apically with two complex teeth, an outermost simple and smooth spine, and one unipinnate strong seta inserted somewhere between the complex teeth; innermost complex tooth with strong spinules along outer margin and with three-dimensional but smooth outer margin; middle complex tooth bifurcate, with several long spinules along longer process.

Maxilliped (fig. 2G) prehensile but robust, four-segmented, composed of syncoxa, basis, and two-segmented endopod. Syncoxa 1.7 times as long as greatest width, unornamented and armed at middle with two equally long, strong setae; one seta bipinnate at distal part, other seta smooth. Basis 1.9 times as long as wide, ornamented with a single cuticular pore and armed, similar to syncoxa, with two equally long, strong setae at middle, both of them strongly bipinnate. First endopodal segment unornamented and unarmed, much smaller than basis and 0.6 times as long as wide. Second endopodal segment also very small, 0.7 times as long as wide, unornamented but armed with seven setae; two apicalmost setae very strong and pinnate, other setae smooth and slender. Longest seta on endopod 1.2 times as long as basal setae and 0.9 times as long as longer seta on syncoxa.

All swimming legs (fig. 3A, B, C, D) with three-segmented exopod and endopod; their armature formula as follows (legend: inner/outer spine or seta; inner/terminal/outer):

	Exopod			Endopod		
Segments	1	2	3	1	2	3
First leg	0/1	1/1	4/2/2	1/0	1/0	3/2/1
Second leg	0/1	1/1	5/2/2	1/0	2/0	3/2/1
Third leg	0/1	1/1	5/2/2	1/0	2/0	3/2/1
Fourth leg	0/1	1/1	5/2/1	1/0	2/0	2/2/1

Fig. 3. *Hemicyclops nichollsi* n. sp., holotype female. A, first swimming leg; B, second swimming leg; C, third exopodal segment of third swimming leg; D, fourth swimming leg; E, fifth leg. Scales = 0.1 mm.

Third exopodal segment spine formula 2.3.3.2 and third endopodal segment spine formula 1.3.3.3. All intercoxal sclerites with concave distal margin, ornamented with hair-like spinules in first leg and with very strong and closely set spinules in other legs. Coxae ornamented with spinules along outer margin (that of fourth leg with additional posterior row of minute spinules) and armed with strong and long, plumose seta at inner distal corner. Spine inserted at inner distal corner of basis of first leg reaching halfway the second endopodal segment, stout and bipinnate. Basis of each leg armed with slender seta at distal outer corner and ornamented with cuticular pore basally near outer margin, as well as with row of strong spinules along distal margin between exopod and endopod; basis of fourth swimming leg additionally ornamented with row of spinules along outer margin, while basis of other legs with additional ornamentation along inner margin. Most exopodal and endopodal setae slender and plumose; inner apical armature element on third exopodal segment of second, third, and fourth leg a transition between a spine and a seta, with serrated outer margin and plumose inner one; both inner setae on third endopodal segment of fourth leg modified into very strong, spiniform armature elements, with a few hair-like spinules at base and sharply serrated distal lateral margins. All exopodal and endopodal spines stout and bipinnate; those on exopod of first leg with additional long spinule near distal tip. All endopodal segments with hair-like spinules along outer margins, which are also present on inner margin of second exopodal segment; most exopodal segments with short spinules along outer margin. Cuticular pores visible on most endopodal and exopodal segments, but sometimes hard to observe. Third endopodal segment of fourth swimming leg about 1.9 times as long as wide; inner apical spine about as long as segment and twice as long as outer apical spine.

Fifth leg (figs. 1B, 3E) without intercoxal sclerite, attached to ventrolateral corner of fifth pedigerous somite, two-segmented (although first segment fused to somite dorsally), comprising ovoid coxobasis and one-segmented exopod. Coxobasis armed with slender and unipinnate outer basal seta and ornamented with three cuticular pores and a row of strong spinules along outer margin distally. Exopod nearly 1.9 times as long as its greatest width, much narrower at basal part, armed with three spines and a single seta, and ornamented with three cuticular pores and strong spinules along inner and outer margins, as well as at base of each spine and seta; seta slender, bipinnate at distal part, positioned centrally and as the most apical element, 0.6 times as long as exopod, 1.4 times as long as inner apical spine, about

1.5 times as long as outer apical (actually subapical) spine, 1.4 times as long as lateral spine and as long as basal seta; all spines strong and with sharply serrated lateral margins.

Sixth leg (fig. 2B) not clearly distinct, cuticular plate, strongly chitinized and curved, without any surface armature or ornamentation.

Male (allotype). Body length, excluding caudal setae, 1.523 mm. Habitus (fig. 4A) less robust than in female, with prosome/urosome ratio 1.4 and greatest width near posterior end of cephalothorax. Body length/width ratio about 2.4; cephalothorax 2.2 times as wide as genital somite. Prosome more ellipsoid than in female, with posterior lateral corners of fourth pedigerous somite less laterally expanded.

Cephalothorax (fig. 4A) 0.8 times as long as its greatest width. Ornamentation of prosomites, colour, and nauplius eye similar to those of female. Hyaline fringes of all somites smooth and lightly frilled.

Genital somite (fig. 4A, B) almost twice as wide as long (dorsal view), with two completely formed spermatophores visible inside, ornamented with numerous pores dorsally and laterally; third and fourth urosomites ornamented with six pores and two sensilla ventrally, four pores dorsally, and two pores laterally on each side; fifth urosomite with only two ventral and two dorsal pores and one lateral pore on each side. Anal somite similar to that of female.

Caudal rami (fig. 4A, B) about 1.8 times as long as wide, divergent, with space between them about half ramus' width. Armature and ornamentation similar as in female.

Antennula (fig. 4C, D) not geniculate, with proportions of segments and armature exactly the same as in female.

Antenna (fig. 4E) similar to that of female, except coxobasis armed with slender plumose seta at distal inner corner, which is slightly shorter than the segment.

Labrum, postoral plate, mandibula, maxillula, and maxilla same as in female.

Maxilliped (fig. 4F) modified, very strong, prehensile appendage; syncoxa smaller than in female, 1.5 times as long as wide, unarmed and unornamented; basis much stronger than in female, heart-shaped, with inflated proximal part (especially proximal inner corner), 1.6 times as long as wide, ornamented with two parallel rows of strong spinules along inner margin and armed with one pinnate seta at 2/3 and one smooth and minute seta at distal inner corner; endopod transformed into strong, smooth, and geniculate

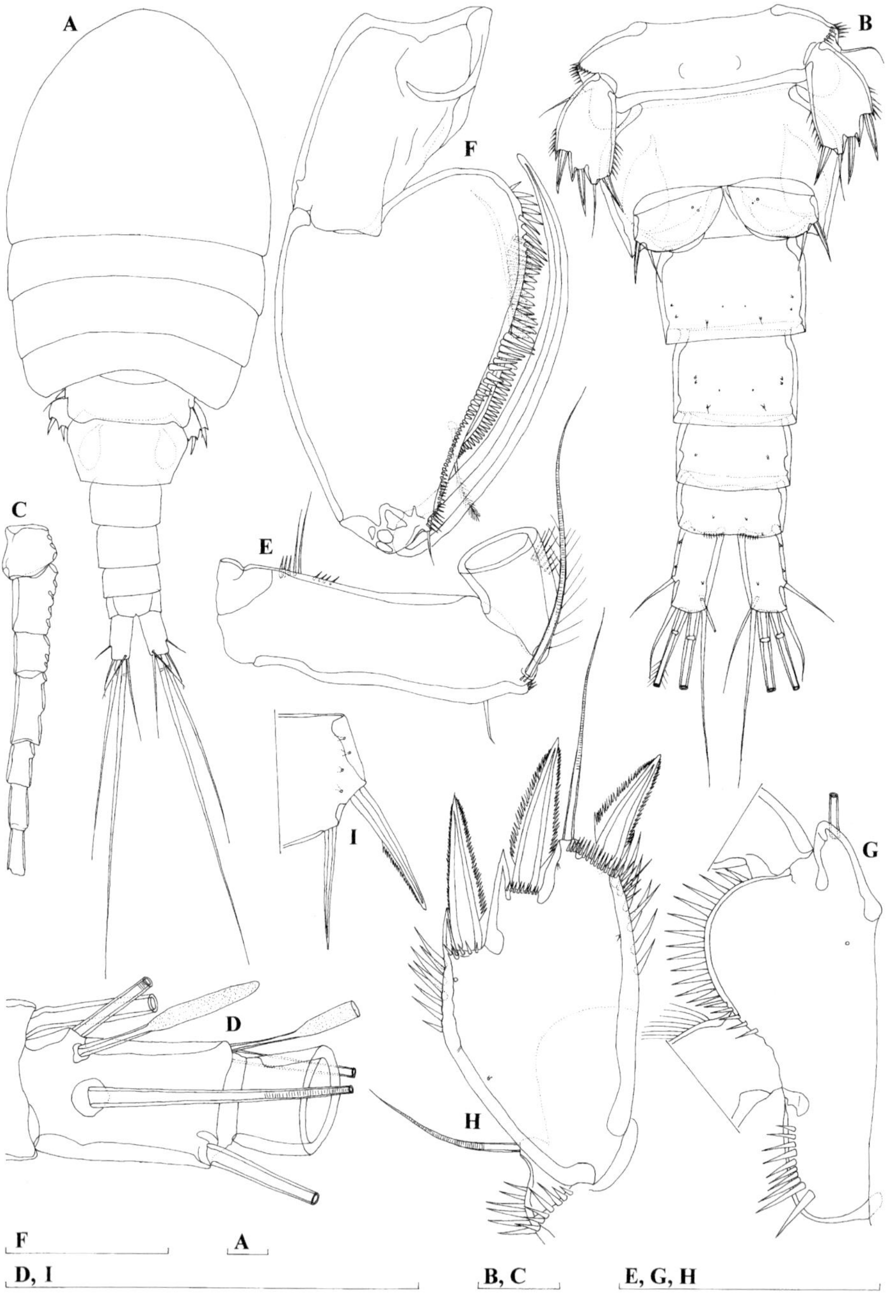

Fig. 4. *Hemicyclops nichollsi* n. sp., allotype male. A, habitus, dorsal view; B, urosome, ventral view; C, antennula without armature; D, fifth antennular segment; E, proximal part of antenna; F, maxilliped; G, basis of first swimming leg; H, left fifth leg, ventral view; I, part of left sixth leg. Scales = 0.1 mm.

spine, slightly longer than basis and forming strong, prehensile organ with inner margin of basis.

Swimming legs similar to those of female, except first swimming leg without spine at distal inner corner of basis (fig. 4G).

Fifth leg (fig. 4B, H) relatively similar to that of female, but coxobasis completely fused to somite; exopod 1.7 times as long as wide, exopodal seta slender and smooth, 0.7 times as long as exopod, 1.9 times as long as inner apical spine, about 1.5 times as long as outer, actually subapical spine and lateral spine, and 1.4 times as long as basal seta.

Sixth leg (fig. 4B, I) distinct, large plate, ornamented with six cuticular pores (two close to inner margin and four close to outer margin) and armed with two strong elements (outer one serrated along inner margin and 1.1 times as long as inner, smooth element).

Variability. — Body length of females ranging from 1.627 mm to 1.729 mm (1.692 mm average; n = 10), while in males it ranges from 1.523 mm to 1.708 mm (1.589 mm average; n = 6). No other type of variability was observed.

Etymology. — The species is named in honour of the late Dr. A. G. Nicholls, who described the first and so far only Australian *Hemicyclops* Boeck and who was also the first to suggest that the genus *Saphirella* T. Scott represents "the young form of *Hemicyclops*" (Nicholls, 1944). The name is a noun in the genitive singular.

Distribution. — *Hemicyclops nichollsi* n. sp. seems to be restricted to southern Queensland and northern New South Wales.

Discussion. — The family Clausidiidae Embleton, 1901 currently comprises nearly 80 species (Ho & Kim, 2004) in eight valid genera: *Clausidium* Kossmann, 1874; *Conchyliurus* Bocquet & Stock, 1957; *Giardella* Canu, 1888; *Hemadona* Ho & Kim, 2004; *Hemicyclops* Boeck, 1873; *Hersiliodes* Canu, 1888; *Hippomolgus* G. O. Sars, 1917; and *Hyphalion* Humes, 1987 (see Boxshall & Halsey, 2004; Ho & Kim, 2004). The generic name *Saphirella* T. Scott, 1894 was widely used for juvenile copepodid stages of *Hemicyclops*, a situation recognized by Nicholls (1944) and later confirmed through rearing experiments by Itoh & Nishida (1991) and Kim & Ho (1992). Many other genera were considered as members of the Clausidiidae but were later either synonymized with existing Clausidiidae genera or transferred to other families; see summary in Boxshall & Halsey (2004).

Hemicyclops is the largest genus in the family Clausidiidae and contains more species than the other seven genera combined. It is also the most diverse ecologically, with free-living representatives in the marine interstitial and littoral, but also in association with a great variety of hosts (often inhabiting their burrows), including sponges, scleractinian corals, gastropods and bivalve molluscs, polychaetes, crustaceans, and echiurans (Humes, 1984; Boxshall & Humes, 1987). All species appear to be intertidal forms or inhabitants of shallow waters, found predominantly in temperate and tropical regions of the world oceans (Mulyadi, 2005). The most recent revisionary work on *Hemicyclops* taxonomy was that of Vervoort & Ramirez (1966), who recognized 23 valid species: *H. aberdonensis* (T. & A. Scott, 1892); *H. acanthosquillae* Humes, 1965; *H. adhaerens* (Williams, 1907); *H. amplicaudatus* Humes, 1965; *H. arenicolae* Gooding, 1960; *H. australis* Nicholls, 1944; *H. axiophilus* Humes, 1965; *H. bacescui* (Serban, 1956); *H. biflagellatus* Humes, 1965; *H. carinifer* Humes, 1965; *H. cylindraceus* (Pelseneer, 1929); *H. diremptus* Humes, 1965; *H. elongatus* Wilson, 1937; *H. intermedius* Ummerkutty, 1961; *H. kombensis* Humes, 1965; *H. leggii* (Thompson & Scott, 1903); *H. livingstoni* (T. Scott, 1894); *H. purpureus* Boeck, 1872; *H. subadhaerens* Gooding, 1960; *H. tamilensis* (Thompson & Scott, 1903); *H. thalassius* Vervoort & Ramirez, 1966; *H. thysanotus* Wilson, 1935; and *H. visendus* Humes, Cressey & Gooding, 1958. They synonymized some (following Gooding, 1960), excluded a number of insufficiently described species or those based on immature stages, and provided a key to 22 species (*H. leggii* was excluded from the key as it was known only from males, but was recognized as a valid species nevertheless). Since then, another 18 species have been described in the genus *Hemicyclops*: *H. acanthophorus* Humes, 1995; *H. apiculus* Humes, 1995; *H. caissarum* Kihara & Rocha, 1993; *H. columnaris* Humes, 1984; *H. ctenidis* Ho & Kim, 1990; *H. geminatus* Stock, 1992; *H. gomsoensis* Ho & Kim, 1991; *H. japonicus* Itoh & Nishida, 1993; *H. javaensis* Mulyadi, 2005; *H. minutus* Mulyadi, 2005; *H. mortoni* Boxshall & Humes, 1987; *H. perinsignis* Humes, 1973; *H. saxatilis* Ho & Kim, 1991; *H. sebastiani* Kihara & Rocha, 1993; *H. spinulosus* Itoh & Nishida, 1998; *H. tanakai* Itoh & Nishida, 2002; *H. ventriplanus* Kim, 2000; and *H. vicinalis* Humes, 1995. Also, Hamond (1968) reported and illustrated a *Hemicyclops* male that he could not identify to species level (as *Hemicyclops* sp.), and Yeatman (1982) described the male of *H. adhaerens* for the first time. This puts the number of valid species to 41 and it would be highly desirable to present a key to aid in their identification. Unfortunately, such a key could not be constructed today without excluding a few more species with incomplete descriptions and this would make any future revision and/or subsequent descriptions of other new species more complicated. A full revision of *Hemicyclops* is needed and it would have to involve redescriptions of most of the species described in the nineteenth and the first half of the twentieth century, a task that is well beyond the scope of this monograph.

Hemicyclops nichollsi n. sp. is characterized by a plesiomorphic armature of the male sixth leg, which bears two elements, unlike most *Hemicyclops* species that only have one element on that appendage. Only three other species share this character: *H. ventriplanus*, described from the washings of a thalassinidean crustacean in Korea by Kim (2000), *H. gomsoensis*, described from crab burrows in Korea by Ho & Kim (1991), and *H. minutus*, collected from the littoral of the northern shore of Java (Indonesia) by Mulyadi (2005). The first species is only remotely related to *H. nichollsi* and could be distinguished by a number of morphological characters, including the shape of the genital double somite, and of the caudal rami and labrum, but also by details in the armature and ornamentation of the antennula, antenna, and maxilla in the female, as well as by the shape and ornamentation of the maxilla and maxilliped in the male. *Hemicyclops gomsoensis* is very close to *H. ventriplanus* and could be distinguished from *H. nichollsi* also by a number of characters, most importantly by the shape of the genital double somite and the caudal rami. *Hemicyclops minutus*, on the other hand, seems to be very close to *H. nichollsi*, but unfortunately is known only from a single male. However, it differs from *H. nichollsi* by the armature of the antennula, maxilla (syncoxa with three setae), maxilliped (one seta on the syncoxa and no setae on the basis), as well as by the relative length of the third endopodal segment of the fourth swimming leg (which is less slender). Some other minor differences include the ornamentation of the mandibula, maxilla, and fifth leg.

Further support for the specific status of *H. nichollsi* is provided by its geographic isolation from the Javanese representative, as it was not recorded in tropical Queensland or in the tropical part of Western Australia (which is much closer to Java). It is interesting to note that all four species (*H. nichollsi*, *H. ventriplanus*, *H. gomsoensis*, and *H. minutus*) are distributed in the West Pacific, which may suggest a closer phylogenetic relationship than becomes apparent from their morphology.

H. nichollsi could be distinguished from its congeners most easily by the shape of the genital double somite, caudal rami, labrum, antenna, or fifth leg, but usually by more than one character. It is the only known species of *Hemicyclops* with sexual dimorphism of the antenna, as the female coxobasis lacks any armature (fig. 2D), while that of the male is normally armed with one seta at the inner distal corner (fig. 4E). The only other Australian representative known so far is *H. australis*, described from the littoral of South Australia by Nicholls (1944). It differs from *H. nichollsi* by the much shorter caudal rami, but also in details in the armature and ornamentation of the labrum, maxilla, and maxilliped, in addition to its armed female antennal coxobasis, and the single seta only on the male sixth leg.

Other *Hemicyclops* species that somewhat resemble *H. nichollsi* are: *H. axiophilus*, described from burrows of a thalassinidean crustacean in Madagascar by Humes (1965); *H. bacescui*, from the interstitial of the Black Sea (Serban, 1956); *H. biflagellatus*, from burrows of unknown origin in Madagascar (Humes, 1965); *H. thysanotus*, associated with several hosts in the United States (Wilson, 1935b; Gooding, 1960); and *H. visendus*, associated with mud shrimps in Madagascar (Humes et al., 1958). However, though they all have an armed female antennal coxobasis and only one seta on the male sixth leg (the male of *H. bacescui* is still unknown), they can still be distinguished from *H. nichollsi* by a few other characters. *Hemicyclops axiophilus*, *H. biflagellatus*, *H. thysanotus*, and *H. visendus* all have the first antennular segment with only four setae, but also an additional small seta on the maxillar syncoxa, the labrum without strongly produced lateral corners, and an armed syncoxa of the male maxilliped. The genital double somite of *H. axiophilus* and *H. visendus* is expanded in the posterior half, while that of *H. biflagellatus* and *H. thysanotus* has a much more rounded anterior expansion than the genital double somite of *H. nichollsi*. The fifth leg of *H. axiophilus* and *H. visendus* has a somewhat shorter and more ovoid exopod than that of *H. nichollsi*, while the fifth leg exopod of *H. biflagellatus* and *H. thysanotus* is more elongated. The caudal rami of *H. nichollsi* are somewhat longer than those of *H. axiophilus*, but shorter than those of *H. thysanotus*; the caudal rami of *H. biflagellatus* are all covered with hair-like spinules. *Hemicyclops visendus* also has a somewhat longer third antennal segment. *Hemicyclops bacescui*, on the other hand, is incompletely described and only known after females. Its first antennular segment is illustrated with only three setae, but that could be an oversight. However, unlike *H. nichollsi*, this species has an additional small seta on the maxillar syncoxa, a somewhat shorter and at midlength constricted genital double somite, a more elongated exopod of the fifth leg, and somewhat longer spiniform setae on the third endopodal segment of the fourth leg.

Genus **Cemihyclops** n. g.

Diagnosis. — Moderately large Clausidiidae, with cyclopiform and relatively slender habitus and first pedigerous somite completely fused to cephalothorax. Hyaline fringes of all somites narrow and smooth. Sclerotized joint between fifth pedigerous somite and genital double somite well developed. Female urosome five-segmented. Genital double somite with anterior lateral corners slightly expanded, longer than wide; copulatory pores and ovipores lie together within lateral genital apertures, covered by reduced sixth legs. Anal somite with very broad and short anal operculum. Caudal rami parallel, 5.5 times as long as wide, armed with one lateral, one dorsal, and four apical setae; lateral seta arising at anterior part; principal apical setae without breaking planes. Antennula long and seven-segmented, without aesthetasc on fifth segment. Antenna four-segmented, with very short fourth segment and armature formula 1.1.4.7. Labrum simple and trapezoidal. Mandibula without palp, armed with two foliaceous spines and two bipinnate setae. Maxillula trilobate, armed with three elements on inner lobe, two on middle, and three on outer one. Maxilla two-segmented, armed with three strong setae on first segment and four elements on second segment (two complex teeth and two simple elements). Female maxilliped large and four-segmented, with armature formula 2.2.0.5; apicalmost element transformed into smooth and prehensile claw. All swimming legs with three-segmented exopod and endopod and without sucking discs; third exopodal segment spine formula 4.4.4.3 and third endopodal segment spine formula 2.3.3.5; seta on first endopodal segment of second, third, and fourth leg plumose, but with sharply serrated outer distal tip; second leg with peculiar pincers formed by transformed proximalmost spine on third exopodal segment, and only spine on first exopodal segment. Fifth leg comprising ovoid coxobasis and elongated one-segmented exopod; coxobasis armed with slender outer basal seta; exopod 4.5 times as long as wide, curved inwards, ornamented with spinules along both inner and outer margins and with one very strong spinule at inner distal corner, armed with three spines and single spiniform seta; innermost apical spine longest and positioned as most apical element. Female sixth leg small cuticular plate, armed with one smooth seta and two minute spines.

Etymology. — The generic name is an anagram of the existing generic name *Hemicyclops*. Gender masculine.

Type and only species. — *Cemihyclops cedunensis* n. sp.

2. **Cemihyclops cedunensis** n. sp.
(figs. 5-7)

Material examined. — Types only: Holotype female dissected on 2 slides (WAM C37205); paratype female on 1 slide in toto (TMAG G5773): Australia, South Australia, Ceduna, 24 December 2002, leg. T. Karanovic, 32°06.800′S 133°40.319′E.

Description. — Female (holotype). Body length, excluding caudal setae, 1.508 mm. Habitus (fig. 5A) not very robust, with prosome/urosome ratio 1.2 and greatest width at second pedigerous (first free) prosomite. Body length/width ratio about 3.1; prosome elliptical in dorsal view, but somewhat constricted between second and third pedigerous somites, about 1.7 times as long as wide and 2.3 times as wide as genital double somite. Posterior corners of fourth pedigerous somite only slightly expanded posterolaterally, other prosomites without pronounced lateral expansions. Fifth pedigerous somite narrower than genital double somite. Colour of preserved specimen milky white; integument weakly chitinized. Nauplius eye not visible. Rostrum broadly rounded and furnished with two large sensilla.

Cephalothorax (fig. 5A) almost triangular, as long as its greatest width, representing 31% of total body length. Surface of cephalic shield with a few small sensilla and cuticular pores, and the same on other prosomites. Hyaline fringe of prosomites narrow and smooth. First pedigerous somite completely fused to cephalosome. Arthrodial membrane between fourth and fifth pedigerous somites well developed, and sclerotized joint between fifth pedigerous somite and genital double somite visible (fig. 5A, B).

Genital double somite (fig. 5B) with anterior lateral corners somewhat expanded and gradually tapering towards posterior margin, about 1.2 times as long as wide, ornamented with only two sensillla ventrally near posterior border. Hyaline fringe of genital double somite and of two succeeding somites narrow and smooth. No cuticular recesses or windows visible. Copulatory pores and ovipores lie together within lateral genital apertures, covered by reduced sixth legs. Copulatory ducts and seminal receptacles not clearly visible inside. Third urosomite also ornamented with only two ventral sensilla, fourth urosomite with two ventral pores. Anal somite with smooth, broad, very short, and almost straight anal operculum, which represents 83% of the somite's width and reaches to only 1/5 of somite's length; ornamented with two large sensilla and two pores dorsally, one pore laterally on each side, and two pores ventrally; its posterior margin finely serrated. Anal sinus smooth and widely open dorsally.

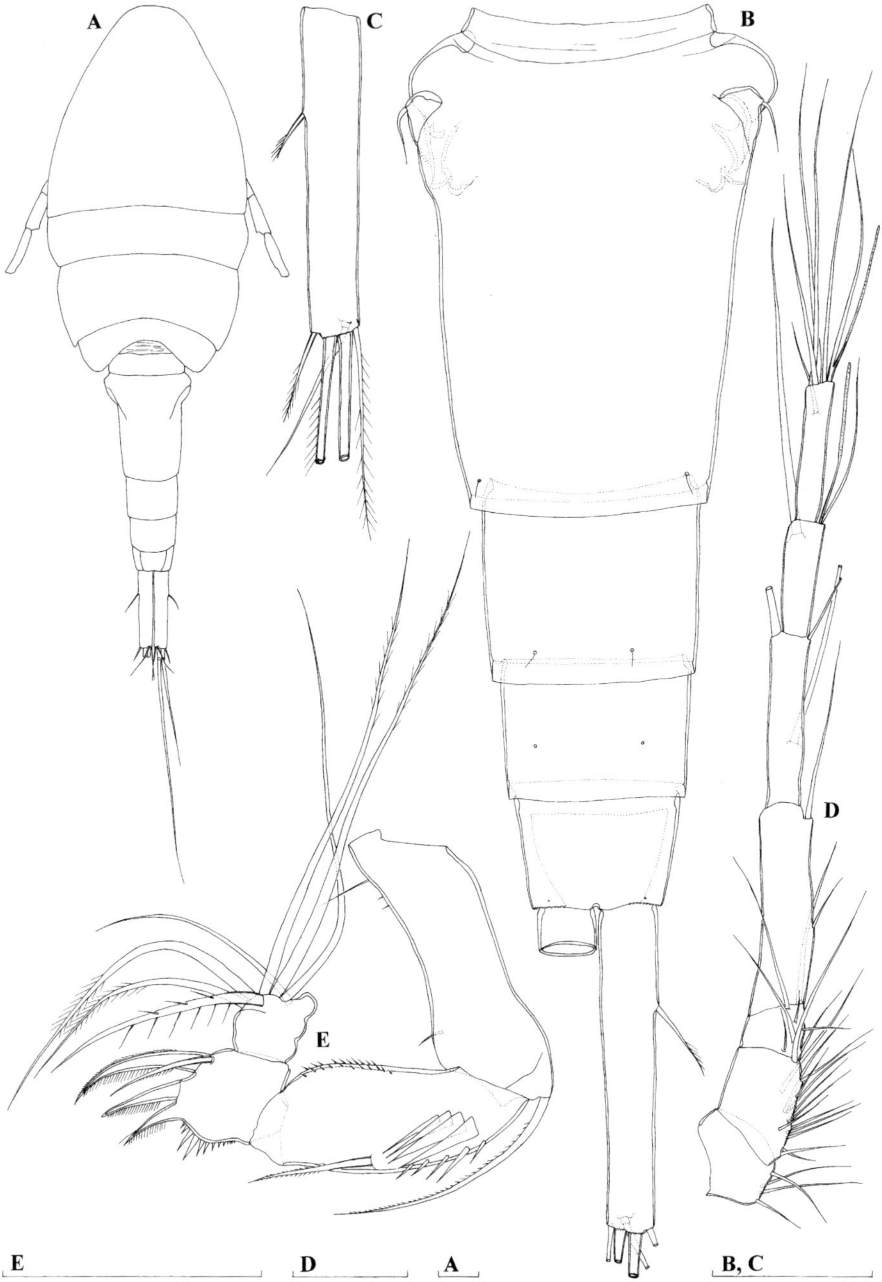

Fig. 5. *Cemihyclops cedunensis* n. gen. et sp., holotype female. A, habitus, dorsal view; B, abdomen, ventral view; C, right caudal ramus, ventral view; D, antennula; E, antenna. Scales = 0.1 mm.

Caudal rami (fig. 5B, C) cylindrical, unornamented, 5.5 times as long as wide (ventral view), parallel, with almost no space between them; they represent 12% of total body length. Distal margin ventrally serrated and straight, with medial cuticular pore not visible. Dorsal seta smooth, about 0.5 times as long as ramus, inserted on small protuberance close to posterior border, uniarticulate at base. Lateral seta smallest, arising at 1/3, smooth and slightly shorter than ramus' width. Outermost apical seta slender and bipinnate, 0.3 times as long as ramus and 0.4 times as long as innermost apical seta, which is also slender and bipinnate. Principal apical setae without breaking planes and bipinnate; inner seta about twice as long as outer one and 0.4 times as long as body.

Antennula (fig. 5D) slender, reaching beyond posterior margin of second pedigerous somite (first free prosomite), seven-segmented, unornamented, with one slender aesthetasc on sixth segment subapically and one on seventh segment apically, and with setal formula: 4.14.7.2.3.2.6. No setae with breaking planes or articulating on basal part, and all setae smooth; one apical seta on seventh segment fused basally to aesthetasc. Length ratio of antennular segments, from proximal end and along caudal margins, 1 : 2.8 : 1.4 : 9.1 : 7.7 : 5 : 6.2.

Antenna (fig. 5E) four-segmented, comprising coxobasis and three-segmented endopod, strongly geniculate with first and last segment pointing in opposite directions; length ratio of antennal segments, 1 : 1.1 : 0.4 : 0.3. Coxobasis about 2.3 times as long as wide, armed with a single unipinnate seta at distal inner corner, ornamented with a few spinules along outer margin; seta representing exopod absent. First endopodal segment 2.6 times as long as wide, armed with one pinnate seta at middle, ornamented with two parallel longitudinal rows of long spinules along inner margin proximally and two rows of small spinules along outer margin distally; seta 0.4 times as long as segment. Second endopodal segment almost triangular, with very short outer margin, about 0.8 times as long as wide, ornamented with strong spinules along inner margin, armed with four setae subapically (three setae very strong, geniculate, and pinnate; one seta slender and smooth). Third endopodal segment quadrangular, unornamented, as long as wide, with produced, blunt outer apical corner, armed with seven strong and long apical setae (five of these pinnate, two smooth).

Labrum simple and large, trapezoidal, ornamented with numerous spinules near and along posterior margin, but not mounted sufficiently properly to allow detailed examination.

Postoral plate destroyed during dissection and not observed.

Mandibula (fig. 6A, B) very simple, unornamented, slightly geniculate, about 3.4 times as long as wide and without any trace of a palp. Coxal gnathobase cutting edge armed with four clearly distinct armature elements: two foliaceous spines and two bipinnate setae. Outer foliaceous spine (fig. 6B) with both margins serrated, while middle spine with very strong spinules along both margins and all over its outer surface. Innermost seta 0.8 times as long as middle seta and as long as middle spine; outermost spine 1.5 times as long as middle spine.

Maxillula (fig. 6C) also very simple, unornamented, unsegmented but trilobate. Inner lobe armed apically with three bipinnate elements, middle lobe armed with two bipinnate elements, and outer lobe with three bipinnate elements. All armature elements strong, of about same length, and those on inner lobe fused to it basally.

Maxilla (fig. 6D, E) unornamented, two-segmented, comprising large and quadriform syncoxa and smaller and more elongate basis. Syncoxa 0.6 times as long as wide, armed with three subequal strong and bipinnate setae on presumably proximal endite; distal endite reduced and unarmed. Basis about 1.2 times as long as wide, armed apically with two complex teeth, an outermost simple and unipinnate spine, and one unipinnate strong seta inserted somewhere between the complex teeth; innermost complex tooth with distal end rounded and wider than its base, armed with 11 very strong spinules along distal margin, fanlike; middle complex tooth (fig. 6E) three-dimensional, with strong base and foliaceous distal part, which is spiral-shaped and deeply serrated along outer margin, its inner margin smooth.

Maxilliped (fig. 6F) prehensile but robust, unornamented, four-segmented, composed of syncoxa, basis, and two-segmented endopod, but first endopodal segment partly fused to basis. Syncoxa large, 1.1 times as long as its greatest width, armed at middle with two equally long, strong setae; one seta strongly bipinnate at distal part, other seta smooth. Basis 2.4 times as long as wide, armed similarly as syncoxa, with two subequal, strong setae at middle, both of them strongly bipinnate. First endopodal segment unornamented and unarmed, largely fused to basis. Second endopodal segment also very small, as long as wide, armed with five setae (three apical and two lateral); innermost apical seta transformed into strong claw, smooth, and forming prehensile organ with basis, about 0.7 times as long as basis and 8 times as long as second endopodal segment; middle apical seta very short and broad, with seven long spinules along apical margin, fanlike, 0.7 times

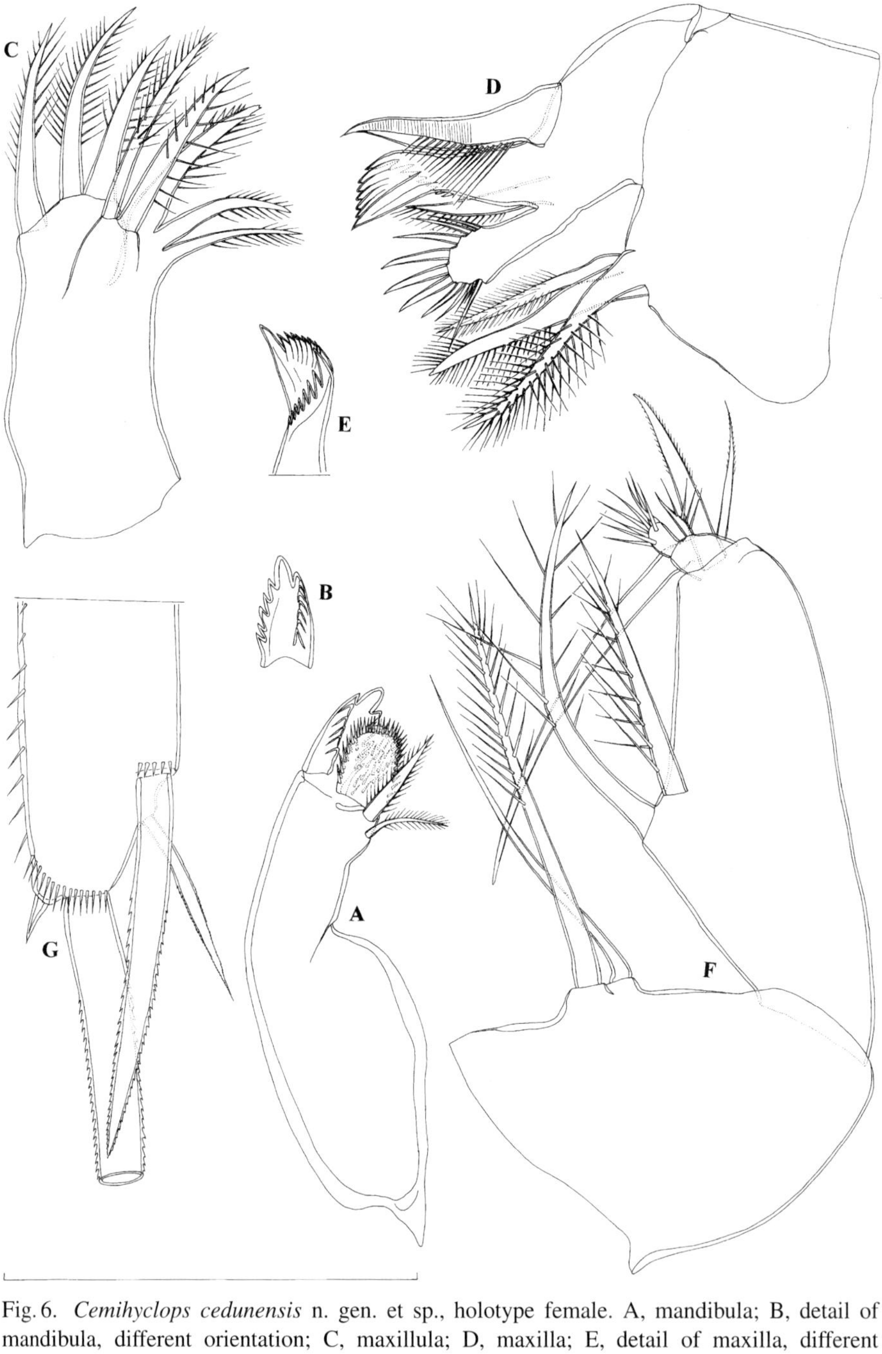

Fig. 6. *Cemihyclops cedunensis* n. gen. et sp., holotype female. A, mandibula; B, detail of mandibula, different orientation; C, maxillula; D, maxilla; E, detail of maxilla, different orientation; F, maxilliped; G, detail of left fifth leg, ventral view. Scale = 0.1 mm.

as long as segment; outer apical seta strong but narrow, spiniform, about as long as segment, ornamented with four long spinules along outer margin; two lateral setae ornamented, with minute spinules along both outer and inner margins, proximal seta much stronger and 1.4 times as long as distal one. Longest seta on endopod 1.1 times as long as the longer basal seta, and 0.9 times as long as the longer seta on the syncoxa.

All swimming legs (fig. 7A, B, C, D, E, F) with three-segmented exopod and endopod; their armature formula as follows (legend: inner/outer spine or seta; inner/terminal/outer):

	Exopod			Endopod		
Segments	1	2	3	1	2	3
First leg	0/1	1/1	4/2/2	1/0	1/0	3/2/1
Second leg	0/1	1/1	5/2/2	1/0	2/0	3/2/1
Third leg	0/1	1/1	5/2/2	1/0	2/0	3/2/1
Fourth leg	0/1	1/1	5/2/1	1/0	2/0	2/2/1

Third exopodal segment spine formula 3.3.3.2 and third endopodal segment spine formula 2.3.3.3. Intercoxal sclerites of first, second, and third leg with concave distal margin, that of fourth leg with convex margin; sclerites of second, third, and fourth legs ornamented with eight to ten spinules distally, that of first leg unornamented. Coxae ornamented with a single spinule at outer distal corner and armed with strong and long, plumose seta at inner distal corner. Spine inserted at inner distal corner of basis of first leg reaching 2/3 of second endopodal segment, with both margins finely serrated. Basis of each leg armed with slender seta at distal outer corner (that of first leg exceptionally long) and ornamented with a row of spinules along distal margin between exopod and endopod; basis of first swimming leg additionally ornamented with row of spinules at base of inner spine; inner distal corners of second, third, and fourth leg roundly pronounced. Most exopodal and endopodal setae slender and plumose; inner apical armature element on third exopodal segment of all legs a transition between a spine and a seta, with serrated outer margin and plumose inner one; both inner setae on third endopodal segment of fourth leg modified into very strong, spiniform armature elements, with smooth proximal part and sharply serrated distal lateral margins; strong seta on first endopodal segment of second, third, and fourth legs with sharply and characteristically serrated outer distal tip, otherwise normally plumose. All exopodal and endopodal spines stout and with finely serrated outer and inner margins; second leg with two characteristically modified spines: the proximal spine on the third exopodal segment, and the spine

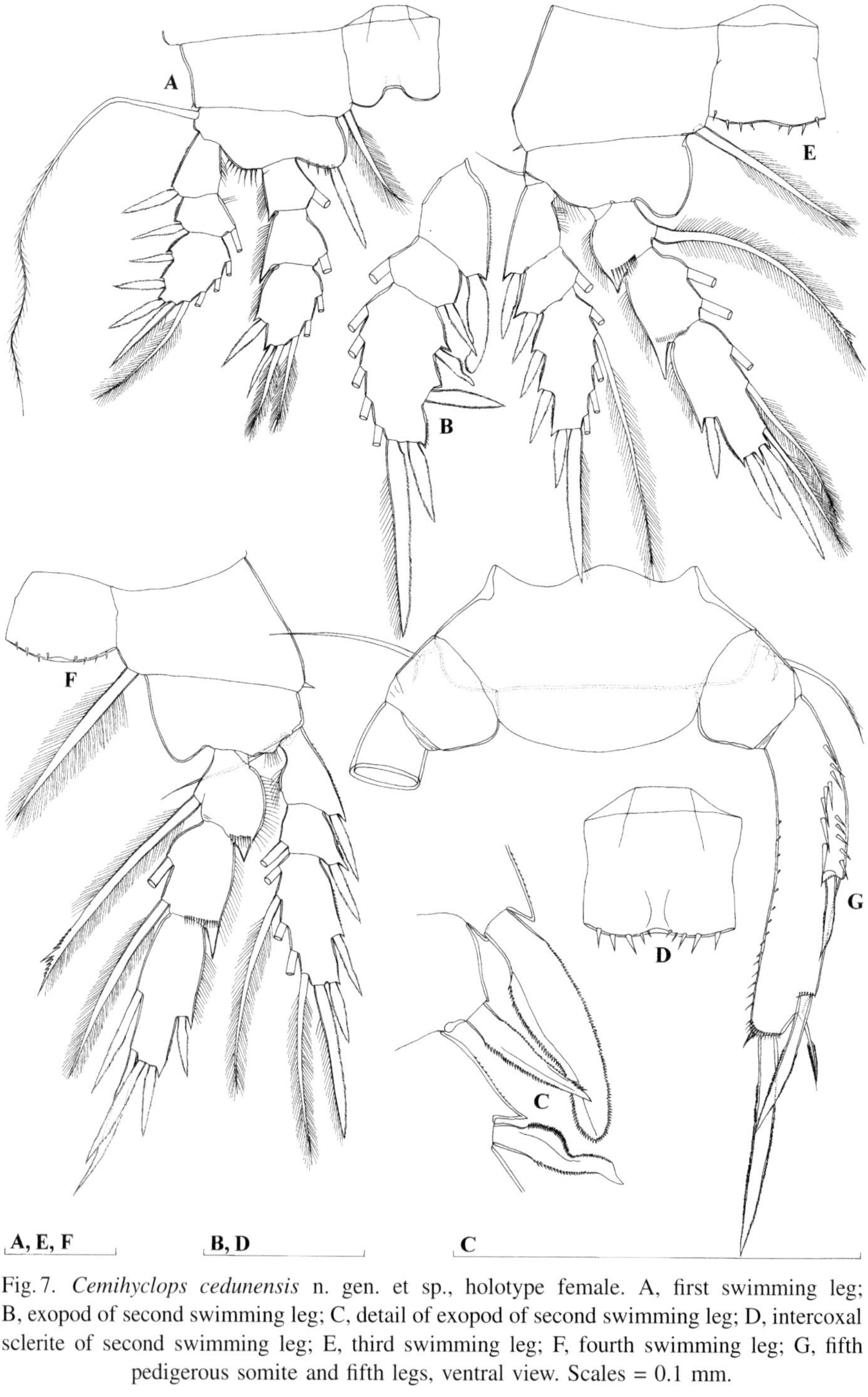

Fig. 7. *Cemihyclops cedunensis* n. gen. et sp., holotype female. A, first swimming leg; B, exopod of second swimming leg; C, detail of exopod of second swimming leg; D, intercoxal sclerite of second swimming leg; E, third swimming leg; F, fourth swimming leg; G, fifth pedigerous somite and fifth legs, ventral view. Scales = 0.1 mm.

on first exopodal segment, which together form a peculiar pair of pincers (fig. 7C). All endopodal segments with hair-like spinules along outer margins, which are also present on inner margin of first exopodal segment; most exopodal segments with minute spinules along outer margin. Cuticular pores not visible on any endopodal or exopodal segment, probably because of very weakly chitinized integument. Third endopodal segment of fourth swimming leg about 2.4 times as long as wide; inner apical spine about 0.8 times as long as segment and 1.7 times as long as outer apical spine.

Fifth leg (figs. 6G, 7G) without intercoxal sclerite, attached to ventrolateral corner of fifth pedigerous somite, two-segmented (although first segment partly fused to somite dorsally), comprising ovoid coxobasis and one-segmented, elongate exopod. Coxobasis unornamented and armed with slender and unipinnate outer basal seta, about 0.8 times as long as wide (ventral view). Exopod nearly 4.5 times as long as its greatest width, slightly wider at basal part than distally, although cylindrical and somewhat curved invards, armed with three spines and a single seta, and ornamented with strong spinules along outer margin proximally, and smaller ones along inner margin, as well as at base of each spine; one strong spinule at inner distal corner; innermost apical spine longest and strongest, positioned as most apical element, 0.7 times as long as exopod, 1.5 times as long as outer subapical spine, 2.6 times as long as lateral spine, and 2.9 times as long as exopodal seta; all spines strong and with sharply serrated lateral margins, exopodal seta bipinnate distally.

Sixth leg (fig. 5B) not clearly distinct, cuticular plate, strongly chitinized, without any surface ornamentation, armed with one short and smooth seta and two minute spines; both spines fused basally to plate.

Male. Unknown.

Variability. — Body length of females ranging from 1.508 mm to 1.562 mm (1.535 mm average; n = 2). No other type of variability observed.

Etymology. — The species is named after the type locality, the city of Ceduna in South Australia. The name is an adjective for place, made with the Latin suffix "-ensis".

Distribution. — *Cemihyclops cedunensis* n. g., n. sp. is known from the type locality only.

Discussion. — Boxshall & Halsey (2004) recognized only seven valid genera in the family Clausidiidae Embleton, 1901: *Clausidium* Kossmann,

1874 (with 11 valid species); *Conchyliurus* Bocquet & Stock, 1957 (nine species); *Giardella* Canu, 1888 (monospecific); *Hemicyclops* Boeck, 1873 (39 species); *Hersiliodes* Canu, 1888 (two species); *Hippomolgus* G. O. Sars, 1917 (three species); and *Hyphalion* Humes, 1987 (also with three valid species). They excluded three previously accepted genera (*Foliomolgus* Kim, 2001; *Leptinogaster* Pelseneer, 1929; and *Pholadicola* Ho & Warde, 1992) from the family and placed these in a generic grouping around the genus *Teredicola* Wilson, 1929, characterized by the derived form of the antenna and mandibula. Unfortunately, Boxshall & Halsey (2004) did not describe a separate family for these four genera, but merely listed them as "*Teredicola*-group". Almost at the same time, Ho & Kim (2004) described a new genus of Clausidiidae (*Hemadona* Ho & Kim, 2004) and discussed the phylogenetic relationships in the family, based on a cladistic analysis of morphological characters. Unaware of Boxshall & Halsey's (2004) paper, Ho & Kim (2004) treated *Foliomolgus*, *Leptinogaster*, and *Pholadicola* as valid Clausidiidae genera, but did not mention the genus *Giardella*, which by some copepodologists was considered a synonym of *Hemicyclops* (see Gooding, 1960; Ho & Wardle, 1992; and Kim, 2001), although not by Nicholls (1944), Vervoort & Ramirez (1966), or Stock (1973). I here accept the narrower concept of the family as advocated by Boxshall & Halsey (2004), with the addition of the subsequently described genus *Hemadona*. This leaves eight valid genera, as mentioned in the Discussion section of the previous species here treated.

Peculiar pincers, formed by the exopodal spines of the second swimming leg, are an autapomorphy that distinguishes *Cemihyclops* n. g. easily from the other eight genera of the Clausidiidae. It can also be distinguished from each of these genera by a number of other characters. Its seven-segmented antennula differs from that found in *Conchyliurus*, *Hemadona*, *Hersiliodes*, *Hippomolgus*, and *Hyphalion* (in which it is at most six-segmented), and the absence of sucking discs on the swimming legs easily distinguishes *Cemihyclops* from species of the genus *Clausidium*. The latter also has only three elements on the mandibula, while the other five genera differ from *Cemihyclops* by many modifications of their mouth appendages and body form, resulting from their parasitic or semiparasitic way of life. On the other hand, *Cemihyclops* is very closely similar to the remaining two clausidiid genera (*Giardella* and *Hemicyclops*), although not as much as those are to each other. *Cemihyclops* could be distinguished from both by the plesiomorphic armature of the third endopodal segment of the first swimming leg, which bears two spines and four setae (vs. one spine and five setae in *Giardella* and *Hemicyclops*). This character is also found in the previously mentioned genera *Conchyliurus*, *Hemadona*, and *Hersiliodes*, which are not closely similar to *Cemihyclops*. *Giardella*, just like *Cemihyclops*, does not have an aesthetasc on the fifth antennular segment (always present in *Hemicyclops*) and has very long caudal rami, but differs additionally from the new genus by the position of the lateral seta on the caudal rami (inserted posteriorly in *Giardella*), by the armature of the mandibula (only three elements in *Giardella*), and by the armature of the endopod of the fourth leg (with only one seta on the second, and only four elements on the third segment in *Giardella*). The position of the lateral seta on the caudal rami also distinguishes *Cemihyclops* from *Hemicyclops*, from which it can be additionally separated by the presence of three strong setae on the syncoxa of the maxilla (third seta either completely reduced or very small in *Hemicyclops*). Even the structure of the fifth leg is somewhat different in these two genera, as its only seta is never spiniform in *Hemicyclops* and the innermost apical spine is not positioned significantly more apically than the seta. Two other peculiar features of *Cemihyclops*, the first endopodal seta of the second, third, and fourth legs having a serrated distal tip, and the remarkable, fanlike innermost complex tooth on the maxillular basis, can be found in some *Hemicyclops* species, thus further supporting a close phylogenetic relationship between these two genera. The former character was reported for *Hemicyclops spinulosus* Itoh & Nishida, 1998, described from crab burrows in Japan (Itoh & Nishida, 1998), and for *H. ctenidis* Ho & Kim, 1990, associated with a polychaete in Korea (Ho & Kim, 1990; Kim & Ho, 1992). The latter character, i.e., the fanlike tooth, was found in the following six species: *Hemicyclops* sp., reported from the United Kingdom by Hamond (1968); *H. apiculatus* Humes, 1995, described from a scleractinian coral in Madagascar by Humes (1995); *H. australis* Nicholls, 1944, from the littoral of South Australia (Nicholls, 1944); *H. intermedius* Ummerkutty, 1961, from weed washings in India (Ummerkutty, 1961); *H. leggii* (Thompson & Scott, 1903), from weed washings in Sri Lanka (Thompson & Scott, 1903); and *H. saxatilis* Ho & Kim, 1991, from crab burrows in Korea (Ho & Kim, 1991).

Its close relationship with *Hemicyclops* and *Giardella* leaves no doubt about placing this new genus in the family Clausidiidae, while the combination of apomorphic and plesiomorphic character states by which it is distinguished from these close relatives (as well as from more distantly related genera of this family) supports its separate generic status. Unfortunately, the males of *Cemihyclops* are still unknown so male characters could not be compared, but since the main distinguishing characters of this family are usually not sexually dimorphic (caudal rami, antennula, second swimming leg, and fifth leg), the generic diagnosis probably would not change appreciably with the discovery of the males.

Family POLYANKYLIIDAE Ho & Kim, 1997
Genus **Polyankylis** Ho & Kim, 1997

3. **Polyankylis australis** n. sp.
(figs. 8-10)

Material examined. — Types only: Holotype ovigerous female dissected on 2 slides (WAM C37207); paratypes, 2 females (1 dissected on 1 slide (TMAG G5774), other in alcohol (WAM C37209)): Australia, South Australia, Robe, 27 December 2002, leg. T. Karanovic, 37°09.725′S 139°45.153′E.

Description. — Ovigerous female (holotype). Body length, excluding caudal setae, 0.612 mm. Habitus (fig. 8A) very robust, with prosome/urosome ratio 1.5 and greatest width at first pedigerous prosomite, which is fused to cephalosome but with visible dorsal suture marking original segmentation. Body length/width ratio about 2.4; prosome elliptical in dorsal view, about 1.4 times as long as wide and 3.5 times as wide as genital double somite. Posterior corners of second (first free), third, and fourth pedigerous somites expanded laterally. Fifth pedigerous somite narrower than genital double somite. Colour of preserved specimen yellowish. Nauplius eye not visible. Rostrum broadly rounded and furnished with two large sensilla.

Cephalothorax (fig. 8A) almost triangular, about as long as its greatest width, representing 40% of total body length, with dorsal suture marking original segmentation of first pedigerous somite, which is otherwise completely fused to the cephalosome. Surface of cephalic shield with numerous small sensilla and cuticular pores, as are the surfaces of the other prosomites. Hyaline fringe of prosomites very narrow and smooth. Arthrodial membrane between fourth and fifth pedigerous somites not visible from dorsal view, and no sclerotized joint between fifth pedigerous somite and genital double somite (figs. 8A, B, 9A). Fifth pedigerous somite ornamented with four pores ventrally and two sensilla dorsally (figs. 8B, 9A).

Genital double somite (figs. 8B, 9B) with laterally expanded anterior part, 1.3 times as long as wide, ornamented with four pores and one sensillum

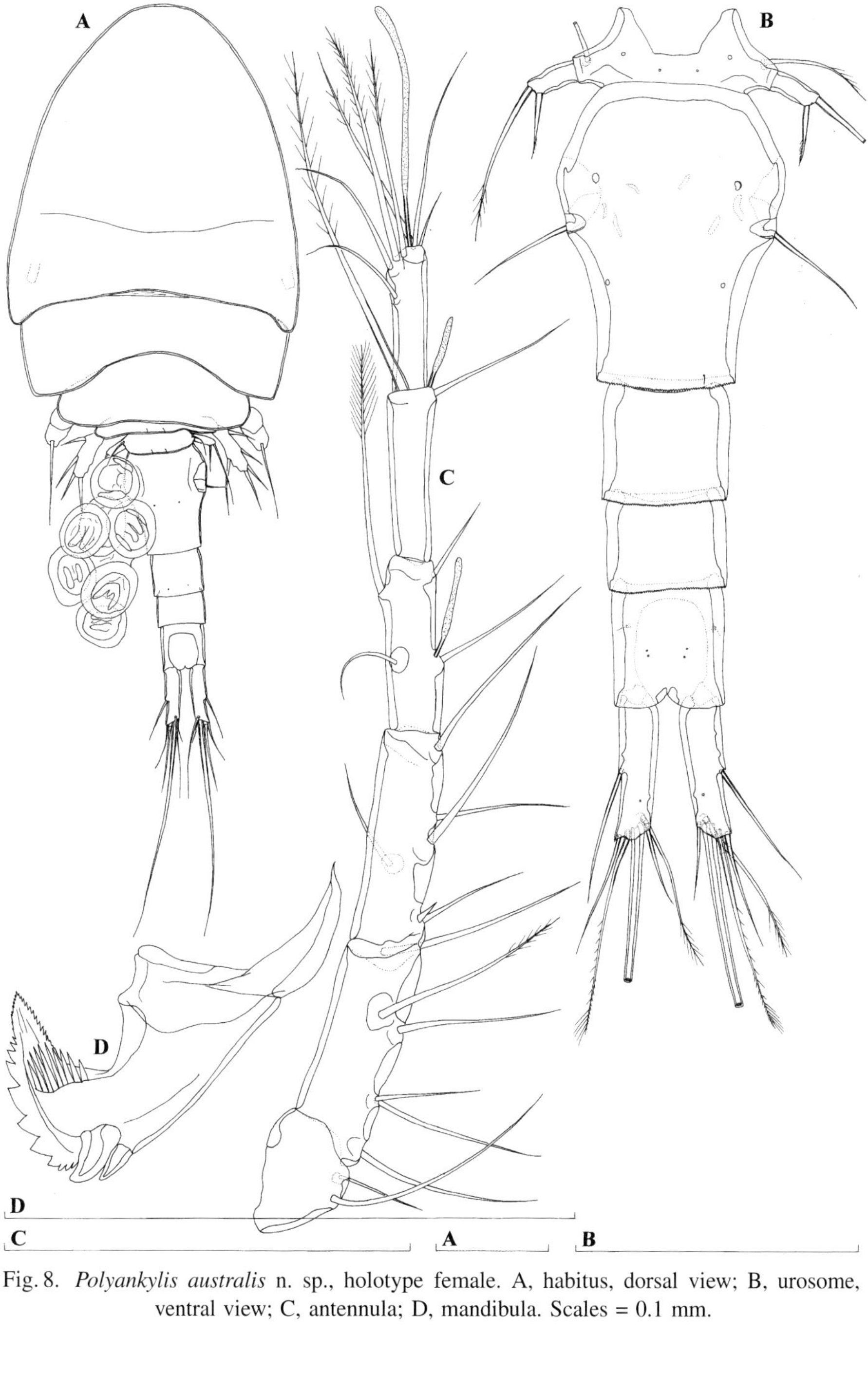

Fig. 8. *Polyankylis australis* n. sp., holotype female. A, habitus, dorsal view; B, urosome, ventral view; C, antennula; D, mandibula. Scales = 0.1 mm.

ventrally and ten pores dorsally. Hyaline fringe of genital double somite and two succeeding somites narrow and finely serrated. No cuticular recesses or windows visible, although some pores very large and almost window-like. Copulatory pores and ovipores lie together within lateral genital apertures, covered by reduced sixth legs. Copulatory ducts and seminal receptacles not clearly visible, although several short internal cuticular ribs are visible from ventral view, that may be part of the reproductive system. One egg sac attached to left anterior part of genital double somite, with eight large eggs with almost completely developed nauplii visible inside, and reaching slightly beyond posterior margin of fourth urosomite. Third urosomite ornamented with only two dorsal pores, fourth urosomite without any ornamentation visible. Anal somite (figs. 8B, 9A) with smooth, broad, very short, and slightly concave anal operculum, which represents 62% of the somite's width and reaches to only 2/5 of the somite's length; ornamented with two large sensilla dorsally and four pores ventrally; its posterior margin smooth both ventrally and dorsally. Anal sinus widely open dorsally and without any ornamentation visible.

Caudal rami (figs. 8A, B, 9A) cylindrical, three times as long as wide (ventral view), very little divergent, with space between them less than one ramus' width; they represent 8% of total body length. Distal margin smooth and slanting, almost diagonal (with outer corner positioned more posteriorly than inner one) both ventrally and dorsally, with ventral medial cuticular pore clearly visible. The ornamentation consists of one ventral, one lateral, and one dorsal pore; no lateral sensillum visible. Dorsal seta about 1.2 times as long as ramus, inserted on small protuberance close to posterior border and inner dorsal corner, uniarticulate at base and bipinnate distally. Lateral seta long and relatively strong, arising at middle, smooth, and about as long as ramus. Outermost apical seta also smooth, 0.8 times as long as ramus and 1.5 times as long as innermost apical seta, which is also smooth. Principal apical setae without breaking planes and bipinnate; inner seta about 3.3 times as long as outer one and 0.3 times as long as body.

Antennula (fig. 8C) relatively slender, not visible from dorsal view but not reaching posterior margin of cephalothorax, six-segmented, unornamented, with one short aesthetasc on fourth segment laterally, one short aesthetasc on fifth segment subapically, one long aesthetasc on sixth segment apically, and setal formula: 2.5.7.4.2.7. No setae with breaking planes or articulating on basal part, and most setae slender and smooth; only one seta on third segment very short and spiniform, and only one seta on second, one on fourth, one on fifth, and three setae on sixth segment pinnate at distal part. Length ratio of antennular segments, from proximal end and along caudal margins, 1 : 3.6 : 4.4 : 3.5 : 3.3 : 3.

Antenna (fig. 9B, C) S-shaped and slender, four-segmented, comprising coxobasis and three-segmented endopod; length ratio of antennal segments, 1 : 1.5 : 0.3 : 0.7. Coxobasis almost three times as long as wide, unornamented and armed with a single, slender and smooth seta at inner distal corner, which is half as long as coxobasis; seta representing exopod absent. First endopodal segment 4.3 times as long as wide, armed with smooth and slender medial seta, and ornamented with longitudinal row of outer spinules at distal part; seta 0.3 times as long as segment. Second endopodal segment small, about 1.2 times as long as wide, unornamented, and armed with two setae laterally and one stout aesthetasc subapically that reaches posterior margin of appendage. Third endopodal segment 3.5 times as long as wide, unornamented, and armed apically with three short, very strong claws and two smooth setae; longer seta only slightly longer than longest claw, and 0.6 times as long as segment.

Labrum relatively large, smooth, and bilobate, but unfortunately not mounted properly on slide to allow detailed examination.

Mandibula (fig. 8D) very simple, with unornamented proximal part, strongly geniculate, about twice as long as wide, and without any trace of a palp. Coxal gnathobase cutting edge armed with two distinct, smooth and short outer spines, and one large and foliaceous inner spine with both margins serrated, and with a row of nine long spinules at its base, which are completely fused at their base. Middle spine about as long as longest spinules and 1.6 times as long as outermost spine.

Maxillula (fig. 9D) also very simple and very small, unornamented and unsegmented, about 1.4 times as long as wide, armed apically with four strong and unipinnate elements (probably all setae); longest armature element about as long as segment.

Maxilla (fig. 9E, F) two-segmented, comprising large, ovoid syncoxa, and much smaller and more slender basis. Syncoxa unarmed and unornamented, about as long as wide and almost three times as wide as basis. Basis about twice as long as wide, also unornamented, armed laterally with strongly unipinnate seta, and apically with one seta and one complex tooth; apical seta with very strong and heavily armed basal part and with slender and finely bipinnate distal part, about 1.8 times as long as basis; apical spine fused at base, making basis look like being drawn into a tapering blade,

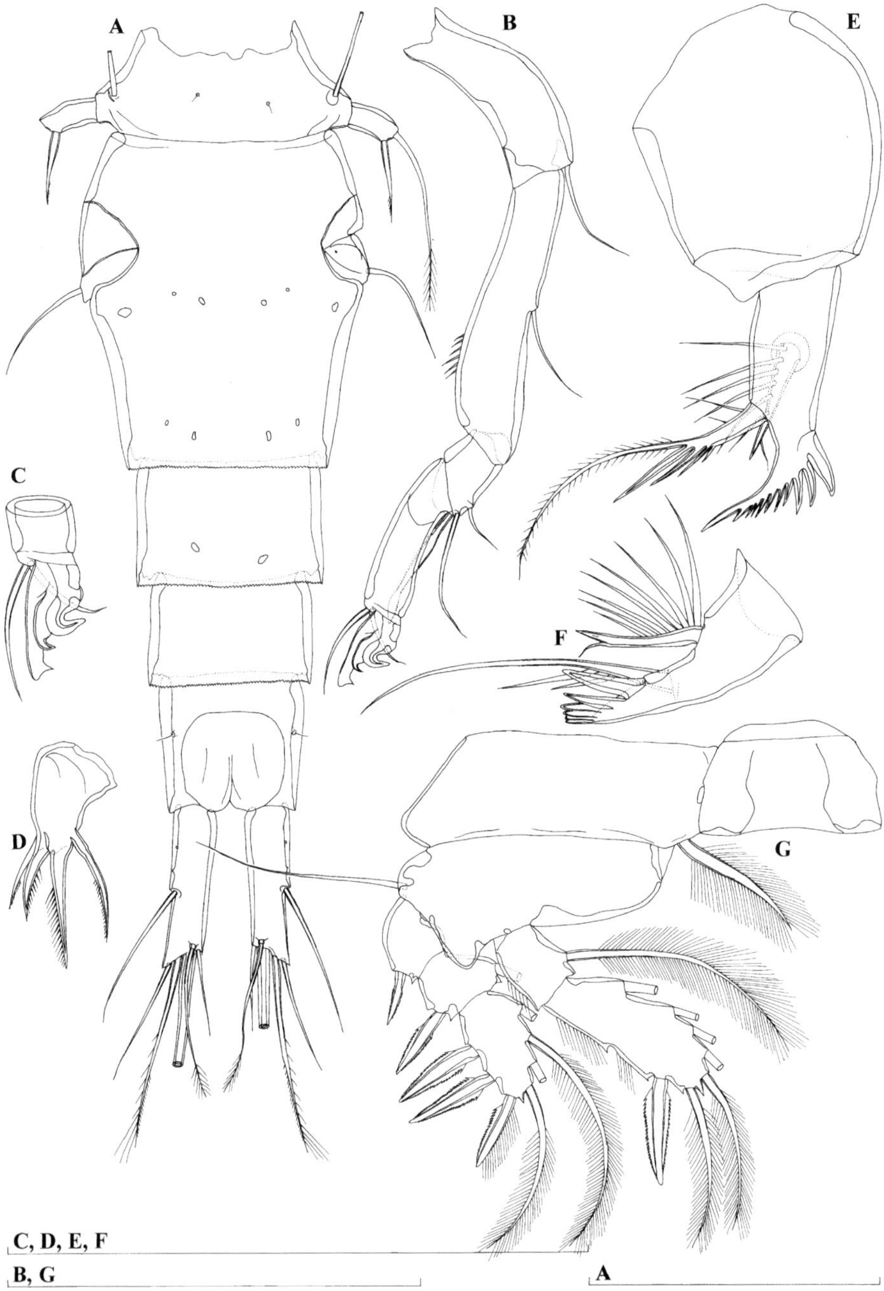

Fig. 9. *Polyankylis australis* n. sp., holotype female. A, urosome, dorsal view; B, antenna; C, distal end of antenna; D, maxillula; E, maxilla; F, detail of maxilla, different orientation; G, first swimming leg. Scales = 0.1 mm.

armed with three large teeth followed by a row of seven smaller ones, only slightly shorter than basis.

Maxilliped (fig. 10A) prehensile but not very large, three-segmented, composed of syncoxa, basis, and one-segmented endopod. Syncoxa trapezoidal, 1.2 times as long as its greatest width, unornamented and unarmed. Basis with inflated proximal inner corner, 1.5 times as long as wide, unornamented and armed with one heavily armed seta laterally, which is as long as the maxilliped, and one finely bipinnate, strong seta at distal inner corner, which is about as long as the basis. Endopodal segment small and conical, unornamented, and armed with one slender and two strong setae; slender seta and one strong seta smooth, other strong seta finely bipinnate. Longest seta on endopod 3.4 times as long as segment and about as long as shorter seta on basis.

First, second, and third swimming legs (figs. 9G, 10B, C) with three-segmented exopod and two-segmented endopod (although its ancestral three-segmented state could still be recognized); fourth swimming leg (fig. 10D) with two-segmented exopod and the endopod absent; their armature formula as follows (legend: inner/outer spine or seta; inner/terminal/outer):

	Exopod			Endopod	
Segments	1	2	3	1	2
First leg	0/1	1/1	3/2/2	1/0	4/2/1
Second leg	0/1	1/1	4/2/2	1/0	4/2/1
Third leg	0/1	1/1	5/2/2	1/0	3/2/1
Fourth leg	0/1	2/2/2	-	-	-

Third exopodal segment spine formula 3.3.3.4 and third endopodal segment spine formula 1.3.3.0. All intercoxal sclerites without surface ornamentation and with almost straight distal margin. Coxae also unornamented; those of first, second, and third legs armed with strong and long, plumose seta at inner distal corner; coxa of fourth leg unarmed. Basis of each leg armed with slender seta at distal outer corner and also unornamented; long and curved inner margin makes large space between inner coxal seta and base of endopod. All exopodal and endopodal setae slender and plumose; those on fourth leg plumose only at distal third. Most exopodal and endopodal spines on first, second, and third leg with both margins sharply serrated; outer apical spine on third exopodal segment serrated along outer margin and pinnate along inner margin; all spines on these three legs short and stout. All exopodal spines on fourth leg long, unipinnate, and relatively slender. All

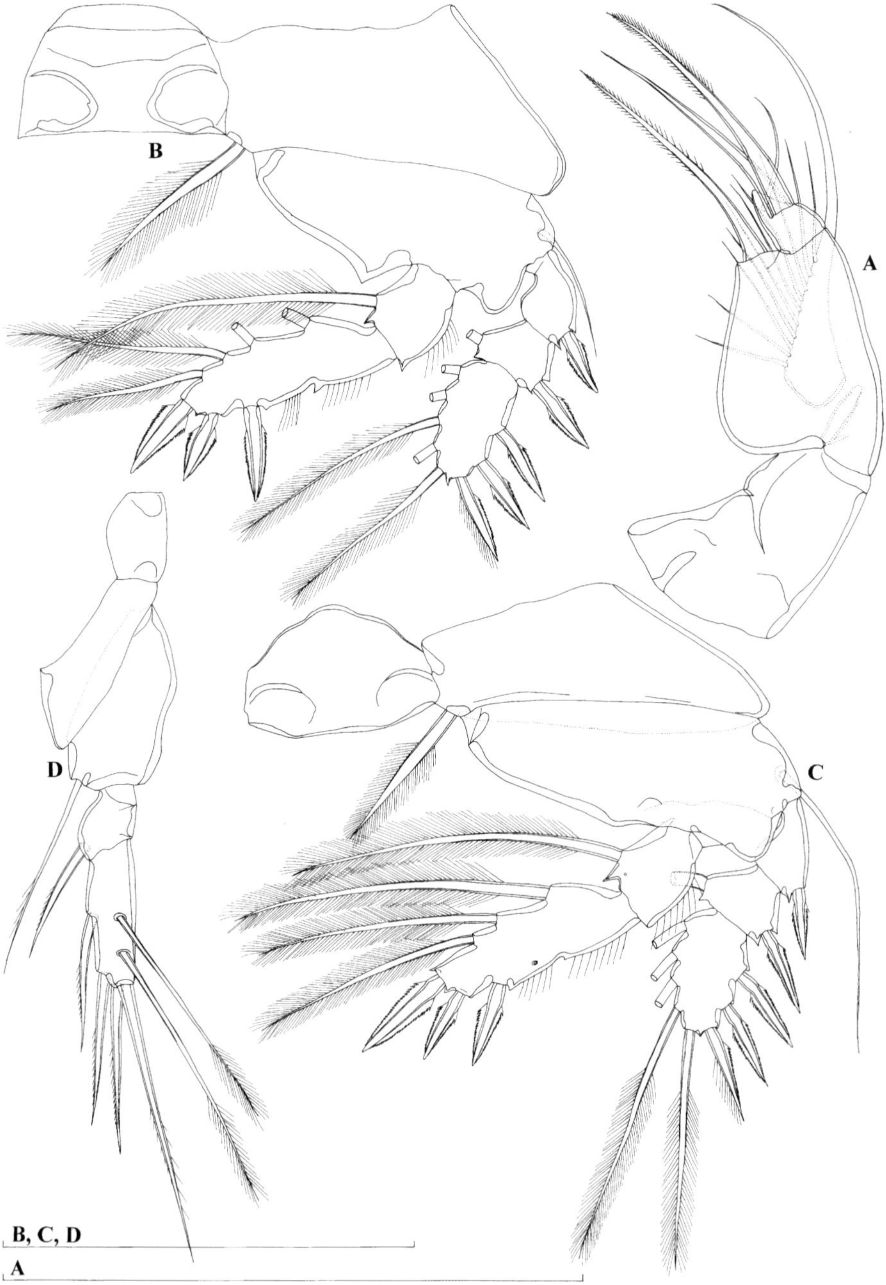

Fig. 10. *Polyankylis australis* n. sp., holotype female. A, maxilliped; B, second swimming leg; C, third swimming leg; D, fourth swimming leg. Scales = 0.1 mm.

endopodal segments with hair-like spinules along outer margins; endopod of third leg additionally ornamented with two pores; exopods unornamented.

Fifth leg (figs. 8B, 9A) without intercoxal sclerite, attached laterally to fifth pedigerous somite, two-segmented, but basal segment completely fused to somite, and one seta attached dorsally to somite remains as its only evidence. Second (the only free) segment small, 1.8 times as long as wide, unornamented, and armed apically with inner spine and outer seta; seta 2.2 times as long as spine and 3.3 times as long as segment.

Sixth leg (fig. 9A) not clearly distinct, cuticular plate, strongly chitinized, ornamented with a single pore, and armed with a single smooth and relatively long seta.

Male. Unknown.

Variability. — Body length of females ranging from 0.608 mm to 0.654 mm (0.625 mm average; n = 3). No other type of variability was observed and males were not collected.

Etymology. — The species is named after the Latin adjective "australis", meaning "southern" and referring both to the country (Australia) and the state (South Australia) of its currently known distribution.

Distribution. — At present, *Polyankylis australis* n. sp. is known from the type locality only.

Discussion. — *Polyankylis australis* n. sp. is only the second representative of the family Polyankyliidae Ho & Kim, 1997 described so far. The first was *Polyankylis orientalis* Ho & Kim, 1997, reported from an intertidal pool in Korea. For that species, Ho & Kim (1997) established a new genus and a new family in the lichomolgoid complex of the poecilostomatoid copepods (sometimes regarded as the superfamily Lichomolgoidea Humes & Stock, 1972). The family was accepted as valid by Boxshall & Halsey (2004), but it is fair to note that all ten currently recognized lichomolgoid families are pretty closely similar, and are chiefly found as symbionts of other marine invertebrates.

Polyankylis australis n. sp. fits into the generic diagnosis of *Polyankylis* Ho & Kim, 1997 perfectly, and could be distinguished from its type species (*P. orientalis*) only by a few morphological characters. These are as follows: more robust habitus from dorsal view; somewhat longer dorsal and innermost apical setae on the caudal rami; longer ultimate segments of the antennula and antenna; absence of a recurved hook on the first antennal segment; two large elements on the ultimate maxillipedal segment in the female (vs. one only in *P. orientalis*); and a completely divided exopod of the fourth swimming leg, which also bears two additional inner setae on the second segment. Unfortunately, males of *P. australis* are still unknown, so the male characters could not be compared.

Although the second representative of the Polyankyliidae was found on a different continent and even in a different hemisphere, it is interesting to note that both species were found at similar latitudes (at around 37°), which may suggest their preference for moderate to cold regions. However, these are probably some of the last refuges for this rare group of poecilostomatoid copepods, allegedly displaced from tropical and temperate regions by more recent and possibly more adaptive forms.

Family CORYCAEIDAE Dana, 1852
Genus **Onychocorycaeus** M. Dahl, 1912

4. **Onychocorycaeus pacificus** (Dahl, 1894)
(figs. 11-13)

Synonymy. —
Corycaeus pacificus Dahl, 1894: 73 (fide Dahl, 1912); Farran, 1936: 136; Wilson, 1950: 195.
Corycaeus (*Onychocorycaeus*) *pacificus* - Dahl, 1912: 103, pl. 14 figs. 1-10; Sewell, 1947: 285; Tanaka, 1957: 95, pl. 9 figs. 13-29; Kang et al., 1990: 53, pl. 4 figs. A-D.
Onychocorycaeus pacificus - Boxshall & Halsey, 2004: 494.
Corycaeus obtusus Farran, 1911: 291, pl. 12 figs. 4-6.
Corycaeus catus Mori, 1937: 134, pl. 74 figs. 1-7.

Material examined. — One female dissected on 2 slides (WAM C37210), Australia, Western Australia, Port Hedland, 4 August 2003, leg. T. Karanovic, 20°18.553′S 118°35.359′E.

Supplementary description. — Female from Port Hedland. Body length, excluding caudal setae, 1.121 mm. Habitus (fig. 11A) very robust, with prosome/urosome ratio 2.3 and greatest width at posterior end of cephalothorax. Body length/width ratio about 2.5; prosome rocket-shaped in dorsal view, about 1.9 times as long as wide and 2.9 times as wide as genital double somite. Specimen transparent, collected long dead, with all soft tissues disintegrated. Nauplius eye not visible, but large, paired cuticular lenses located on anterior part of cephalothorax. Rostrum broadly rounded and furnished with several large pores.

Cephalothorax (fig. 11A) almost triangular, about as long as its greatest width, representing 41% of total body length, with posterior corners only

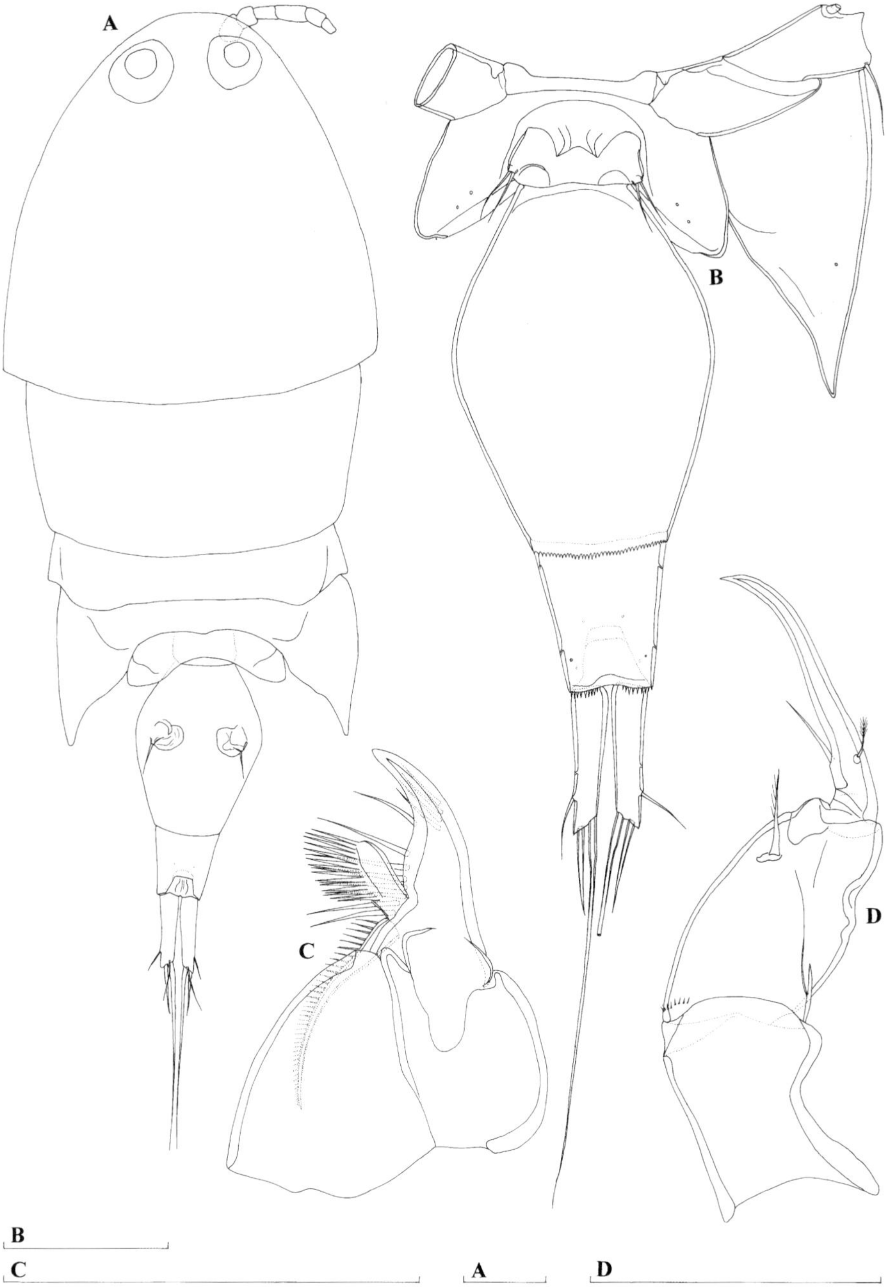

Fig. 11. *Onychocorycaeus pacificus* (Dahl, 1894), female from Port Hedland (1.121 mm). A, habitus, dorsal view; B, urosome, ventral view; C, maxilla; D, maxilliped. Scales = 0.1 mm.

slightly expanded laterally. Surface of cephalic shield with numerous small sensilla and cuticular pores, as also applies for the other prosomites. Hyaline fringe of prosomites narrow and smooth. First pedigerous somite not fused to cephalosome, without any lateral expansions, longer than other four pedigerous somites combined, and 0.4 times as long as cephalothorax. Second pedigerous somite with only slightly expanded and flattened lateral corners, looking as very small wings. Posterior corners of third pedigerous somite expanded laterally and posteriorly as robust epimeral plates, that terminate very sharply and reach the genital apertures when considered along the length axis of the animal. Fourth pedigerous somite with much less expanded, and rounded posterior lateral corners. Epimeral plates of third and fourth pedigerous somites ornamented with pores even ventrally (fig. 11B), two on fourth and one on third somite. Fifth pedigerous somite very small, unornamented, and without pronounced lateral expansions, much narrower than genital double somite and partly fused dorsally to fourth pedigerous somite. No arthrodial membrane between fourth and fifth pedigerous somites, or sclerotized joint between fifth pedigerous somite and genital double somite (fig. 11A, B).

Genital double somite (figs. 11B, 12A) laterally expanded at middle, 1.4 times as long as wide, ornamented with only four dorsal pores, with hyaline fringe smooth dorsally and finely serrated ventrally. No cuticular recesses, windows, or sensilla visible. Copulatory pores and ovipores lie together within dorsolateral genital apertures, which are situated at middle of genital double somite and covered by reduced sixth legs. Copulatory duct short and seminal receptacle small, kidney-like in shape, and heavily sclerotized. The anal somite (figs. 11B, 12A) is the only remaining free abdominal somite, it tapers posteriorly so that the anterior part is 1.8 times as wide as the posterior margin, with smooth, narrow, very short, and almost straight anal operculum, which represents 31% of the somite's width and reaches to 2/3 the somite's length; ornamented with two large sensilla and four pores dorsally, two pores laterally on each side, two pores ventrally, and a transverse row of spinules on the posterior margin ventrally; its posterior margin finely serrated laterally and dorsally. Anal sinus smooth and widely open dorsally. Anal somite 0.4 times as long as genital double somite and only slightly longer than wide.

Caudal rami (figs. 11A, B, 12A) cylindrical, but with "diagonally chopped" posterior margin (outer corner reaches further than inner one), 4.6 times as long as wide (dorsal view), slightly divergent, with very small space between them; they represent almost 8% of the total body length. Distal margin dorsally serrated, ventrally smooth, and without visible medial cuticular pore. Ornamentation consisting of a single lateral cuticular pore; no lateral sensillum visible. Dorsal seta about 0.7 times as long as ramus, inserted on small protuberance that reaches posterior border and is situated close to inner dorsal corner; this seta uniarticulate at base and smooth. Lateral seta the shortest one, about 1.7 times as long as ramus' width, inserted at 4/5 of ramus' length. Only two apical setae present, both smooth and without breaking planes; inner apical seta 4.4 times as long as outer one and 2.7 times as long as ramus.

Antennula (fig. 12B) very short when compared to cephalothorax, six-segmented, unornamented, with setal formula: 2.7.3.3.3.6. No setae with breaking planes or articulating on basal part, and all setae smooth; each segment with one very slender and relatively short seta, but two apical setae about as long as antennula; no aesthetascs visible. Length ratio of antennular segments, from proximal end, and along caudal margins, 1 : 1.3 : 1.6 : 2.4 : 1.2 : 1.

Antenna (fig. 12C, D) four-segmented, comprising coxobasis and three-segmented endopod, strongly geniculate; length ratio of antennal segments, 1 : 1.9 : 0.5 : 0.7. Coxobasis quadriform, slightly wider than long, unornamented and armed with robust, prehensile, and smooth spine, that reaches beyond posterior tip of antenna. First endopodal segment twice as long as wide, armed with one smooth and short spine basally, and with two strong teeth on inner margin distally. Second endopodal segment almost triangular, very short, unornamented, and armed with one large and one small claw. Third endopodal segment almost quadriform, 1.6 times as long as wide, unornamented, armed with two small claws laterally and one large claw apically, and one minute smooth seta; only apical claw distinct at base. Apical claw on third endopodal segment only slightly longer than larger claw on second segment and 1.1 times as long as basal spine on first endopodal segment.

Labrum, mandibula, and maxillula not mounted satisfactorily to allow detailed examination.

Maxilla (fig. 11C) conical and two-segmented, comprising large syncoxa and much smaller basis. Syncoxa unarmed and unornamented, about 1.3 times as wide as long. Basis produced into geniculate, spiniform, and spinulose process, armed with two also very spinulose elements on inner margin; proximal element more slender and about 2.5 times as long as distal one.

Maxilliped (fig. 11D) prehensile and not very robust, three-segmented, composed of syncoxa, basis, and one-segmented endopod. Syncoxa quadri-

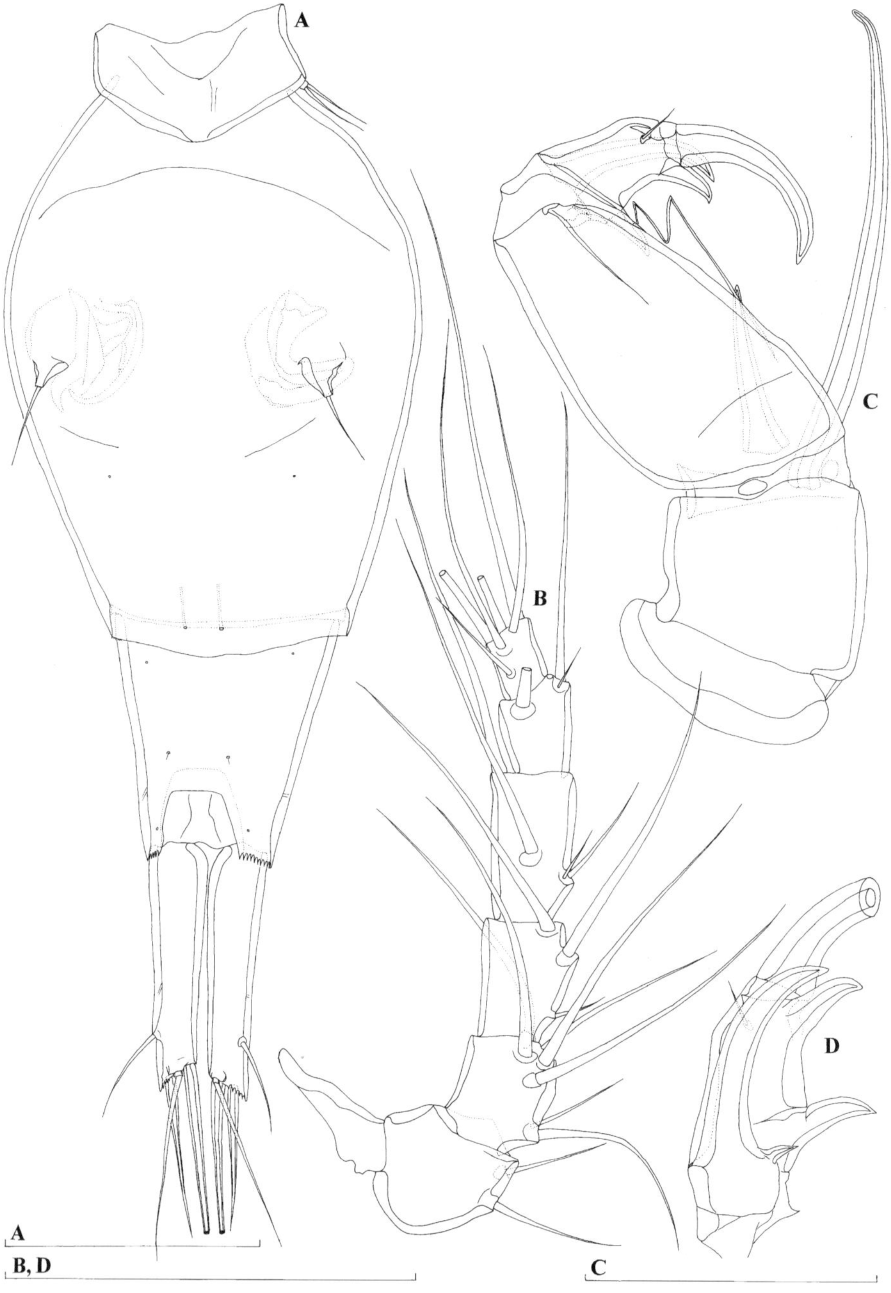

Fig. 12. *Onychocorycaeus pacificus* (Dahl, 1894), female from Port Hedland (1.121 mm). A, urosome, dorsal view; B, antennula; C, antenna; D, detail of antenna, different orientation. Scales = 0.1 mm.

form, 1.3 times as long as greatest width, unornamented and unarmed. Basis 1.9 times as long as wide and 1.3 times as long as syncoxa, ornamented with a short row of small spinules basally, and armed with a single unipinnate inner seta at 2/3. Endopod completely incorporated in terminal smooth and geniculate claw, which is about as long as basis and armed with two short setae; inner seta smooth, about twice as long as outer plumose seta, and somewhat shorter than only seta on basis.

All swimming legs (fig. 13A, B, C, D, E) with three-segmented exopod; endopod of first, second, and third leg also three-segmented, while endopod of fourth leg one-segmented; their armature formula is as follows (legend: inner/outer spine or seta; inner/terminal/outer):

	Exopod			Endopod		
Segments	1	2	3	1	2	3
First leg	0/1	1/1	4/2/2	1/0	1/0	3/2/0
Second leg	0/1	1/1	5/2/2	1/0	2/0	3/1/0
Third leg	0/1	1/1	5(3)/2/2	1/0	2/0	2/0/0
Fourth leg	0/0	1/1	5/1/1	0/1/0	-	-

Third exopodal segment spine formula 4.4.4.2 and third endopodal segment spine formula 0.1.0.0. All intercoxal sclerites with concave distal margin and without any surface ornamentation; intercoxal sclerite of fourth leg very wide. Coxae also unornamented; those of first and second leg armed with plumose seta at inner distal corner; those of third and fourth leg unarmed. Basis of each leg also unornamented; basis of first, third, and fourth legs armed with slender seta at distal outer corner, basis of second leg unarmed; basis of fourth leg inflated and very large as compared to other segments; no spine at distal inner corner of first leg. All exopodal and endopodal setae slender and plumose, and all spines stout and sharply serrated. Endopodal segments of first, second, and third legs with hair-like spinules along outer margins, which are also present on inner margin of first exopodal segment; no other ornamentation visible. Endopod of fourth leg minute, about twice as long as wide, unornamented, and armed with a single apical seta. Third endopodal segment of third leg without apical armature but with rounded spinulose tip.

Fifth leg (fig. 11B) reduced to minute segment, basally largely fused to somite, attached ventrally to its lateral corner, and armed apically with two equally long, smooth setae.

Fig. 13. *Onychocorycaeus pacificus* (Dahl, 1894), female from Port Hedland (1.121 mm). A, first swimming leg; B, second swimming leg; C, basis and rami of third swimming leg; D, coxa and intercoxal sclerite of third swimming leg; E, fourth swimming leg. Scales = 0.1 mm.

Sixth leg (fig. 12A) a not clearly distinct, triangular cuticular plate, strongly chitinized, unornamented, and armed apically with a single smooth seta, which is about as long as the setae on the fifth leg.

Variability. — Only one female was collected and examined. One of the third swimming legs of that specimen has only three setae along the inner margin of the third exopodal segment (fig. 13C), while the opposite appendage is normally built, with five inner setae. No other type of asymmetry was observed.

Distribution. — *Onychocorycaeus pacificus* (Dahl, 1894) is widely distributed in the Pacific and Indian oceans. It was recorded in Australia by Farran (1936) from four stations outside the Great Barrier Reef.

Discussion. — The family Corycaeidae Dana, 1852 was last revised by Dahl (1912), who recognized a single genus and seven subgenera, and provided illustrations and keys to all species. Seven subgenera were accorded full generic status by Boxshall & Halsey (2004), "pending phylogenetic revision of the whole family at the generic level". They also provided a key to all 39 valid species of this family. Corycaeidae are marine planktonic forms that typically inhabit the epipelagic zone of the world's oceans. The two species presented in this monograph were almost certainly buried in the sand by the wave action, as they were collected dead, with all soft tissues completely disintegrated.

The genus *Onychocorycaeus* M. Dahl, 1912 still numbers only seven species, as it did when it was first established as a subgenus of *Corycaeus* Dana, 1846 by Dahl (1912): *O. agilis* (Dana, 1849); *O. catus* (Dahl, 1894); *O. giesbrechti* (Dahl, 1894); *O. latus* (Dana, 1849); *O. ovalis* (Claus, 1863); *O. pacificus* (Dahl, 1894); and *O. pumilus* (Dahl, 1912). They are small forms (up to 1.18 mm long), with a large prosome, short caudal rami, and a one-segmented endopod of the fourth swimming leg, armed with a single apical seta.

Onychocorycaeus pacificus (Dahl, 1894) could easily be distinguished from its congeners by the rounded (not pointed) posterolateral corners of its fourth pedigerous somite. It differs from *O. pumilus* additionally by its size and prosome/urosome proportions, from *O. agilis* and *O. giesbrechti* by a much shorter anal somite, and from *O. latus* and *O. catus* by the length and proportions of the caudal rami and the genital double somite. *Onychocorycaeus ovalis* seems to be most closely related to *O. pacificus* from all congeners, as was already noted by Sewell (1947), who also speculated about their possible synonymy. However, he did not formally synonymize them, nor did Boxshall & Halsey (2004), so I treat them as two valid species. None of the synonyms of *O. pacificus* given here are new: *Corycaeus obtusus* Farran was synonymized by Dahl (1912) and *Corycaeus catus* Mori was recognized as a synonym of *O. pacificus* by Tanaka (1957).

5. **Onychocorycaeus catus** (Dahl, 1894) (figs. 14-16)

Synonymy. —

Corycaeus catus Dahl, 1894: 72 (fide Dahl, 1912: 99); Farran, 1911: 290, pl. 12 figs. 1-3; Farran, 1936: 138.

Corycaeus (*Onychocorycaeus*) *catus* - Dahl, 1912: 99, pl. 13 figs. 17-24; Sewell, 1947: 284; Tanaka, 1957: 94; ? Vilela, 1968: 40, pl. 15 fig. 3a-c; Kang et al., 1990: 54, pl. 4 figs. E-I; ? Campos-Hernández & Suárez-Morales, 1993: 122, fig. 3G-J.

Onychocorycaeus catus - Boxshall & Halsey, 2004: 494.

Corycaeus latus - Mori, 1937: 136, pl. 74 figs. 8-10.

[non] *Corycaeus catus* Mori, 1937: 134, pl. 74 figs. 1-7.

Material examined. — One female dissected on 1 slide (WAM C37211), Australia, Queensland, Alva (near Ayr), 12 August 2003, leg. T. Karanovic, 19°27.232′S 147°28.924′E.

Supplementary description. — Female from Alva. Body length, excluding caudal setae, 1.060 mm. Habitus (fig. 14A) very robust and similar to previous species. Specimen transparent, collected long dead (with all soft tissues disintegrated) and damaged, so complete habitus could not be drawn, nor body proportions measured. Nauplius eye not visible, but large, paired cuticular lenses located on anterior part of cephalothorax. Integument not strongly chitinized. Rostrum (fig. 14D) broadly rounded and furnished with nine large pores.

Cephalothorax about as long as its greatest width, with posterior corners not expanded laterally, damaged (torn) dorsally. Surface of cephalic shield with numerous small sensilla and cuticular pores, as also applies to the other prosomites (fig. 15A). Hyaline fringe of prosomites narrow and smooth. First pedigerous somite not fused to cephalosome, without any lateral expansions, shorter than in previous species, about half as long as other four pedigerous somites combined. Second pedigerous somite with only slightly expanded and pointed lateral corners (fig. 14A). Posterior corners of third pedigerous somite expanded laterally and posteriorly as robust epimeral plates, that terminate very sharply and reach the level of the genital apertures. Fourth pedigerous somite with much less expanded and shorter, but also sharply

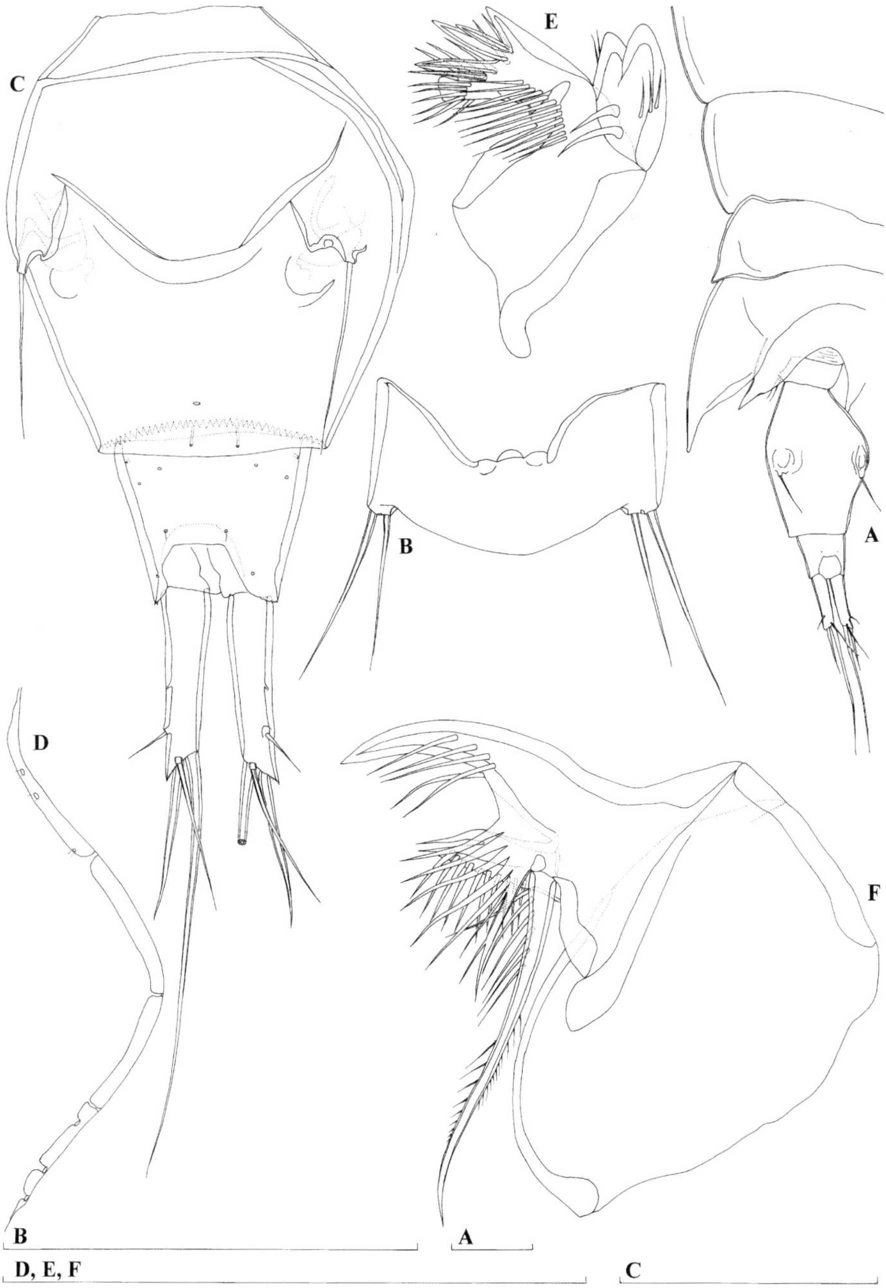

Fig. 14. *Onychocorycaeus catus* (Dahl, 1894), female from Alva (1.060 mm). A, urosome and left part of prosome, dorso-lateral view; B, fifth pedigerous somite and fifth legs, ventral view; C, abdomen, dorsal view; D, rostrum; E, mandibula; F, maxilla. Scales = 0.1 mm.

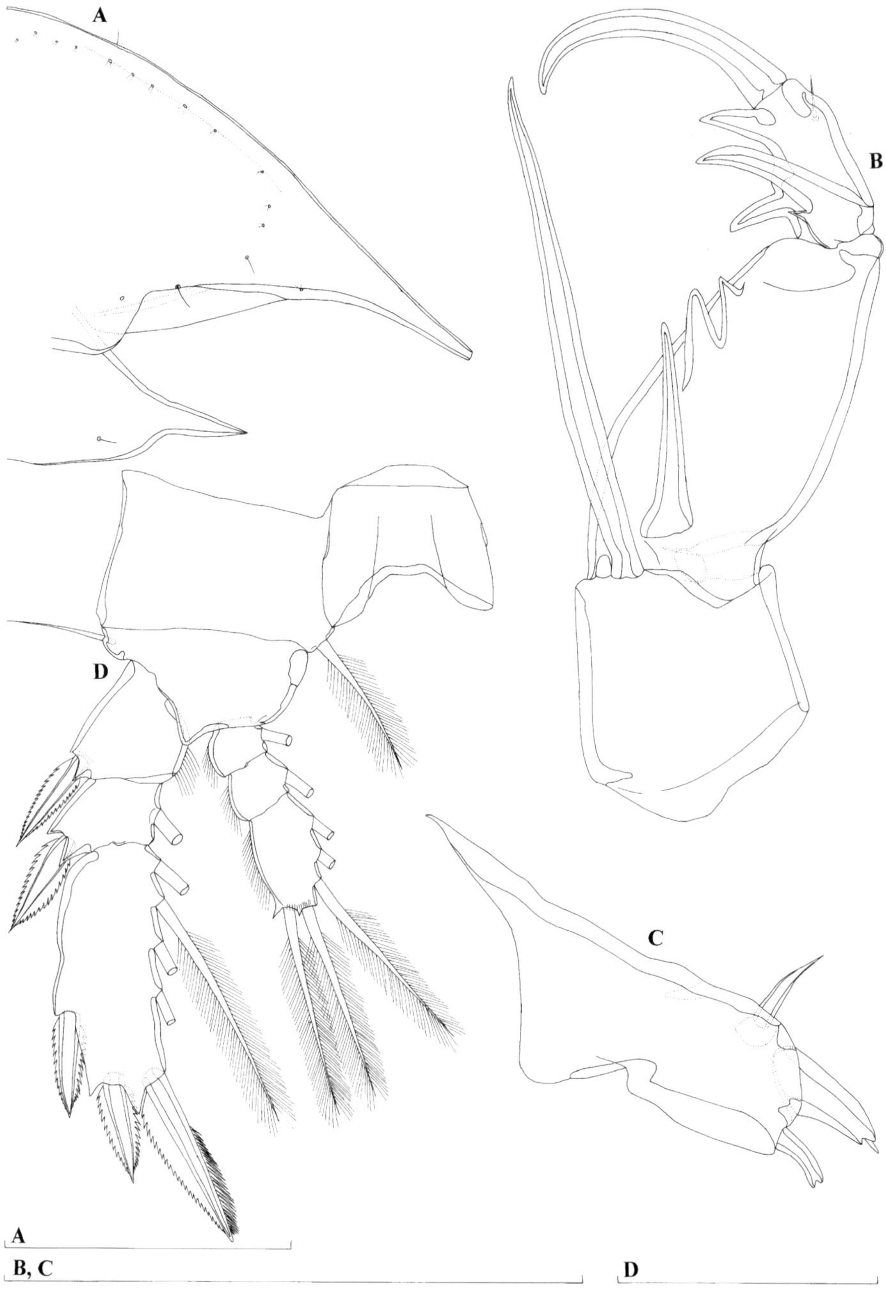

Fig. 15. *Onychocorycaeus catus* (Dahl, 1894), female from Alva (1.060 mm). A, pleural corners of fourth and fifth pedigerous somites; B, antenna; C, maxillula; D, first swimming leg. Scales = 0.1 mm.

pointed lateral corners (fig. 15A). Epimeral plates of third and fourth pedigerous somites ornamented with numerous pores and several sensilla (fig. 15A). Fifth pedigerous somite very small, unornamented, and without pronounced lateral expansions, much narrower than genital double somite, but not partly fused to fourth pedigerous somite. Arthrodial membrane between fourth and fifth pedigerous somites visible from dorsal view. No sclerotized joint between fifth pedigerous somite and genital double somite.

Genital double somite (fig. 14C) laterally expanded in middle, only slightly longer than wide, ornamented with only three dorsal pores, with hyaline fringe smooth dorsally and serrated ventrally. No cuticular recesses, windows, or sensilla visible. Copulatory pores and ovipores lie together inside dorsolateral genital apertures, which are situated at middle of genital double somite and covered by reduced sixth legs. Copulatory duct short and seminal receptacle small, but not all clearly visible. Anal somite (fig. 14C), the only remaining free abdominal somite, tapers posteriorly so that anterior part is 1.6 times as wide as posterior margin, with smooth, narrow, very short, and almost straight anal operculum, which represents 33% of the somite's width and reaches to 2/3 of the somite's length; ornamented with two sensilla and six pores dorsally, two pores laterally on each side, two pores ventrally, and a transverse row of spinules on posterior margin ventrally; the posterior margin finely serrated laterally and dorsally. Anal sinus smooth and widely open dorsally. Anal somite shorter than in previous species, about 0.8 times as long as wide and 0.4 times as long as genital double somite.

Caudal rami (fig. 14C) cylindrical but with "diagonally chopped" posterior margin (outer corner reaches further than inner one), four times as long as wide (dorsal view), slightly divergent, with very small space between them; they represent almost 8% of total body length. Distal margin dorsally serrated, ventrally smooth, and without visible medial cuticular pore. The ornamentation consists of a single lateral cuticular pore; no lateral sensillum visible. Dorsal seta about 0.8 times as long as ramus, inserted almost on posterior border and close to inner dorsal corner; uniarticulate at base, and smooth. Lateral seta shortest one, about as long as ramus' width, inserted at 3/4 of ramus' length. Only two apical setae present, both very stout, smooth, and without breaking planes; inner apical seta 2.7 times as long as outer one and 2.2 times as long as ramus.

Antennula same as in previous species.

Antenna (fig. 15B) also very similar to that of previous species, with only slightly longer apical claw on third endopodal segment.

Labrum not mounted satisfactorily to allow detailed examination.

Mandibula (fig. 14E) small, very simple, and about 1.6 times as long as wide, with two smooth and short setae as well as two large, bifurcate, and heavily spinulose blades on distal margin. Ornamentation consisting of a diagonal row of 12 large spinules at base of the larger and less chitinized blade.

Maxillula (fig. 15C) also very small and simple, unornamented, about 2.5 times as long as wide, armed with three strong and smooth elements on distal margin, two of which with bifurcate tip.

Maxilla (fig. 14F) similar to that of previous species, conical and two-segmented, comprising large syncoxa and much smaller basis. Syncoxa unarmed and unornamented, almost 1.7 times as wide as long. Basis produced into geniculate, spiniform, and spinulose process, armed with two also very spinulose elements on inner margin; proximal element simple, more slender, and about 2.4 times as long as distal one, which has a bifurcate tip.

Maxilliped damaged and not observed completely, but similar to that of previous species.

All swimming legs (figs. 15D, 16A, B, C) with three-segmented exopod; endopod of first, second, and third leg also three-segmented, that of fourth leg one-segmented; their armature formula as follows (legend: inner/outer spine or seta; inner/terminal/outer):

		Exopod			Endopod	
Segments	1	2	3	1	2	3
First leg	0/1	1/1	4/2/1	1/0	1/0	3/2/0
Second leg	0/1	1/1	5/2/1	1/0	2/0	3/1/0
Third leg	0/1	1/1	5/2/1	1/0	2/0	2/0/0
Fourth leg	0/1	1/0	5/1/1	0/1/0	-	-

Third exopodal segment spine formula 3.3.3.2 and third endopodal segment spine formula 0.1.0.0. All intercoxal sclerites with concave distal margin and without any surface ornamentation; intercoxal sclerite of fourth leg very wide. Coxae also unornamented; those of first and second leg armed with plumose seta at inner distal corner; those of third and fourth leg unarmed. Basis of each leg also unornamented, except for one pore on third leg; basis of first, third, and fourth legs armed with slender seta at distal outer corner, basis of second leg unarmed; basis of fourth leg inflated and large when compared to other segments; no spine at distal inner corner of first leg. All exopodal and endopodal setae slender and plumose and all spines stout and sharply

Fig. 16. *Onychocorycaeus catus* (Dahl, 1894), female from Alva (1.060 mm). A, second swimming leg; B, third swimming leg; C, fourth swimming leg. Scale = 0.1 mm.

serrated. Endopodal segments of first, second, and third leg with hair-like spinules along outer margins, which are also present on inner margin of first exopodal segment; no other ornamentation visible. Endopod of fourth leg minute, about twice as long as wide, unornamented and armed with a single apical seta. Third endopodal segment of third leg without apical armature but with rounded spinulose tip.

Fifth leg (fig. 14B) reduced to minute segment, basally largely fused to somite and attached ventrally to its lateral corner, armed apically with two smooth setae; outer seta about 1.3 times as long as inner one.

Sixth leg (fig. 14C) not clearly distinct, triangular cuticular plate, strongly chitinized, unornamented, and armed apically with a single smooth seta, which reaches the posterior margin of the genital double somite.

Variability. — Only one female was collected and examined, and no asymmetry was observed.

Distribution. — This is also an Indo-Pacific species and it was recorded in Australian waters from the Great Barrier Reef and from Christmas Island (Sewell, 1947). Reports of this species from the Atlantic Ocean (Vilela, 1968; Campos-Hernández & Suárez-Morales, 1993) have yet to be verified.

Discussion. — *Onychocorycaeus catus* (Dahl, 1894) is another planktonic species, buried in the sand by the wave action. Although the only female specimen was partly damaged and had an abnormal armature on the swimming legs, it was easily identified as *O. catus* due to its short anal somite (which is shorter than the caudal rami), long genital double somite (about 3.5 times as long as the anal somite), the very long and sharply pointed epimeral plates of the third and fourth pedigerous somites, and the very long seta on the sixth leg. It differs from the previous species (*O. pacificus* (Dahl, 1894)) also by the length of the apical spine on the third exopodal segment of the fourth swimming leg, which is much longer than the segment (while being shorter than the segment in *O. pacificus*).

Onychocorycaeus catus is morphologically very similar to the Mediterranean *O. ovalis* (Claus, 1863) and the Indo-Pacfic *O. pumilus* (Dahl, 1912), as also noted by Sewell (1947). However, the latter species is much smaller in size (females are up to 0.75 mm long) and has a much more robust prosome, while the former has somewhat shorter caudal rami and a shorter (single) seta on the female sixth leg, as well as a longer anal somite, and less pointed epimeral plates of the third and fourth pedigerous somites. It is interesting to note that Dahl (1912, pl. 12 fig. 28) reported *O. pumilus* with an abnormal armature of the second swimming leg, which is very similar to that of my specimen of *O. catus*, as it only has three spines on the third exopodal segment. This certainly is an abnormality, also occurring on the first and third legs of my specimen. The armature of the first three swimming legs is very conservative in the family Corycaeidae Dana, 1852 and thus useless for species identification, although anomalies seem to be more frequent than usually reported (see also the variability section of the previous species). Certainly, the reduced number of exopodal spines is not a specific character of *O. pumilus*, as Sewell (1947) reported two females of this species from the Arabian Sea with normal armature on the swimming legs.

The specimen of *O. catus* studied in this monograph is the largest ever reported (1.06 mm), although only slightly outside the known specific range. Dahl (1912) reported a female length of 0.95 mm, while Farran (1936) reported 0.92-1.01 mm, Sewell (1947) 0.80-0.867 mm, Tanaka (1957) 0.93-1.00 mm, Vilela (1968) 0.89-0.96 mm, and Kang et al. (1990) reported female specimens ranging between 0.87 and 0.95 mm in length. Interestingly, the largest previously reported specimen (1.01 mm) was recorded from the Great Barrier Reef (Farran, 1936), which is also the previously known locality of this species closest to the one reported in this monograph.

TAXONOMIC ACCOUNT:
ORDER CYCLOPOIDA RAFINESQUE, 1815

Family OITHONIDAE Dana, 1853
Genus **Oithona** Baird, 1843

6. **Oithona nana** Giesbrecht, 1893
(figs. 17-19)

Synonymy. —

Oithona nana Giesbrecht, 1893: 538, pl. 4 fig. 8, pl. 34 figs. 10, 11, 20, 24-26, 34, 35, pl. 44 figs. 2, 4, 6 (fide Nishida, 1985: 52); Esterly, 1905: 209, fig. 51a-c; Farran, 1913: 186; Rosendorn, 1917: 40, fig. 24a-d; Sewell, 1924: 791; Früchtl, 1924: 92; Pesta, 1928: 76, fig. 58; Kiefer, 1929b: 9; Mori, 1937: 113, pl. 63 figs. 1-8; Nicholls, 1944: 15; Sewell, 1947: 254; Gonzalez & Bowman, 1965: 272, fig. 20c-g; Vilela, 1965: 12; Bossanyi & Bull, 1971: 15; Marcus, 1971: 71; Hamond, 1973: 339; Chen et al., 1974: 33, pl. 2 figs. 6-12; Marcus, 1975: 72; Nishida et al., 1977: 138, figs. 11a-h, 12a-h; Nishida, 1985: 52, figs. 27-32; Holmes, 2000: 4.

Oithonina nana - Sars, 1913: 5; Wilson, 1932: 316, fig. 190c, d.

Oithona attenuata Farran, 1913: 187, pl. 30 figs. 3-7; Roserdorn, 1917: 42, fig. 25a-h; Früchtl, 1924: 92; Kiefer, 1929b: 8; Farran, 1936: 124; Nicholls, 1944: 15; Sewell, 1947: 253; Lindberg, 1953: 6; Chen et al., 1974: 34, pl. 2 figs. 13-15, pl. 3 figs. 1-3; Nishida et al., 1977: 128, fig. 3a-e; Nishida, 1985: 21, figs. 5-10.

Oithona attenuata var. *latithoracica* Früchtl, 1923: 454; Früchtl, 1924: 93, figs. 47, 48.

Oithona plumosa Lindberg, 1947: 50, fig. 1e; Lindberg, 1950: 265, fig. 5a-h.

Oithona plumosa pseudoplumosa Lindberg, 1950: 265, figs. 5i, 6a-g.

Oithona ovata Lindberg, 1950: 261, figs. 3a-f, 4a-e.

Material examined. — One female dissected on 1 slide (WAM C37212), Australia, South Australia, Smoky Bay, 24 December 2002, leg. T. Karanovic, 32°22.735′S 133°56.006′E.

Supplementary description. — Female from Smoky Bay. Body length, excluding caudal setae, 0.638 mm. Specimen transparent, collected long dead (with almost all soft tissues disintegrated) and with both antennulae broken. Integument not strongly chitinized. Habitus (fig. 17A) of typical cyclopoid shape, slender, with prosome/urosome ratio 1.1 and greatest width at posterior end of cephalothorax. Body length/width ratio about 3.9; cephalothorax about 3.2 times as wide as genital double somite. Posterior corners of cephalothorax and pedigerous somites without pronounced lateral expansions. Fifth pedigerous somite narrower than genital double somite. Nauplius eye not visible.

Rostral expansion not very large, ventrally directed, and broadly rounded; rostrum reduced.

Prosome (fig. 17A) comprising cephalothorax and four free pedigerous somites. Cephalothorax slightly longer than its greatest width, representing 28% of total body length. Surface of cephalic shield with several pores and sensilla; no other ornamentation visible. Hyaline fringe of prosomites narrow and smooth. Fifth pedigerous somite ornamented with two dorsal pores only (fig. 19D). Sclerotized joint between fifth pedigerous somite and genital double somite well developed, almost as a pseudosomite (fig. 17B).

Genital double somite (fig. 17B) laterally expanded at middle, 1.8 times as long as wide, 1.8 times as long as third urosomite, twice as long as fourth urosomite, and almost three times as long as anal somite, ornamented with only one dorsal and two ventral pores, with hyaline fringe smooth both dorsally and ventrally. No cuticular recesses, windows, or sensilla visible. Copulatory pores and ovipores lie together inside lateral genital apertures, which are situated at 2/5 of genital double somite's length and covered by reduced sixth legs. Copulatory duct short and seminal receptacle large and ovoid; internal space between two seminal receptacles equal to one receptacle's width. Third urosomite ornamented with only two ventral pores, about 1.6 times as long as wide, with smooth hyaline fringe. Fourth urosomite 1.5 times as long as wide, also ornamented with only two ventral pores and with smooth hyaline fringe. Anal somite (fig. 17B) with smooth, relatively narrow, and slightly convex anal operculum, which represents 44% of the somite's width and reaches to its middle; ornamented with two large sensilla, two pores ventrally, and a transverse row of minute spinules on posterior margin. Anal sinus smooth and widely open dorsally.

Caudal rami (fig. 17B) cylindrical, 2.7 times as long as wide (ventral view), divergent, with very small space between them; they represent 6% of total body length. Distal margin ventrally with a row of minute spinules, with medial cuticular pore hardly visible. No other ornamentation visible. Dorsal seta broken on both rami, inserted on small protuberance reaching posterior border of ramus. Lateral seta arising at 2/5, smooth, and slightly longer than the ramus' width. Outermost apical seta slender and bipinnate distally, about 0.9 times as long as ramus and 0.6 times as long as innermost apical seta, which is also bipinnate and slender. Principal apical setae with breaking planes, and both broken.

Antennula (fig. 17C) very long and robust, but broken on both sides and only proximal six segments with incomplete armature observed. Their setal

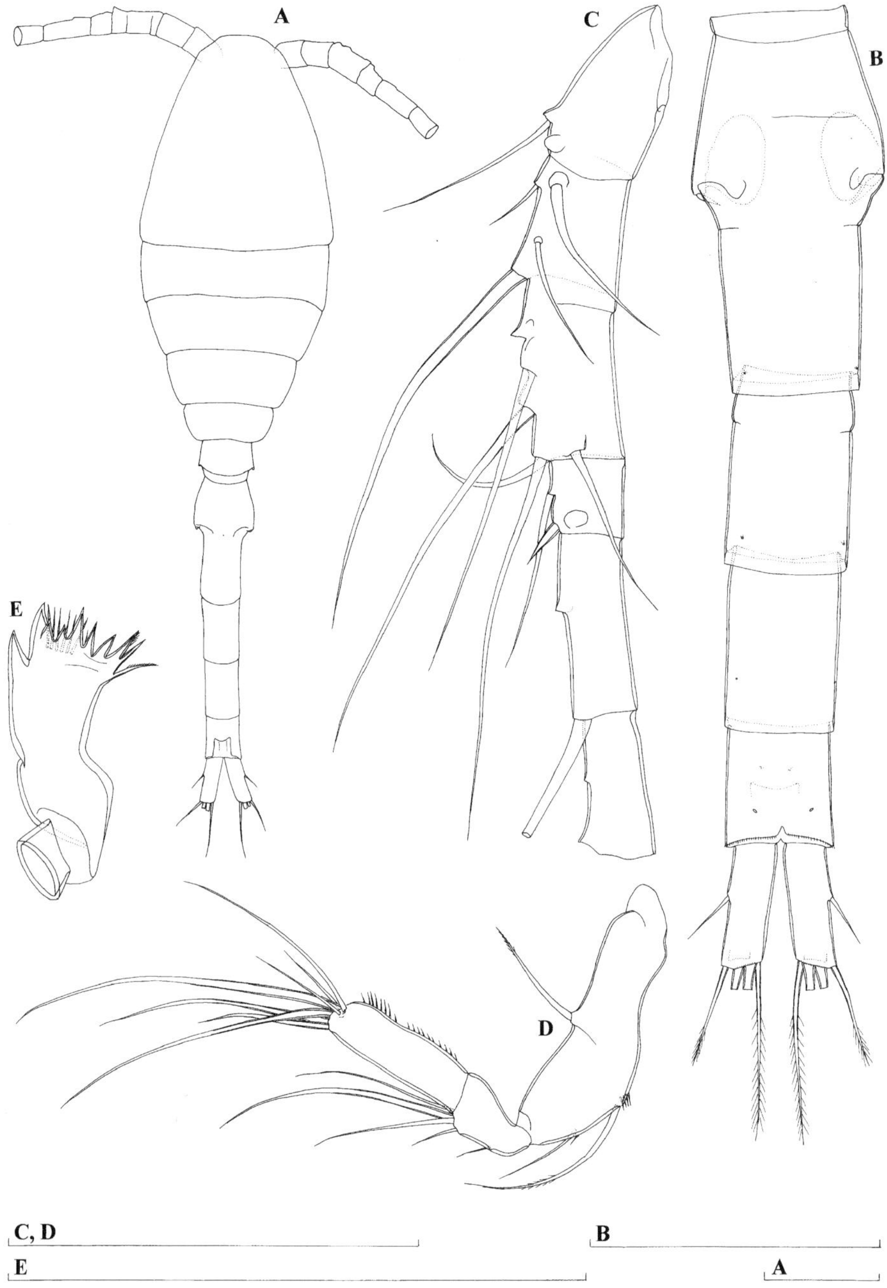

Fig. 17. *Oithona nana* Giesbrecht, 1893, female from Smoky Bay (0.638 mm). A, habitus, dorsal view; B, abdomen, ventral view; C, proximal part of antennula; D, antenna; E, coxa of mandibula. Scales = 0.1 mm.

formula (that also includes broken elements): 2.4.6.3.2.2. One seta on fourth segment short and spiniform, all other setae slender, and all setae smooth. Length ratio of first six antennular segments, from proximal end and along caudal margins, 1 : 1 : 1.1 : 0.6 : 1.3 : 1.1.

Antenna (fig. 17D) four-segmented, comprising coxobasis and three-segmented endopod, although coxobasis and first endopodal segment partly fused; antenna constricted and geniculate between first and second endopodal segment. Coxobasis slightly longer than first endopodal segment and about twice as long as wide, unornamented and armed with a single exopodal seta, which is bipinnate at the distal end and about as long as the coxobasis. First endopodal segment conical, 1.3 times as long as wide, armed with one bipinnate seta at middle, ornamented with a few spinules at base of seta. Second endopodal segment smallest, 1.5 times as long as wide, unornamented and armed with five smooth setae (three lateral, two subapical; one subapical seta geniculate, longer, and much stronger than any other seta on that segment). Third endopodal segment the longest, 3.1 times as long as wide and 2.5 times as long as second endopodal segment, armed with seven smooth setae apically (four of which strong and geniculate) and ornamented with a longitudinal row of spinules along outer (caudal) margin.

Labrum not mounted satisfactorily to allow detailed examination.

Mandibula (figs. 17E, 18A) comprising coxa with well developed gnathobase and distal palp; palp consisting of allobasis (incorporating first endopodal segment), one free endopodal segment, and a four-segmented exopod, although exopodal segments largely fused. Coxal gnathobase cutting edge clearly divided into three regions: inner group of four complex and strong teeth and six long spinules, innermost tooth shortest; middle group of three simple teeth, middle one unipinnate, other two smooth, and four spinules; and outermost, short and unipinnate seta. Allobasis large, unornamented and armed with one smooth seta laterally and one smooth and one strong and heavily armed geniculate element apically. Free endopodal segment minute, armed with four slender apical setae. Exopod half as long as allobasis, unornamented and armed with long, slender, and plumose setae; setal formula 1.1.1.2.

Maxillula (fig. 18B) composed of syncoxa with well developed arthrite, basis, and one-segmented endopod and exopod. Arthrite of syncoxa with eight elements; four apical elements strong, pinnate, and geniculate, proximalmost element also strong and pinnate, but not geniculate; other three lateral elements slender, one of them bipinnate at distal end, two smooth.

Fig. 18. *Oithona nana* Giesbrecht, 1893, female from Smoky Bay (0.638 mm). A, mandibula; B, maxillula; C, maxilla; D, maxilliped. Scale = 0.1 mm.

Coxa unornamented and armed with one smooth epipodal seta; coxal endite unarmed. Basis elongated, much longer than arthrite of syncoxa, about 2.6 times as long as wide, unornamented, and armed with three setae on proximal and one seta on distal endite; two setae on proximal endite very strong and geniculate, one of these also pinnate, third seta slender and smooth; only seta on distal endite slender and bipinnate, somewhat shorter than slender seta on proximal endite. Endopod small and short, unornamented, and armed with four short, slender, and bipinnate setae apically. Exopod larger than endopod, half as long as basis, and twice as long as endopod, unornamented, and armed apically with four slender, pinnate, and very long setae.

Maxilla (fig. 18C) large and five-segmented, comprising praecoxa, coxa, basis, and two-segmented endopod. Praecoxa quadriform, 1.5 times as wide as long, unornamented, with only one seta on distal endite, and unarmed proximal endite. Coxa twice as long as praecoxa, 1.5 times as long as wide, also unornamented, armed with three setae on proximal endite, two of which strong, and with long spinules along both margins; distal endite highly mobile and bearing two slender as well as one completely fused and very stout setae apically. Basis short, expanded into spinulose, smooth claw, and armed with one strong and one small setae; strong seta the longest on the maxilla, 2.4 times as long as basal claw, with long spinules proximally and much smaller spinules distally; small seta smooth and slender, 0.3 times as long as basal claw. First endopodal segment slightly shorter than basis, armed with three robust and heavily armed, geniculate setae. Second endopodal segment about as long as basis, 1.6 times as long as wide, armed with three strong and two slender setae.

Maxilliped (fig. 18D) slender, four-segmented, composed of syncoxa, basis, and two-segmented endopod; length ratios of segments, from proximal end, 1 : 0.6 : 0.4 : 0.1. Syncoxa 4.1 times as long as wide, unornamented, with setal formula on its endites 0.1.2.2; all setae short, strong, and heavily ornamented. Basis 3.3 times as long as wide, ornamented with longitudinal row of six strong spinules along inner margin, armed distally with two long and very robust setae. First endopodal segment 2.4 times as long as wide, unornamented, and armed with three setae along inner margin; distal one the strongest and longest on the maxilliped, 3.4 times as long as middle seta, 2.8 times as long as proximalmost seta, and twice as long as the segment. Second endopodal segment very small, slightly wider than long, unornamented, but armed with three setae; innermost seta ornamented with four long spinules, very long and strong, almost as long as longest seta on first endopodal segment, 1.7 times as long as middle seta, which is also strong but smooth, and almost 12 times as long as the outermost seta, which is smooth and slender.

All swimming legs (fig. 19A, B, C) with three-segmented exopod and endopod; their armature formula as follows (legend: inner/outer spine or seta; inner/terminal/outer):

	Exopod			Endopod		
Segments	1	2	3	1	2	3
First leg	1/1	1/1	3/2/3	1/0	1/0	3/2/1
Second leg	1/1	1/1	4/2/3	1/0	2/0	3/2/1
Third leg	1/1	1/1	4/2/3	1/0	2/0	3/2/1
Fourth leg	1/1	1/1	4/2/2	1/0	2/0	2/2/1

Third exopodal segment spine formula 4.4.4.3; no spines on endopods. All intercoxal sclerites without surface ornamentation and with rounded protrusions at both sides of distal margin, making it concave. Coxae also unornamented, armed with plumose seta at inner distal corner. Basis of first leg with spine inserted at inner, protruded corner, which does not reach distal margin of first endopodal segment, stout and bipinnate, ornamented with several long spinules along inner margin; basis of other legs unornamented, with inner distal corner pronounced as blunt spiniform process; each basis armed with slender seta on outer margin, which is exceptionally long on the first leg. All exopodal and endopodal setae slender and plumose. Most exopodal spines short and stout, with both margins sharply serrated; outer apical spine on third exopodal segment of each leg very long, with smooth inner and sharply serrated outer margin. All endopodal segments with hair-like spinules along outer margins; exopodal segments of second, third, and fourth leg with minute spinules along outer margins; no other ornamentation visible. All endopods shorter than exopods. Third endopodal segment of fourth swimming leg about 1.8 times as long as wide.

Fifth leg (fig. 19D) with proximal segment completely fused to somite, of which one lateral seta on fifth pedigerous somite its the only remnant (in the present specimen broken on both sides). The only free segment attached ventrally to lateral corner of somite, small and more than 3 times as long as wide, ornamented with four long spinules along inner margin distally, and armed with a single long and distally plumose seta apically.

Sixth leg (fig. 17B) not clearly distinct, cuticular plate, armed with a single smooth spine.

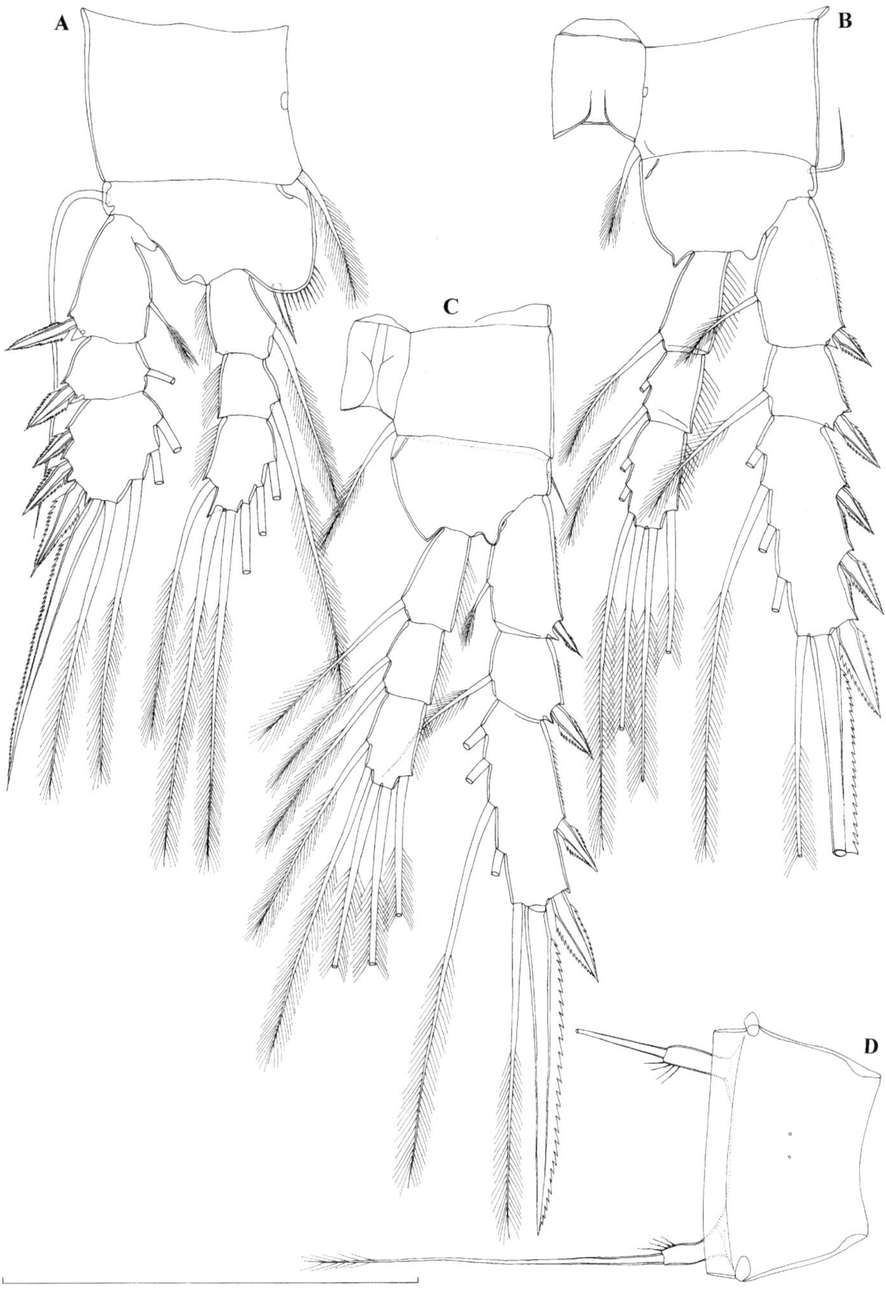

Fig. 19. *Oithona nana* Giesbrecht, 1893, female from Smoky Bay (0.638 mm). A, first swimming leg; B, third swimming leg; C, fourth swimming leg; D, fifth pedigerous somite and fifth legs, dorsal view. Scale = 0.1 mm.

Variability. — Only one female was collected and examined and no asymmetry was observed.

Distribution. — *Oithona nana* Giesbrecht, 1893 is widely distributed in the tropical and temperate regions of the Indian, Pacific, and Atlantic oceans, as well as in the Mediterranean Sea. It was recorded many times from Australian waters (see Farran, 1913; Nicholls, 1944; Lindberg, 1953; Nishida, 1985; McKinnon, 2000) and was reported from the littoral of South Australia by Nicholls (1944).

Discussion. — Of some 92 named species and subspecies of the family Oithonidae Dana, 1853, around 51 are considered valid species (Boxshall & Halsey, 2004) and these are classified into three accepted genera: *Dioithona* Kiefer, 1935 (with three valid species); *Limnoithona* Burckhardt, 1913 (with only two representatives); and *Oithona* Baird, 1843 (comprising the majority of the known species). The genus *Paroithona* Farran, 1908, that comprised oithonids with two-segmented endopods of the swimming legs, was rejected by Boxshall & Halsey (2004), based on an unpublished Ph.D. thesis of Abihay (fide Boxshall & Halsey, 2004). Oithonids are abundant members of the marine planktonic communities worldwide, and are very common in epipelagic and mesopelagic zones of the open ocean, in neritic waters, and in estuaries. Because of their abundance and high metabolic rates, oithonids are considered to play important roles in the biological production of the ocean, linking the lower trophic levels to higher ones (Nishida, 1985). Not a single species was reported from the interstitial, and the specimen presented here was certainly buried in the sand by the wave action, as it was collected with almost all soft tissues disintegrated and with both antennulae broken.

In Australian waters, the identification of oithonids is relatively easy, thanks to an excellent study of this family in the Indian and Pacific oceans by Nishida (1985), but estuaries and near-coastal waters may still contain some undescribed species as shown by McKinnon (2000), who described two new species from mangrove waters of Queensland. The present finding of *O. nana* Giesbrecht, 1893 in Smoky Bay is hardly surprising, as it was already reported from the littoral of South Australia by Nicholls (1944), who found this species sympatrically with *O. attenuata* Farran, 1913. Reports of these two species living together are too frequent to be coincidental. For example, Nishida (1985: 153) in his detailed study of nine inlets or embayment waters in the Indian and the Pacific oceans, always found them together if either one was present. This raised my suspicion about the specific status of

O. attenuata and prompted careful examination of my only specimen of *O. nana*.

Oithona nana stands well apart from other congeners, and G. O. Sars even tried to establish a new genus for it, as he wrote (Sars, 1913: 4): "One of the species described by Dr. Giesbrecht, *O. nana*, seems in some respects to differ more conspicuously from the others, and may perhaps be regarded as the type of a separate, though nearly allied genus, for which the name *Oithonina* may be proposed". The genus *Oithonina* G. O. Sars, 1913 was supported by Wilson (1932), but most copepodologists preferred to retain *O. nana* in the genus *Oithona*, together with its closely related congener, *O. attenuata*. Farran (1913) in his description of *O. attenuata* stated: "This species is closely allied to *O. nana* and differs mainly in the long drawn-out abdomen and furca and in the longer 1st antenna". Rosendorn (1917) stated that *O. attenuata* was closely allied to *O. nana* and was distinguishable from the latter only by the longer prosome and caudal rami. Later on, the prosome to urosome ratio was used as a main distinguishing character between the two species (see Kiefer, 1929b), but Nishida (1985) showed that this character cannot reliably be used. The latter author studied extensive material of both species from the Indian and Pacific oceans and found them also to be dimorphic. He described the "stocky" form of *O. attenuata* and the "plumosa" form of *O. nana*, that bridge the gap in the habitus appearance between the two "typical" forms. He (Nishida, 1985) used the appearance of the maxillula instead to distinguish the two species, and identified all specimens with a short dorsalmost apical element on the praecoxal arthrite as *O. nana*, and those with a long element as *O. attenuata*. McKinnon supported this character, when he wrote (McKinnon, 2000: 105): "The occurrence of this character in *O. attenuata* implies that it has been secondarily derived in this species, given its close relationship to *O. nana* Giesbrecht, 1892 (Nishida, 1985)".

However, as my specimen has the dorsalmost apical element on the praecoxal arthrite of the maxillula shorter than in *O. attenuata* and longer than in *O. nana*, I formally synonymize these species here. Another clue to their synonymy (besides their too frequent sympatric occurrence and transitional stages in body proportions and maxillula ornamentation) actually lies in their size. The larger form (*O. attenuata*) always has the smaller prosome/urosome ratio, which means that its larger size is a consequence of its more elongated urosome; their prosomes are of similar sizes. Not surprisingly, the longer form (with longer urosome) always has longer caudal rami. Farran's (1936) record of *O. attenuata* from the Great Barrier Reef, measuring 1.78 mm, is certainly a printing error and should read 0.78 mm, which is well within the specific boundaries (see Nishida, 1985: 29, but note erroneous column headings in his table 2).

Family CYCLOPINIDAE G. O. Sars, 1913
Genus **Cyclopinella** G. O. Sars, 1913
[= *Muceddina* Jaume & Boxshall, 1996; = *Barathricola* Humes, 1999]

Diagnosis (emended). — Moderately large Cyclopinidae (female from 0.57 to 0.8 mm; males from 0.47 to 0.57 mm), with relatively slender habitus and free first pedigerous somite. Urosome five-segmented in female and six-segmented in male. Female genital double somite with well developed ventral cuticular suture; single copulatory pore situated ventrally; seminal receptacle small and more or less ovoid; paired ovipores located laterally or somewhat dorsolaterally, covered by reduced sixth legs. Anal somite longer than wide, with broad and short anal operculum. Caudal rami cyclindrical, longer than anal somite and from 3.4 to 11 times as long as wide, armed with six setae; lateral seta inserted from 2/5 to 2/3 of ramus' length and always clearly laterally. Female antennula 12- to 15-segmented; male antennula 13- to 17-segmented. Antenna four-segmented, lacking exopodal seta. Mandibula with two- or one-segmented endopod, and exopod reduced in size: the latter can be four-, three-, or one-segmented, or even reduced to a single seta. Maxillula with relatively short basis and one-segmented endopod and exopod. Maxilla large, comprising praecoxa, coxa, basis, and three-segmented endopod. Maxilliped slender, composed of syncoxa, basis, and three- or five-segmented endopod. All swimming legs with three-segmented exopods and endopods and without inner setae on first exopodal segment; all five elements on third endopodal segment of fourth leg transformed into spines, as are at least two apical elements on third endopodal segment of third leg; third exopodal segment spine formula 4.4.4.3 or 4.4.3.3, while setal formula 4.5.5.5. Female fifth leg with small intercoxal sclerite and comprising coxa, basis, and one-segmented exopod; coxa quadriform and unarmed; basis of size and shape similar to coxa, but armed with outer seta; exopod more than three times as long as wide and as long as or longer than coxa and basis combined, armed with one seta as most apical element, and three stout spines (one spine at first third laterally, one at outer distal corner subapically or laterally, and one at inner distal corner subapically). Male fifth leg three- or four-segmented, with one or two additional setae on inner exopodal margin.

Female sixth leg with one or two minute spines and one somewhat longer seta. Male sixth leg with three well developed elements.

Type species. — *Cyclopinella tumidula* G. O. Sars, 1913.

Other species. — *Muceddina multispinosa* Jaume & Boxshall, 1996b; *Barathricola rimensis* Humes, 1999; *Cyclopinella tincanbayensis* n. sp.

7. **Cyclopinella tincanbayensis** n. sp.
(figs. 20-22)

Material examined. — Types only: Holotype female dissected on 1 slide (WAM C37213); paratypes, 4 copepodids in alcohol (WAM C37214): Australia, Queensland, Tin Can Bay, 14 August 2003, leg. T. Karanovic, 25°16.728′S 152°50.400′E.

Description. — Female (holotype). Body length, excluding caudal setae, 0.567 mm. Habitus (fig. 20A) relatively slender, with prosome/urosome ratio 1.3 and greatest width at 2/3 of cephalothorax. Body length/width ratio about 3.4; cephalothorax about 2.7 times as wide as genital double somite. Posterior corners of cephalothorax and free pedigerous somites without pronounced lateral expansions. Urosome narrow and fifth pedigerous somite narrower than genital double somite. Colour of preserved specimen milky white. Nauplius eye not visible. Rostrum broadly rounded and furnished with two large sensilla.

Prosome (fig. 20A) comprising cephalothorax and four free pedigerous somites (first pedigerous somite free), ovoid, about twice as long as wide (dorsal view); length ratio of prosomal somites, beginning with cephalothorax, 100 : 27 : 24 : 23 : 16. Cephalothorax slightly longer than its greatest width, represents 30% of total body length; surface of cephalic shield with several small sensilla and many pores of various sizes. Hyaline fringe of prosomites narrow and smooth. Fifth pedigerous somite ornamented with only two large dorsal sensilla and with smooth fringe dorsally and ventrally. Sclerotized joint between fifth pedigerous somite and genital double somite well developed, almost as a pseudosomite (fig. 20B).

Genital double somite (fig. 20B) with anterior part somewhat inflated laterally, about 1.4 times as long as wide (dorsal view), ornamented with several lateral rows of minute spinules and discontinuous and complex suture line ventrally at first third, which bears four spinules (two on each side). Hyaline fringe of genital double somite and two succeeding somites narrow, not frilled, and completely smooth. No cuticular recesses or sensilla visible. Single copulatory pore relatively large, situated medially just beyond middle;

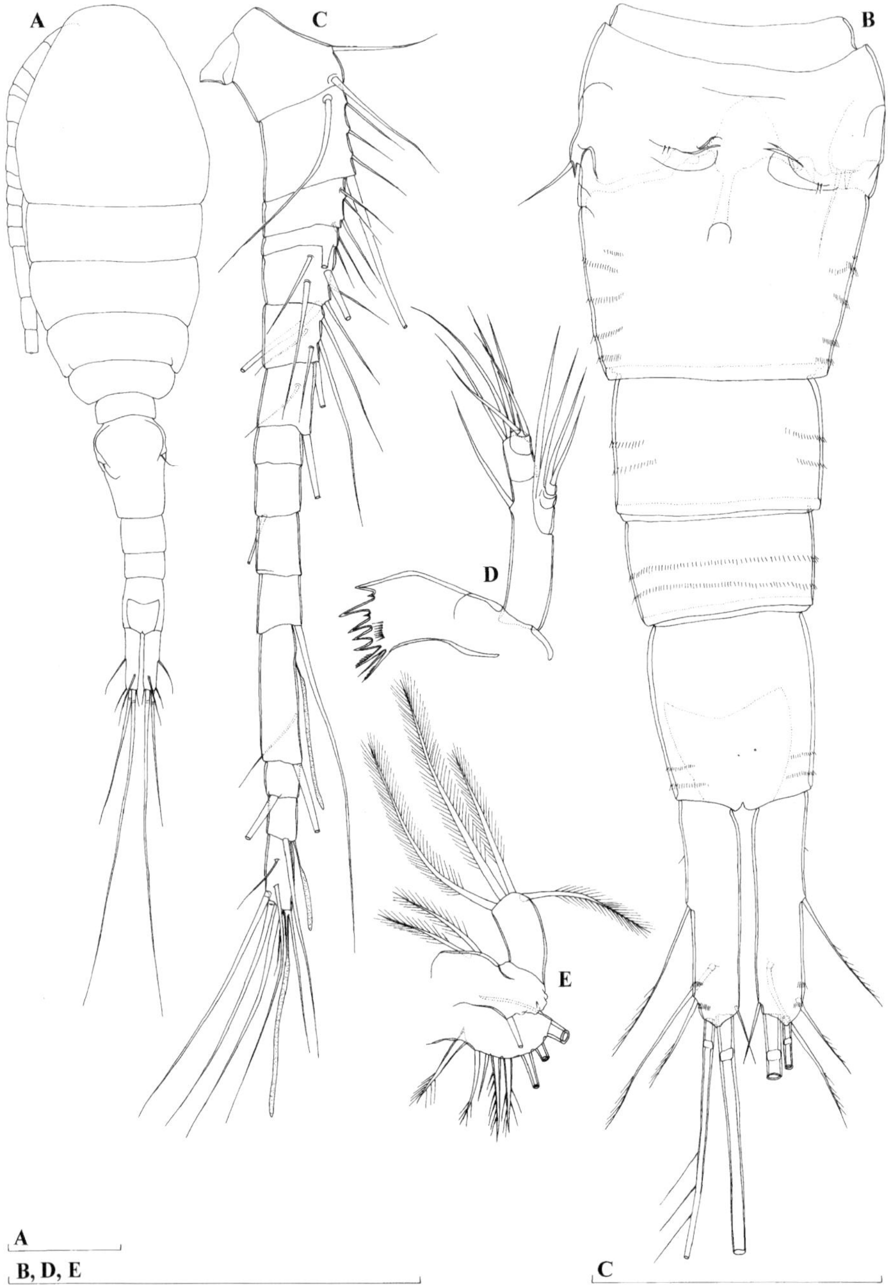

Fig. 20. *Cyclopinella tincanbayensis* n. sp., holotype female. A, habitus, dorsal view; B, abdomen, ventral view; C, antennula; D, mandibula; E, damaged maxillula. Scales = 0.1 mm.

copulatory duct long, narrow, and rigidly sclerotized, directed anteriorly and attached to posterior part of seminal receptacle. Seminal receptacle small, simple, and ovoid, about 1.3 times as long as wide and representing only 22% of double somite's width. Paired ovipores located dorsolaterally at 1/3 of double somite's length, covered by reduced sixth legs and connected to receptacle via long and narrow receptacle ducts. Third urosomite ornamented with two lateral rows of minute spinules on each side, while fourth urosomite with two continuous ventrolateral rows of minute spinules. Anal somite with smooth, broad, and convex anal operculum, which represents 73% of the somite's width and reaches only to 2/5 of the somite's length; ornamented with two large dorsal sensilla and two ventral pores, as well as with two lateral rows of minute spinules on each side. Anal sinus widely open dorsally, ornamented with several diagonal rows of minute spinules. Length ratio of abdominal somites, beginning with genital double somite, 100 : 40 : 29 : 49.

Caudal rami (fig. 20B) cylindrical, four times as long as wide (ventral view) and 1.3 times as long as anal somite, parallel, with very small space between them; they represent 9% of total body length. Distal margin ventrally smooth and convex, with medial cuticular pore clearly visible. Ornamentation consisting of two lateral rows of minute spinules posteriorly and one lateral sensillum at 1/4 of the ramus' length. Dorsal seta relatively long, slightly shorter than ramus, inserted at 3/4 of ramus' length, uniarticulate at base and unipinnate distally. Lateral seta also long and slender, arising at middle, unipinnate, and about 0.8 times as long as ramus. Outermost apical seta slender and unipinnate, less than half the ramus' length and 1.6 times as long as innermost apical seta, which is smooth and also very slender. Principal apical setae with breaking planes and bipinnate; inner seta about 2.4 times as long as outer one and 0.5 times as long as body.

Antennula (fig. 20C) long and slender, almost reaching middle of third pedigerous somite, 15-segmented, unornamented, with one slender aesthetasc on twelfth segment laterally, one on fourteenth segment subapically, and one on fifteenth segment apically, and with setal formula: 2.5.4.2.6.5.2.0.1.0.1.2.1. 1.7. Two apical setae with breaking planes and no setae articulating on basal part; all setae smooth and slender, seta on eleventh segment longest; one apical seta on fifteenth segment fused basally to aesthetasc. Length ratio of antennular segments, from proximal end and along caudal margins, 1 : 1.8 : 0.7 : 0.4 : 1 : 1.4 : 1.3 : 0.8 : 1.1 : 1.2 : 1.3 : 2.8 : 0.8 : 0.8 : 1.6.

Antenna (fig. 21A) four-segmented, comprising coxobasis and three-segmented endopod (although coxobasis and first endopodal segment partly

Fig. 21. *Cyclopinella tincanbayensis* n. sp., holotype female. A, antenna; B, maxilla; C, maxilliped; D, first swimming leg. Scale = 0.1 mm.

fused along caudal margin), with strong geniculation between first and second endopodal segments; length ratio of antennal segments, 1 : 0.8 : 0.8 : 1. Coxobasis about 2.5 times as long as wide, unornamented, and armed with a single smooth seta at inner distal corner; seta representing exopod absent. First endopodal segment 1.9 times as long as wide, armed with one smooth seta at middle, ornamented with longitudinal row of minute spinules along external margin. Second endopodal segment also about 1.9 times as long as wide, ornamented with transverse row of hair-like spinules on external margin, armed with five smooth setae (three lateral, two subapical; one subapical seta geniculate, much longer and stronger than any other seta on that segment). Third endopodal segment 2.6 times as long as wide, armed with six apical setae, two of which strong, geniculate, and bipinnate, ornamented with two transverse rows of spinules on caudal margins, which form central and distal frills.

Labrum not mounted satisfactorily to allow detailed examination.

Mandibula (fig. 20D) comprising coxa with well developed gnathobase and distal palp; palp consisting of basis, two-segmented endopod, and three-segmented exopod. Coxal gnathobase cutting edge divided into three regions: inner group of two complex and strong teeth; middle group of three simple teeth and a row of eight short spinules at their base; and outermost group of three short setae, two of which unipinnate. Basis large, more than three times as long as wide, unornamented, and armed with one smooth seta laterally. Endopod 0.3 times as long as basis, attached almost apically to basis (as its extension), with proximal segment twice as long as distal one, and with three smooth setae on each segment. Exopod about as long as endopod but much narrower, with setal formula 1.1.2.

Maxillula (fig. 20E) damaged and not completely examined, composed of small syncoxa with well developed arthrite, basis, and one-segmented exopod, as well as, probably, a one-segmented endopod (missing as a result of damage). Arthrite of syncoxa with eight elements; four apical elements strong, pinnate, and geniculate, proximalmost element also strong and pinnate, but not geniculate. Coxa largely missing; coxal epipodite with two marginal setae. Basis elongate, about as long as arthrite of syncoxa, but apical part damaged and complete armature unknown; only one apical seta observed. Exopod large, almost twice as long as wide, armed apically with four long, slender, and plumose setae.

Maxilla (fig. 21B) large and six-segmented, comprising praecoxa, coxa, basis, and three-segmented endopod. Praecoxa quadriform, 1.6 times as wide as long, unornamented, with four setae on proximal endite and only one seta on distal endite. Coxa smaller than praecoxa, also unornamented, armed with three pinnate setae on proximal, elongated endite; distal endite highly mobile and bearing three slender setae. Basis shorter than coxa, expanded into spinulose, geniculate, and unipinnate claw, and armed with one strong and one minute setae; strong seta 1.8 times as long as basal claw. First endopodal segment slightly shorter than basis, armed with four robust and geniculate setae. Second endopodal segment about half as long as first endopodal one, armed with two strong setae. Third endopodal segment minute, armed with one strong and three short, slender, and smooth setae.

Maxilliped (fig. 21C) slender, seven-segmented, composed of syncoxa, basis, and five-segmented endopod. Syncoxa 2.6 times as long as greatest width, unornamented, with praecoxal endites weakly discernible and unarmed; two coxal endites well-developed, each with two pinnate setae. Basis 2.1 times as long as wide, ornamented with serrated cuticular plate distally, armed with two subequal setae proximally on inner margin. Endopod slightly shorter than basis, unornamented, with setal formula 1.1.1.1.4; all, except two apical, endopodal setae very strong and all smooth.

All swimming legs (figs. 21D, 22A, B, C) with three-segmented exopod and endopod; their armature formula as follows (legend: inner/outer spine or seta; inner/terminal/outer):

		Exopod			Endopod	
Segments	1	2	3	1	2	3
First leg	0/1	1/1	3/2/3	1/0	1/0	3/2/1
Second leg	0/1	1/1	4/2/3	1/0	2/0	3/2/1
Third leg	0/1	1/1	4/2/2	1/0	2/0	3/2/1
Fourth leg	0/1	1/1	4/2/2	1/0	2/0	2/2/1

Third exopodal segment spine formula 4.4.3.3. All intercoxal sclerites relatively large and wide, with concave distal margin; that of first leg unornamented, others with a row of minute spinules along distal margin. Coxae unornamented, armed with plumose (first leg) of pinnate (other legs) seta at inner distal corner. Basis of each leg ornamented with a row of spinules along distal margin between exopod and endopod, armed with slender seta on outer margin; basis of first leg with additional spine on distal inner corner, which almost reaches posterior margin of second endopodal segment, with strong spinules basally and smaller spinules distally; inner distal corners of second, third, and fourth leg pronounced as sharp and short, spiniform

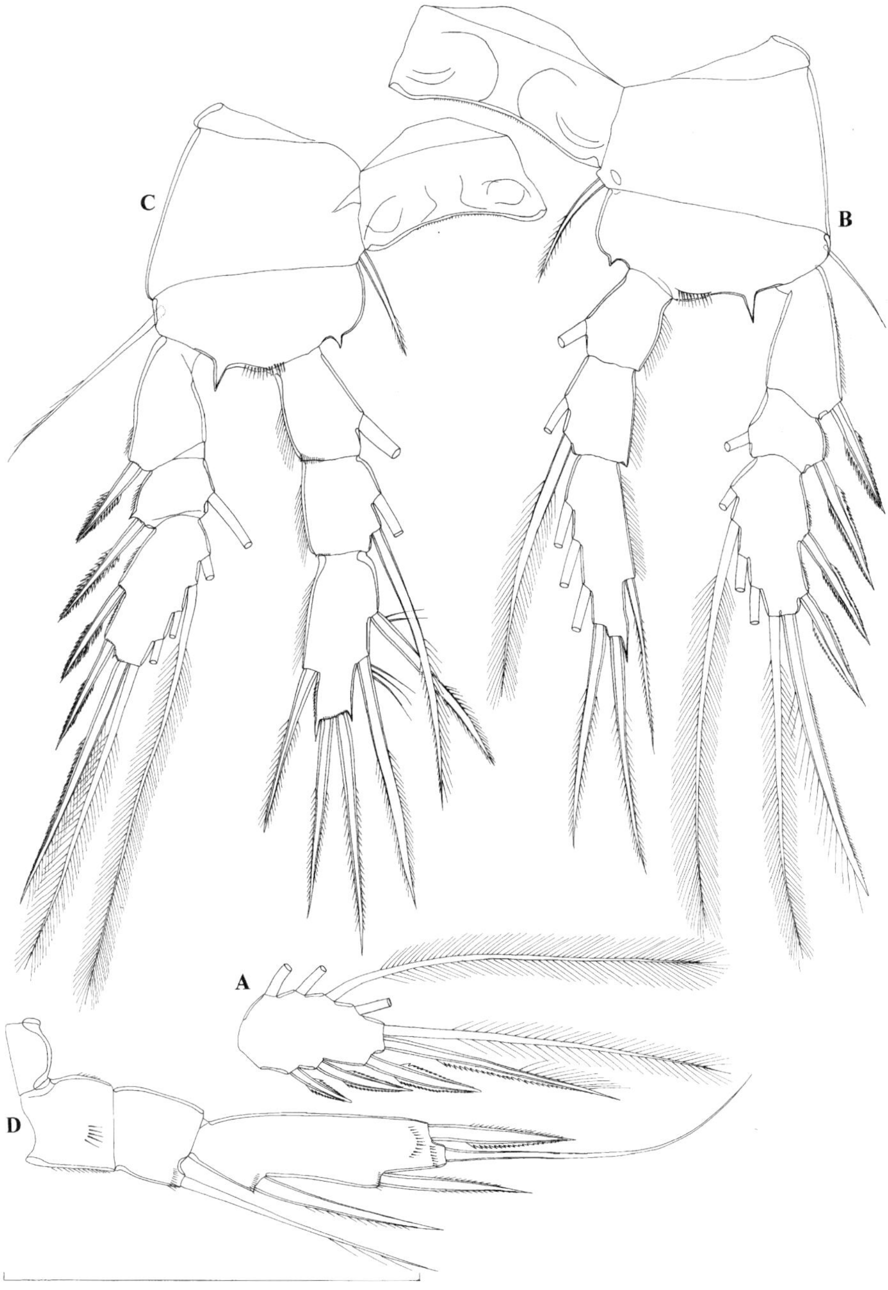

Fig. 22. *Cyclopinella tincanbayensis* n. sp., holotype female. A, third exopodal segment of second swimming leg; B, third swimming leg; C, fourth swimming leg; D, fifth leg. Scale = 0.1 mm.

processes; similar but somewhat larger process also between exopod and endopod. Most exopodal and endopodal setae slender and plumose, but apical and outer setae on third endopodal segment of second, third, and fourth leg strong and spiniform (almost as a transitional stage between a seta and a spine), as well as two inner setae on third endopodal segment of fourth leg and distal inner seta on second endopodal segment of that same leg; inner setae on third endopodal segment of fourth leg with several large spinules basally, unlike the setae of any other swimming leg. Most exopodal spines stout and with both margins sharply serrate; however, outer apical spine on third exopodal segment of each leg with serrated outer and plumose inner margin. All endopodal segments with hair-like spinules along outer margin and all exopodal segments with minute spinules along outer margin. Third endopodal segment of fourth swimming leg about 2.3 times as long as wide; inner apical element 1.2 times as long as outer.

Fifth leg (fig. 22D) with small intercoxal sclerite and three-segmented, comprising coxa, basis, and one-segmented exopod. Intercoxal sclerite quadrangular, with convex distal margin, about twice as wide as long, unornamented, reaching 1/3 of coxa. Coxa about as wide as long, unarmed, ornamented with minute spinules along inner and outer margins, and one arched row of five long spinules at middle. Basis about as large as coxa and only slightly longer than wide, armed with a single long, unipinnate seta on short setophore on outer margin distally; ornamented with short row of minute spinules at base of setophore. Exopod 2.8 times as long as basis and 3.4 times as long as its greatest width, somewhat narrower at basal part, armed with three spines and a single seta, and ornamented with spinules at base of each armature element; seta smooth, positioned centrally and as most apical element, slightly longer than exopod and basis combined, 2.2 times as long as inner apical spine, about 2.1 times as long as outer subapical spine, and 1.7 times as long as lateral spine; all spines strong and pinnate.

Sixth leg (fig. 20B) not clearly distinct, ovoid cuticular plate, armed with inner smooth and minute spine and outer smooth and much longer seta.

Male. Unknown.

Variability. — Only one adult female was collected and examined, and no asymmetry was observed.

Etymology. — The species is named after the type locality, the small city of Tin Can Bay (population approximately 2000) in Queensland. The name is an adjective for place, made with the Latin suffix "-ensis".

Distribution. — *Cyclopinella tincanbayensis* n. sp. is known from the type locality only.

Discussion. — Cyclopinids are widely distributed in the marine environment from polar to tropical seas and from shallow coastal waters to the deep sea, mostly as benthic organisms (Boxshall & Halsey, 2004). As no diagnostic apomorphy for the Cyclopinidae G. O. Sars, 1913 has been recognized, Ho & Thatcher (1989) and Huys & Boxshall (1991) concluded that the family is probably paraphyletic. Martínez Arbizu (2000a, b; 2001a, b) started revising it and proposed four new families, but they were not accepted by Boxshall & Halsey (2004) as they were not described as a result of a comprehensive and parsimony-based phylogenetic analysis of all cyclopinids, including the type genus. Although the groups of genera proposed by Martínez Arbizu (2000a, b; 2001a, b) may be monophyletic, their main distinguishing characters are not sufficient to justify family level separation. Boxshall & Halsey (2004) criticized some of these characters and the cyclopinids presented herein will expose even more of them as unreliable, so here I recognize only the family Cyclopinidae.

The predominance of monotypic genera suggests that the phylogenetic relationships between species and genera are poorly understood (Boxshall & Halsey, 2004). Mouth appendages are very conservative in most copepod families and are only rarely an important part of generic diagnoses. Consequently, many species of copepods have been described without even mentioning the mouthparts. However, in the family Cyclopinidae the mouth appendages are very variable even among closely related species (and sometimes even interspecifically), but never enough to justify a family level taxon. This proved to be very confusing for many copepodologists accustomed to the conservative mouth appendages in the other families or orders they worked on, so they usually resorted to describing new genera. Unfortunately, the mouth appendages can be completely transformed in cyclopoids in a single evolutionary event, as was shown by Boxshall & Strong (2006) for the first known parasitic *Eucyclops* Claus, 1893 from Lake Tanganyika. That is why I think the mouth appendages should not be an important part of generic diagnoses in the family Cyclopinidae, unless they are consistent with other generic characters. Accordingly, I here unite three previously recognized monospecific genera (*Cyclopinella* G. O. Sars, 1913; *Muceddina* Jaume & Boxshall, 1996; and *Barathricola* Humes, 1999) together with my new species into a single genus. Otherwise, using commonly accepted criteria, the new Australian species could be described as a new genus. The emended diagnosis for this now more broadly defined genus *Cyclopinella* is given above.

Martínez Arbizu (2000b: 210) recognized *Cyclopinella*, *Muceddina*, and *Barathricola* as a monophyletic group, when he referred to it as a part of an as yet undiagnosed new family, the Schminkepinellidae. He considered the absence of exopodal setae on the antenna as the main diagnostic character for this group, a character shared by my new species. However, considering how variable the armature of the antennal coxobasis is in other groups of Cyclopinidae, even in a single specimen (see further in this monograph), I contend that the morphology of the fourth and fifth legs constitute more valuable diagnostic characters. Namely, the four species united here share the third endopodal segment of the fourth swimming leg with all armature elements transformed into spines, as well as an almost identically shaped female fifth leg. This fifth leg not only has the same segmentation and armature in all four species, but also has an elongated exopod, more than three times as long as wide, and always as long as or longer than the coxa and basis combined. I consider these two characters as autapomorphies of this group, as it is hard to imagine them arising independently, especially in combination. Other cyclopinid genera (such as *Oromina* Jaume & Boxshall, 1997, *Ginesia* Jaume & Boxshall, 1997, and even *Cyclopinodes* Wilson, 1932) have a relatively similar female fifth leg (which is three-segmented and with an armature formula 0.1.4), but they always have a very short exopod (see Sars, 1913; Wilson, 1932; Jaume & Boxshall, 1997). Besides the above mentioned absence of the antennal exopodal setae, all four *Cyclopinella* species also have relatively similar caudal rami (which are longer than the anal somite and armed with six setae, with the lateral seta inserted at 2/5 to 2/3 of the ramus' length), no inner setae on the first exopodal segment of all swimming legs, as well as a completely free first pedigerous somite (which is almost certainly a plesiomorphic character). Also, the female genital double somite is very similar in *Muceddina*, *Barathricola*, and our new species (unfortunately, not described for *Cyclopinella*), with a ventral cuticular suture, a medially located single copulatory pore, and a small and ovoid seminal receptacle.

Sars (1913) described *Cyclopinella tumidula* as a new genus and new species from "several distant places on the Norvegian coast", but always from "a muddy bottom, partly covered by decaying algae". He found this species at depths from 20 to 40 fathoms (36 to 73 m). Later, Chappuis & Serban (1953) reported it from the Romanian shore of the Black Sea and

Wells (1965) found it in the sub-littoral muds off the coast of Scotland. Unfortunately, these authors did not provide any illustrations and the former record is especially suspicious as the authors admitted that they relied solely on the very short diagnosis of Kiefer (1929b) in identifying their single female specimen ("Tous les caractères de cet exemplaire concordent avec la brève description donnée par Kiefer dans sa Monographie des Cyclopides dans le "Tierreich""), although they did not observe the mandibula. The shape of this appendage, with its reduced exopod, was thought to be a very important generic character of *Cyclopinella* by both Sars (1913) and Lindberg (1952). The latter author gave a very good short diagnosis for this genus, emphasizing the shape of both the fourth and the fifth legs, but in his key to 11 then known cyclopinid genera he used only the shape of the mandibula. The genus was regarded as monospecific so far.

Jaume & Boxshall (1996b) described *Muceddina multispinosa* as a new genus and species from anchialine caves in Sardinia, the Balearic Islands, and the Canary Islands. They interpreted the absence of the exopodal seta on the antenna as a troglomorphic feature and did not notice a close relationship of their species with *Cyclopinella*, which they did not even mention in the paper (although they listed Sars (1913) in the reference section), but rather relied on the generic key provided by Huys & Boxshall (1990) in establishing closer relationships between *Muceddina* and other cyclopinid genera. It is interesting to note that *M. multispinosa* apparently displayed very little variability among three very disjunct island populations, interpreted by Jaume & Boxshall (1996b) as a consequence of their highly conservative morphology over evolutionary time. However, Holmes & Gotto (2000) reported this species from the littoral of Ireland, proving that it is actually more widely distributed and its populations thus less isolated. In the Medditerranean, this species could actually be a glacial relict if it is not present in the littoral.

Humes (1999) described *Barathricola rimensis* as a new genus and species from a depth of 2254 m at a hydrothermal site on the Juan de Fuca Ridge, northeastern Pacific. He correctly recognized that "*Barathricola* and *Cyclopinella* may be closely related, since they share several derived characters", but he did not recognize a close relationship of these two genera with *Muceddina*, although he listed Jaume & Boxshall (1996b) in the reference section.

Cyclopinella tincanbayensis n. sp. is the first representative of this genus from the marine interstitial and can be distinguished from its three congeners by many morphological characters (including the shape of the caudal rami, the segmentation of the antennula and its armature, and even the species' general habitus), but most importantly by the shape of the mandibular palp, which is more reduced than in *Cyclopinella multispinosa* (Jaume & Boxshall, 1996) n. comb., but less reduced than in *Cyclopinella rimensis* (Humes, 1999) n. comb. or *Cyclopinella tumidula* G. O. Sars, 1913. In fact, there is an evolutionary trend towards a more reduced mandibular palp in this genus, which is consistent with reductions in other cephalic appendages and in the maxilliped. The most primitive species is *Cyclopinella multispinosa*, which has a 15-segmented antennula, an armed inner distal corner of the antennal coxobasis, a four-segmented mandibular exopod (armed with five setae), three setae on the maxillular coxal epipodite, and a seven-segmented maxilliped with an armature formula 5.2.1.1.1.1.4. *Cyclopinella tincanbayensis* is the next one in the chain, also with a 15-segmented antennula and an armed inner distal corner of the antennal coxobasis, but with a three-segmented mandibular exopod (armed with four setae), with only two setae on the maxillular coxal epipodite, and with a seven-segmented maxilliped with armature formula 4.2.1.1.1.1.4. *Cyclopinella rimensis* has a 13-segmented antennula, an unarmed antennal coxobasis, a one-segmented mandibular exopod (armed with two setae), an unarmed maxillular coxal epipodite, and a seven-segmented maxilliped with armature formula 2.2.1.1.1.1.2. *Cyclopinella tumidula* is the most reduced member of the genus, with a 12-segmented antennula, an unarmed antennal coxobasis, a completely reduced mandibular exopod comprising one seta as its only remnant, an unarmed maxillular coxal epipodite and a five-segmented maxilliped with armature formula 3.2.1.1.3. It is unfortunate that the type species of the genus is also the most reduced one, but there is nothing we can do about that, as it was designated by Sars (1913). The same evolutionary trend is found in the shape of the male fifth leg in the genus, with *C. multispinosa* having a four-segmented leg and two additional setae on the inner exopodal margin, *C. rimensis* having a three-segmented leg and only one additional seta on the inner exopodal margin, and *C. tumidula* having a male fifth leg that is exactly the same as in the female (i.e., three-segmented and without any setae on the inner exopodal margin); unfortunately, *C. tincanbayensis* is known only after females. All these gradual reductions I interpret as intergeneric evolution and they are not surprising if we consider how widely different the habitats of the four species are. Using the key to genera for this family by Boxshall & Halsey (2004), the new species would key out as *Cyclopinopsis* Smirnov, 1935, but this genus is only remotely related to *Cyclopinella*, and this result is indicative of the inadequate generic diagnoses in this family.

Genus **Hemicyclopina** Herbst, 1953
[= *Glareolina* Huys & Boxshall, 1990]

Diagnosis (emended). — Small Cyclopinidae (females from 0.318 to 0.467 mm; males from 0.271 to 0.310 mm), with very slender habitus (almost harpacticiform) and a free first pedigerous somite. Urosome five-segmented in female and six-segmented in male. Female genital double somite with ventral cuticular suture; a single small copulatory pore situated ventrally; paired seminal receptacles fused medially, expanded posteriorly and laterally; paired ovipores located laterally, covered by reduced sixth legs. Anal somite broader than long, with very broad and short anal operculum. Caudal rami cylindrical, as long as or slightly longer than anal somite and twice as long as wide, armed with six setae; lateral seta inserted at midlength. Female antennula from 11- to 20-segmented, but if each suture could be counted as a real segmental boundary, from 15- to 20-segmented. Male antennula 15-segmented. Antenna four-segmented, with one, two, or no exopodal setae; four apical elements strongly geniculate and relatively short. Mandibula with or without basal seta; endopod two-segmented (although first segment could be partly fused to basis) with two setae on first segment and four or five setae on second; exopod four-segmented, with armature formula 1.1.1.2. Maxillula with or without epipodal seta; basis large, armed with three setae on proximal endite and one or two setae on distal endite; endopod one-segmented, armed with four, five, or six setae; exopod also one-segmented, always armed with four setae. Maxilla robust but not very large, comprising praecoxa, coxa, basis, and three- or two-segmented endopod. Maxilliped long and prehensile, composed of syncoxa, basis, and five- or four-segmented endopod; first endopodal segment much smaller than second, which is elongated; four armature elements originating from last two (in four-segmented endopod) or three (in five-segmented endopod) endopodal segments transformed, strong and prehensile. All swimming legs with three-segmented endopods and exopods, and no spines on endopods; third exopodal segment spine formula 4.4.4.3 or 3.4.4.3, while setal formula could be 4.4.4.4, 4.5.5.5, or 5.5.5.5. Female fifth leg with small intercoxal sclerite, positioned ventrally and comprising coxobasis and one-segmented exopod; coxobasis as wide as long or slightly wider, armed with basal seta on outer margin; exopod somewhat longer than coxobasis and about 1.5 times as long as wide, much narrower at basal part, armed with three spines and a single seta (one spine laterally, two spines apically, and a seta in between); lateral spine always longer and more robust than outer apical spine. Male fifth leg with two additional setae on inner exopodal margin. Female sixth leg semicircular, small plate, armed with dorsally directed inner spine and posteriorly directed, slender and shorter, outer seta. Male sixth leg distinct, large plate, armed with two setae.

Type species. — *Paracyclopina setifera* Herbst, 1952.

Other species. — *Hemicyclopina kliei* Herbst, 1953; *Hemicyclopina begoniae* Martínez Arbizu, 2001; *Hemicyclopina martinezi* n. sp.

8. **Hemicyclopina martinezi** n. sp.
(figs. 23-25)

Material examined. — Types only: Holotype female dissected on 1 slide (WAM C37214); allotype male dissected on 1 slide (WAM C37216); paratypes, 1 male + 12 females (3 ovigerous) + 2 copepodids (1 female dissected on 1 slide (WAM C37217), 2 females on 1 slide in toto (TMAG G5775), others in alcohol (WAM C37218)): Australia, New South Wales, Wollongong, Boat Harbour, 17 August 2003, leg. T. Karanovic, 34°25.326′S 150°54.296′E.

Description. — Ovigerous female (holotype). Body length, excluding caudal setae, 0.454 mm. Habitus (fig. 20A) very slender, harpacticiform, with prosome/urosome ratio 1.7 and greatest width at middle of cephalothorax. Body length/width ratio about 3.5; cephalothorax about 2.1 times as wide as genital double somite. Posterior corners of cephalothorax and first free pedigerous somite without pronounced lateral expansions; second and especially third and fourth pedigerous somites expanded laterally (dorsal view). Urosome cylindrical and fifth pedigerous somite wider than genital double somite. Colour of preserved specimen yellowish. Nauplius eye not visible. Rostrum broadly rounded and furnished with two large sensilla.

Prosome (fig. 23A) comprising cephalothorax and four free pedigerous somites (first pedigerous somite free), cylindrical and constricted at first pedigerous somite, about 2.2 times as long as wide (dorsal view); length ratio of prosomal somites, beginning with cephalothorax, 100 : 21 : 22 : 20 : 11. Arthrodial membranes clearly visible from dorsal view between each prosomal somite, as well as between prosome and urosome. Cephalothorax ovoid, slightly longer than its greatest width, representing 32% of total body length; surface of cephalic shield with several small sensilla and pores. Hyaline fringe of prosomites narrow and smooth. Fifth pedigerous somite ornamented with only two large dorsal sensilla, and with smooth fringe dorsally and ventrally. Sclerotized joint between fifth pedigerous somite and genital double somite well developed, almost as a pseudosomite (fig. 23B).

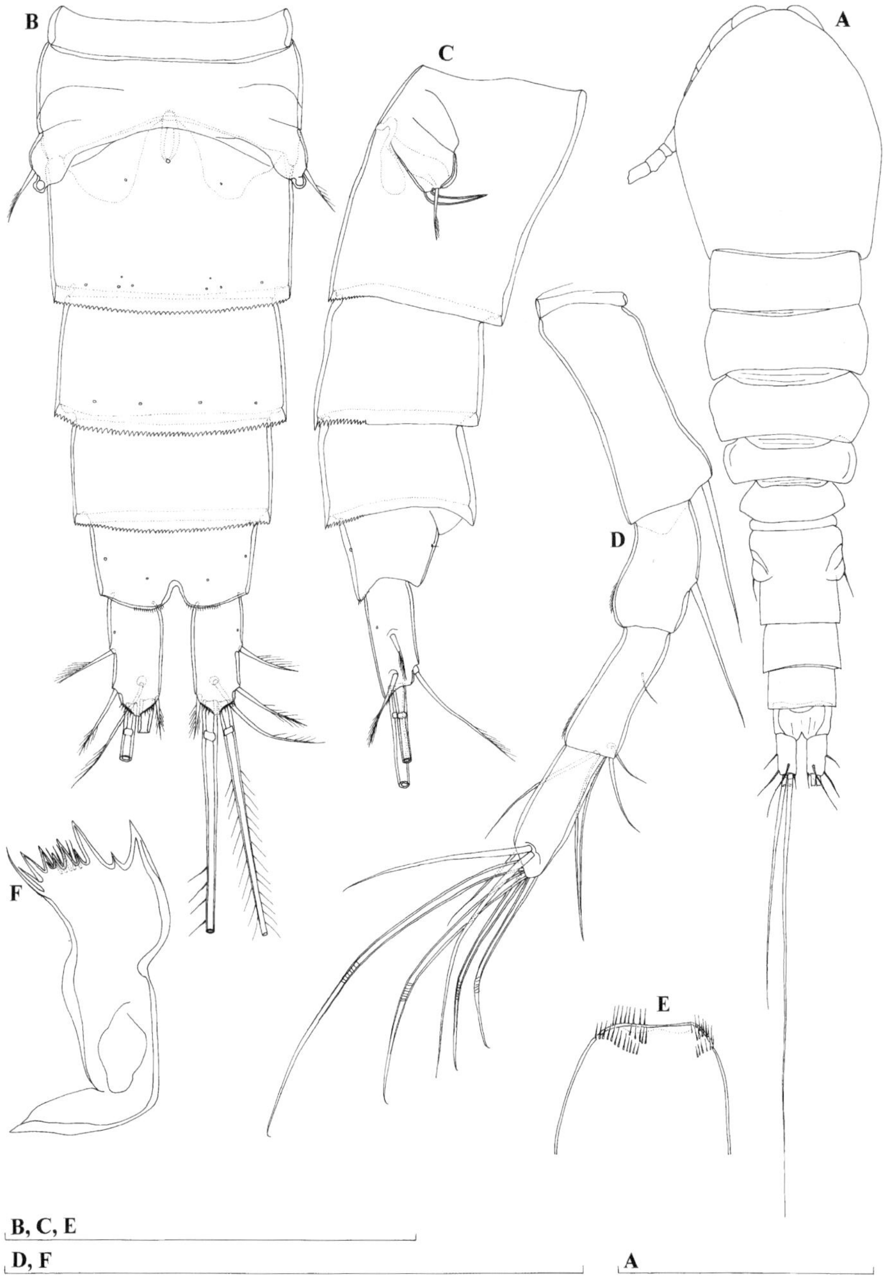

Fig. 23. *Hemicyclopina martinezi* n. sp., holotype female. A, habitus, dorsal view; B, abdomen, ventral view; C, abdomen, lateral view; D, antenna; E, labrum; F, coxa of mandibula. Scales = 0.1 mm.

Genital double somite (fig. 23B, C) almost cylindrical, with anterior part only 1.1 times as wide as posterior, slightly wider than long, ornamented with characteristic pattern of ten pores ventrally. No cuticular recesses or sensilla visible. Hyaline fringe of genital double somite and two succeeding somites narrow, smooth dorsally, and finely serrated ventrally. Copulatory pore very small and ovoid, situated medially at 2/5 of double somite's length; copulatory duct relatively long, bottle-shaped from ventral view, and siphon-like from lateral, much wider than copulatory pore and rigidly sclerotized, attached to narrow bridge between two seminal receptacles. Seminal receptacles large, like hemispheres expanded posteriorly, about 1.6 times as wide as long. Paired ovipores located laterally at middle, covered by reduced sixth legs and connected to receptacles via very short and relatively narrow receptacle ducts. Third urosomite ornamented with four ventral pores posteriorly, while fourth urosomite without any ornamentation. Anal somite with smooth, very broad, and convex anal operculum, which represents 75% of the somite's width and reaches to only 1/5 of the somite's length; ornamented with two large dorsal sensilla and four pores ventrally, as well as with interrupted row of minute spinules along posterior margin ventrally. Anal sinus widely open dorsally, ornamented with several diagonal rows of minute spinules. Length ratio of abdominal somites, beginning with genital double somite, 100 : 54 : 43 : 34.

Caudal rami (fig. 23B, C) cylindrical, twice as long as wide (ventral view) and 1.3 times as long as anal somite, parallel, with space between them about half of one ramus' width; they represent 5% of total body length. Distal margin ventrally convex, with medial cuticular pore clearly visible, ornamented with small spinules. The only additional ornamentation consists of one ventrolateral pore at first third of ramus' length. Dorsal seta relatively long, slightly longer than ramus, inserted in small recess at 4/5 of ramus' length, uniarticulate at base and unipinnate distally. Lateral seta relatively long and slender, arising at middle, unipinnate and about 0.6 times as long as ramus. Outermost apical seta slender and bipinnate, 0.7 times as long as ramus and 1.8 times as long as innermost apical seta, which is also slender and bipinnate. Principal apical setae with breaking planes and bipinnate; inner seta about 1.9 times as long as outer one and 0.6 times as long as body.

Antennula (fig. 24A) short and stout, reaching 3/4 of cephalothorax in length, 13-segmented (although second segment with two subdivisions on caudal margin, indicating ancestral 15-segmented stage), unornamented, with

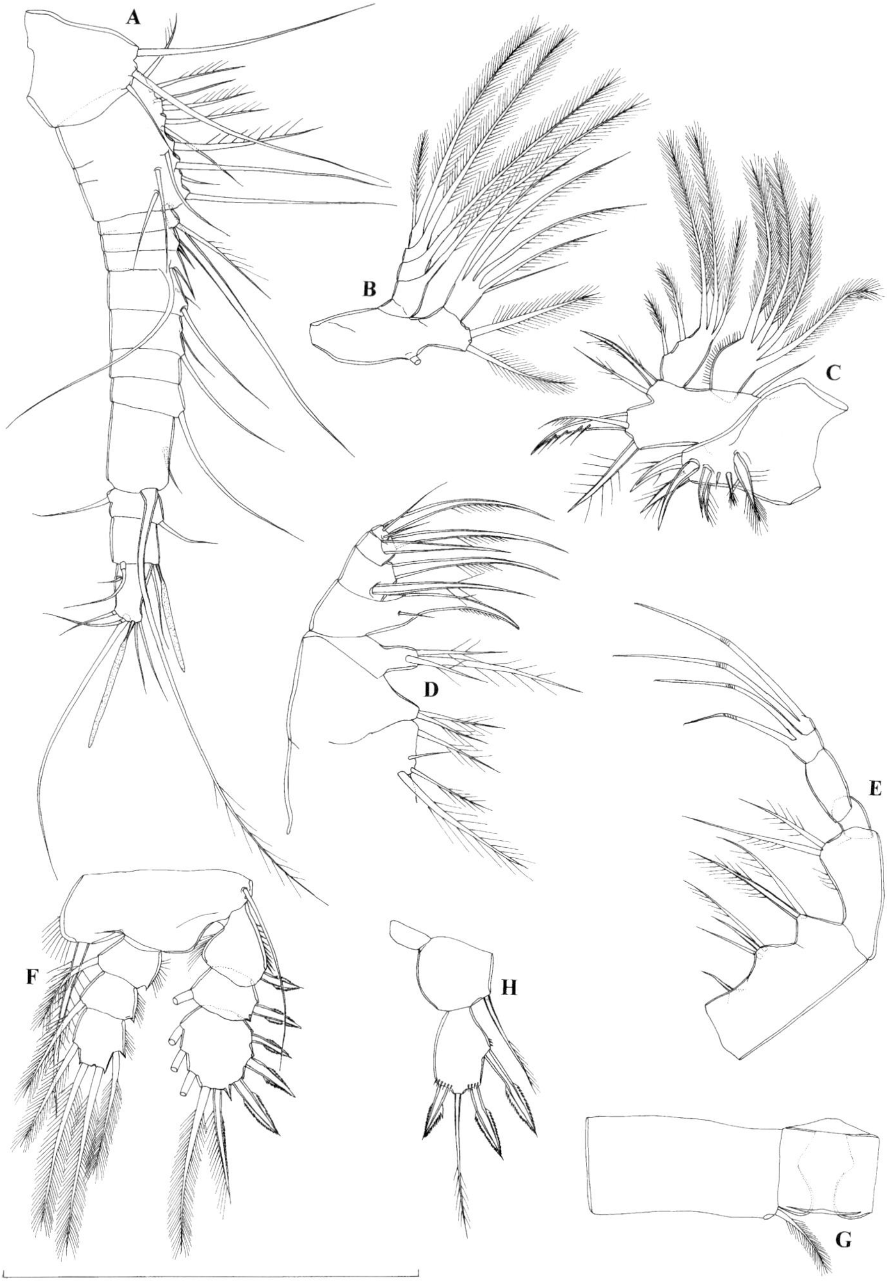

Fig. 24. *Hemicyclopina martinezi* n. sp., holotype female. A, antennula; B, palp of mandibula; C, maxillula; D, maxilla; E, maxilliped; F, basis and rami of first swimming leg; G, coxa and intercoxal sclerite of first swimming leg; H, fifth leg. Scale = 0.1 mm.

one slender aesthetasc on twelfth segment subapically as well as one on thirteenth segment apically, and setal formula: 3.14.1.1.2.2.2.0.1.2.2.2.6. No setae with breaking planes, and one seta on twelfth segment and two apical setae on thirteenth segment articulating on basal part; most setae smooth and slender; subapical seta on tenth segment longest and strongest; one seta on the fifth and the only seta on the fourth segment short and spiniform. Length ratio of antennular segments, from proximal end and along caudal margins, 1 : 2.5 : 0.3 : 0.4 : 0.4 : 1 : 1 : 0.7 : 0.6 : 2.1 : 0.7 : 1.1 : 1.4.

Antenna (fig. 23D) four-segmented, comprising coxobasis and three-segmented endopod, with only small geniculation between coxobasis and first endopodal segment; length ratio of antennal segments, 1 : 0.5 : 0.6 : 0.6. Coxobasis about 2.3 times as long as wide, unornamented, and armed with a single smooth seta at inner distal corner; a seta representing the exopod is absent. First endopodal segment 1.7 times as long as wide, armed with one smooth seta at middle, ornamented with longitudinal row of minute spinules along external margin. Second endopodal segment about 2.3 times as long as wide, ornamented similarly as previous segment, armed with five smooth setae (three lateral, two subapical; one subapical seta geniculate, much longer and stronger than any other seta on that segment). Third endopodal segment 2.7 times as long as wide, unornamented, and armed with seven apical setae, four of which very strong and geniculate.

Labrum (fig. 23E) trapezoidal, large and simple, with smooth and straight distal margin and ornamented with three rows of long spinules on each side at distal lateral corner.

Mandibula (figs. 23F, 24B) comprising coxa with well developed gnathobase and distal palp; palp consisting of basis, one-segmented endopod, and four-segmented exopod. Coxal gnathobase cutting edge divided into three regions: inner group of two complex and strong teeth; middle group of four simple teeth and five long spinules; and outermost group of two short and smooth setae. Basis large, 2.5 times as long as wide, unornamented, and armed with one seta laterally and two slender and plumose setae apically. Endopod small, quadriform, armed apically with four pinnate and one smooth setae. Exopod slender, half as long as basis and more than twice as long as endopod, unornamented, and with setal formula 1.1.1.2.

Maxillula (fig. 24C) composed of syncoxa with well developed arthrite, basis, and one-segmented endopod and exopod. Arthrite of syncoxa with eight elements; four apical elements strong and spiniform, proximalmost element longest and plumose. Coxa unornamented and armed with one epipodal smooth seta, which has actually moved to basal part of basis; coxal endites weakly discernible, unarmed. Basis elongated, much longer than arthrite of syncoxa, almost twice as long as wide, unornamented, and armed with three setae on proximal and two setae on distal endite; two setae on proximal endite very strong and spiniform; all setae on basis pinnate. Endopod as long as basis' width and about as long as exopod but more slender, unornamented, and armed with two short lateral setae on inner margin and three long setae apically; all endopodal setae slender and plumose. Exopod 1.5 times as long as wide, with longitudinal row of hair-like spinules along inner margin, which is also convex, and armed apically with four slender, plumose, and very long setae.

Maxilla (fig. 24D) robust but not very large, six-segmented, comprising praecoxa, coxa, basis, and three-segmented endopod (although praecoxa and coxa partly fused). Praecoxa quadriform, 1.5 times as wide as long, unornamented, with three setae on proximal and only one seta on distal endite. Coxa about as long as praecoxa, also unornamented, armed with two pinnate setae on proximal, elongated endite; distal endite highly mobile and bearing three slender and pinnate setae. Basis much shorter than coxa, expanded into spinulose, geniculate, and unipinnate claw, and armed with a single smooth and minute seta. First endopodal segment shorter than basis, armed with two robust and geniculate setae. Second endopodal segment about half as long as first endopodal segment, armed with two strong setae. Third endopodal segment minute, armed with one strong and two short, slender, and smooth setae.

Maxilliped (fig. 24E) large and slender, prehensile, six-segmented, composed of syncoxa, basis, and four-segmented endopod. Syncoxa 2.1 times as long as greatest width, unornamented, with proximal praecoxal endite weakly discernible and unarmed, and distal praecoxal endite armed with a single smooth seta; two coxal endites well-developed, proximal with three and distal with two setae. Basis twice as long as wide, much narrower at basal part, ornamented with several large spinules along inner margin proximally, and armed with two pinnate setae distally on inner margin. Endopod slightly longer than basis, unornamented, with setal formula 0.0.2.2; all endopodal setae very strong, smooth, and geniculate. Length ratio of maxilipedal segments, from proximal end, 100 : 77 : 31 : 42 : 16 : 8.

All swimming legs (figs. 24F, G, 25A, B) with three-segmented exopod and endopod; their armature formula as follows (legend: inner/outer spine or seta; inner/terminal/outer):

	Exopod			Endopod		
Segments	1	2	3	1	2	3
First leg	0/1	1/1	3/2/3	1/0	1/0	1/2/1
Second leg	0/1	1/1	3/2/3	1/0	1/0	1/2/0
Third leg	0/1	1/1	3/2/3	1/0	1/0	1/2/0
Fourth leg	0/1	1/1	3/2/2	1/0	1/0	1/2/0

Third exopodal segment spine formula 4.4.4.3 and no spines on endopods. All intercoxal sclerites relatively small, quadriform, with straight distal margin, and unornamented. Coxae also unornamented, armed with plumose seta at inner distal corner. Basis of each leg ornamented with a row of hair-like spinules along inner margin, and armed with a slender seta on outer margin, which is very long on first leg; basis of first leg with additional robust spine on distal inner corner, 1.3 times as long as endopod; inner distal corners of second, third, and fourth leg pronounced as short spiniform processes; similar, but somewhat larger and blunt process also between exopod and endopod. All exopodal and endopodal setae slender and plumose. Most exopodal spines stout and with both margins sharply serrated; however, outer apical spine on third exopodal segment of each leg with serrated outer and plumose inner margin. All endopodal segments with hair-like spinules along outer margin; first exopodal segment of each leg with hair-like spinules along inner margin and with normal, short spinules along outer margin; second and third endopodal segment of second and third leg additionally ornamented with a transverse row of spinules. Third endopodal segment of fourth swimming leg about 1.8 times as long as wide; both apical setae of same length, twice as long as segment.

Fifth leg (fig. 24H) with small intercoxal sclerite and two-segmented, comprising coxobasis and a one-segmented exopod. Intercoxal sclerite quadrangular, with convex distal margin, about twice as wide as long, unornamented, reaching 1/4 of coxobasis. Coxobasis about as wide as long, unornamented, armed with a single unipinnate basal seta on outer margin. Exopod 1.2 times as long as coxobasis and 1.5 times as long as its greatest width, much narrower at basal part, armed with three spines and a single seta, and ornamented with spinules at base of each spine; seta bipinnate at distal part, positioned centrally and as the most apical element, almost as long as exopod and coxobasis combined, 2.5 times as long as inner apical spine, about 1.9 times as long as outer apical spine, and 1.5 times as long as lateral spine; all spines strong and with sharply serrated margins.

Sixth leg (fig. 23C) not clearly distinct, but large cuticular plate, armed with inner smooth and dorsally directed spine, and outer unipinnate and slender seta of about the same length.

Male (allotype). Body length, excluding caudal setae, 0.275 mm. Habitus (fig. 25C) even more slender than in female, with prosome/urosome ratio 1.6 and greatest width at 2/3 of cephalothorax. Body length/width ratio about 4.3; cephalothorax 1.7 times as wide as genital somite.

Cephalothorax (fig. 25C) 1.3 times as long as its greatest width, with posterior corners not pronounced laterally. Arthrodial membrane well developed between third and fourth pedigerous somites, as well as between prosome and urosome. Ornamentation of prosomites, colour, and nauplius eye similar to those of female. Hyaline fringes of all prosomites smooth.

Genital somite (fig. 25D) 1.6 times as wide as long (dorsal view), with two completely formed and rather large spermatophores inside, ornamented with only two pores dorsally and with smooth hyaline fringe; third, fourth, and fifth urosomites without any ornamentation, and with serrated hyaline fringe ventrally and smooth dorsally. Anal somite similar to that of female, but without ventral pores visible.

Caudal rami (fig. 25D) about 1.8 times as long as wide, parallel, with space between them less than half of ramus' width. Armature and ornamentation similar to those of female, except anterior ventral pore absent.

Antennula (fig. 25E) large, almost equal in length to cephalothorax, 15-segmented, strongly digeniculate, with geniculations between eighth and ninth and between thirteenth and fourteenth segments. One long and slender aesthetasc on tenth, fourteenth, and fifteenth segment each, and setal formula: 4.7.2.2.2.2.2.2.1.2.1.2.1.1.7. All setae on eleventh and twelfth segments very short and spiniform; other setae slender and most setae smooth. Only spiniform setae on eleventh and twelfth segments, one seta on second, and one seta on third segment pinnate; only three setae on fifteenth segment articulating at basal part.

Antenna, labrum, mandibula, maxillula, maxilla, maxilliped, and swimming legs very similar to those of female.

Fifth leg (fig. 25F) with two additional slender, short and plumose setae on inner margin of exopod; segmentation and remaining armature same as in female. No ornamentation visible.

Sixth leg (fig. 25D) distinct, large plate, unornamented and armed with two pinnate setae; outer seta unipinnate and slightly longer than inner one,

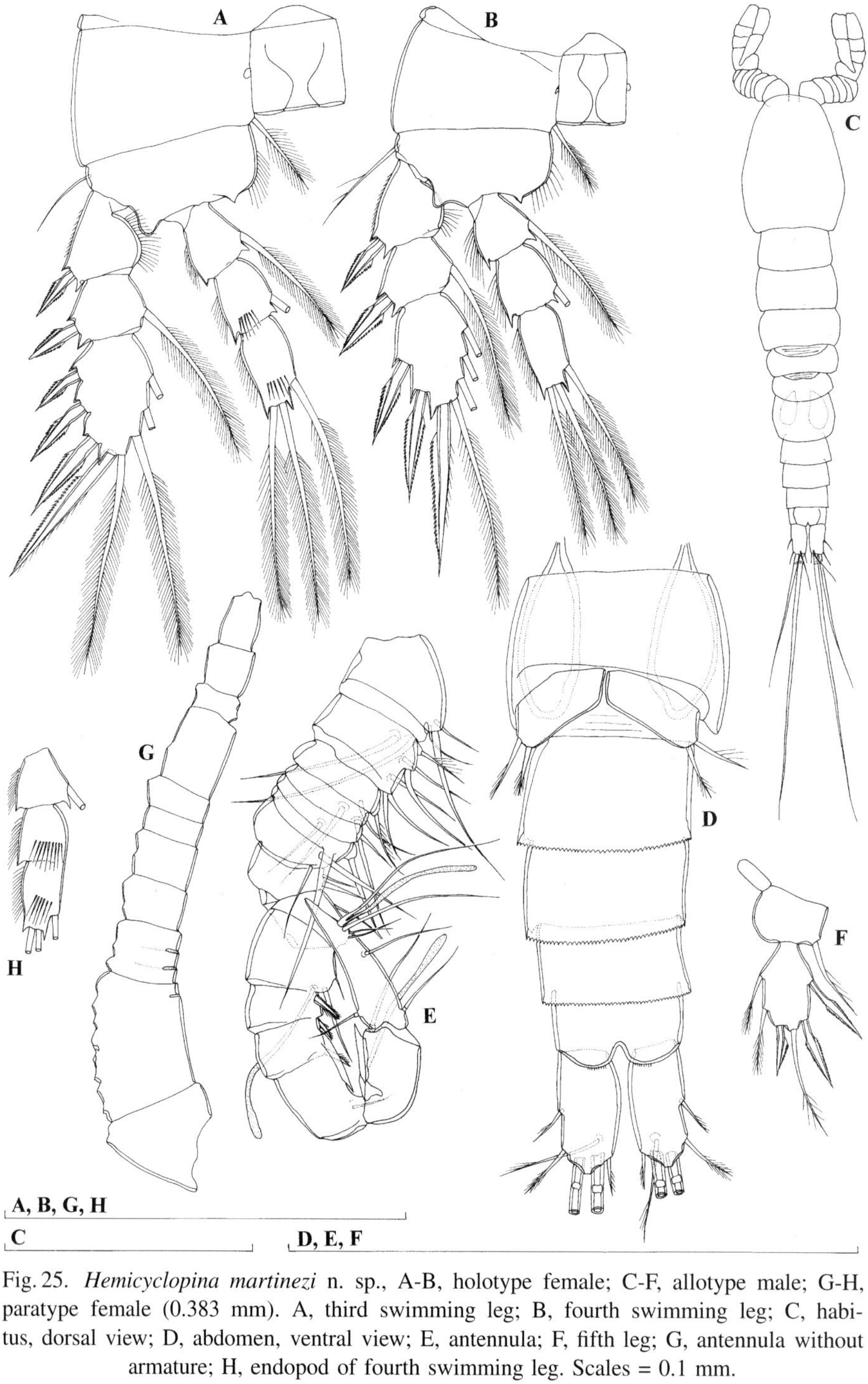

Fig. 25. *Hemicyclopina martinezi* n. sp., A-B, holotype female; C-F, allotype male; G-H, paratype female (0.383 mm). A, third swimming leg; B, fourth swimming leg; C, habitus, dorsal view; D, abdomen, ventral view; E, antennula; F, fifth leg; G, antennula without armature; H, endopod of fourth swimming leg. Scales = 0.1 mm.

which is bipinnate in its distal part; setae hardly reaching beyond middle of third urosomite.

Variability. — Body length of females ranging from 0.346 mm to 0.454 mm (0.383 mm average; n = 10), while in males it ranges from 0.271 mm to 0.275 mm (0.273 mm average; n = 2). One paratype female has an abnormal endopod on one fourth swimming leg (fig. 25H), with the second and third segments partly fused and the armature of the second segment missing, while the opposite fourth leg has a normal endopod. The same female has one antennula with third, fourth, and fifth segments partly fused (fig. 25G).

Etymology. — The species is named in honour of Dr. Pedro Martínez Arbizu, from the DZMB-Forschungsinstitut Senckenberg in Wilhelmshaven, Germany, for his outstanding contribution to the taxonomy of the family Cyclopinidae G. O. Sars, 1913.

Distribution. — *Hemicyclopina martinezi* n. sp. is known from the type locality only.

Discussion. — The genus *Hemicyclopina* was established by Herbst (1953), who designated *Paracyclopina setifera* Herbst, 1952 as its type species. This species was described by him (Herbst, 1952) from the groundwater (interstitial) of a sandy beach at Cannes, France, after several males and females. In the same paper in which he established this genus, Herbst (1953) also described another new species from the French interstitial (near Banyuls) as *Hemicyclopina kliei*, but only after females. Huys & Boxshall (1990) considered *Hemicyclopina* as "another ill-defined genus" and designated *H. kliei* as the type species of their new genus *Glareolina*. The last species described so far in the genus *Hemicyclopina* was *H. begoniae* Martínez Arbizu, 2001, found as a single female in the interstitial of a sandy beach at Valencia, Spain, by Martínez Arbizu (2001b). Surprisingly, Martínez Arbizu (2001b) accepted the genus *Glareolina* as valid, although his new species showed a number of character states transitional between *Hemicyclopina* and *Glareolina*. My new species shows even more character states transitional between these two genera, so the latter one has to be dissolved.

Using the most recent key to the genera of the family Cyclopinidae as provided by Boxshall & Halsey (2004), *Hemicyclopina martinezi* n. sp. could not be assigned to any of the 43 genera recognized and would key out somewhere between the genera *Procyclopina* Herbst, 1955 and *Allocyclopina* Kiefer, 1954 (couplet 28). Unfortunately, it has very little in common with these two genera, showing that the genera in this family are indeed ill-defined. In my opinion, the two closest relatives to the genus *Hemicyclopina* are the genera *Paracyclopina* Smirnov, 1935 and *Arctocyclopina* Mohammed & Neuhof, 1985. The latter genus still contains only one species, *A. pagonasta*, which was described from the annual sea ice in the Arctic Frobisher Bay, Canada, after both sexes by Mohammed & Neuhof (1985). It differs from *Hemicyclopina* by its non-prehensile maxilliped, its anteriorly positioned lateral seta on the caudal rami, but most importantly by its fifth leg, which is placed laterally on the somite and has all the armature elements as slender setae (which was also observed and confirmed by Martínez Arbizu, 2000a). The genus *Paracyclopina* was described by Smirnov (1935) to accommodate his new species *Paracyclopina nana* Smirnov, 1935, which was collected from brackish waters at the mouth of the Suifun River, near Vladivostok. He also recognized the similarities between *P. nana* and two species described by Sewell (1924) from brackish waters of Chilka Lake, India (*Cyclopina intermedia* Sewell, 1924 and *Cyclopina longifurca* Sewell, 1924), but it was Lang (1946) who was the first to formally include these species into the genus *Paracyclopina*, together with another Indian species described by Sewell (1934), i.e., *Cyclopina minuta* Sewell, 1934. Finally, Lindberg (1952) confirmed these four species as valid members of the genus *Paracyclopina* and provided a short generic diagnosis, which has been used ever since. No other species of this genus has been described subsequently, but some of the species have been re-examined and the most important generic characters have been confirmed (see Martínez Arbizu, 2000a). In contrast to *Arctocyclopina*, the fifth leg of *Paracyclopina* has the same armature elements as in *Hemicyclopina*, but in *Paracyclopina* there is no intercoxal sclerite, the coxobasis is fused to the somite, and the whole leg is displaced laterally. Interestingly, Martínez Arbizu (2000a) recognized the close relationship between *Arctocyclopina* and *Paracyclopina* (including them in his new family Cyclopettidae), but he (Martínez Arbizu, 2001b) assigned *Hemicyclopina* to another new family (Hemicyclopinidae). As mentioned in the discussion section for the previous species, these new families are not accepted here.

Hemicyclopina martinezi n. sp. could easily be distinguished from its three congeners by the reduced endopodal armature of its swimming legs, as well as by an unarmed second endopodal segment of the maxilliped and very short innermost apical setae on the caudal rami. These differences, however, are not sufficient to establish a new genus, so the new species is the first representative of the genus *Hemicyclopina* from the Southern Hemisphere.

Interestingly, all four species of this genus were found in the marine interstitial, as are all four species of *Paracyclopina* in brackish waters, and the only species of the closely related genus *Arctocyclopina* was found in Arctic Sea ice (which is an environment that also develops brackish water conditions). Thus, it is hard not to imagine the ancestor of this branch of the family Cyclopinidae as a brackish water dweller in moderate to cold regions of the world.

Genus **Allocyclopina** Kiefer, 1954

9. **Allocyclopina australiensis** n. sp.
(figs. 26-29)

Material examined. — Type: Holotype female dissected on 1 slide (WAM C37219): Australia, South Australia, Yorke Peninsula, Stransbury, 26 December 2002, leg. T. Karanovic, 34°54.524′S 137°47.873′E.

Other material: Australia, South Australia, Eyre Peninsula, Cowell, 25 December 2002, leg. T. Karanovic, 33°41.180′S 136°55.786′E: 1 male dissected on 1 slide (WAM C37220).

Description. — Female (holotype). Body length, excluding caudal setae, 0.509 mm. Habitus (fig. 26A) relatively slender, with prosome/urosome ratio 1.3 and greatest width at posterior end of cephalothorax. Body length/width ratio about 3.2; cephalothorax about 2.5 times as wide as genital double somite. Posterior corners of cephalothorax and second (first free) pedigerous somite without pronounced lateral expansions; third and fourth pedigerous somites slightly expanded laterally (dorsal view). Urosome cylindrical and fifth pedigerous somite somewhat wider than genital double somite. Colour of preserved specimen yellowish. Nauplius eye not visible. Rostrum broadly rounded and furnished with two large sensilla.

Prosome (fig. 26A, B) comprising cephalothorax and three free pedigerous somites (first pedigerous somite fused to cephalothorax, but with dorsal suture marking its position still visible), almost cylindrical and twice as long as wide (dorsal view); length ratio of prosomal somites, beginning with cephalothorax, 100 : 29 : 22 : 20 : 4. Arthrodial membranes clearly visible from dorsal view between third and fourth pedigerous somites, as well as between prosome and urosome. Cephalothorax somewhat longer than its greatest width, representing 33% of total body length; surface of cephalic shield with many sensilla and pores of different sizes. Hyaline fringe of prosomites narrow and smooth. Fifth pedigerous somite ornamented only with two large dorsal sensilla and with smooth fringe dorsally and ventrally. Scle-

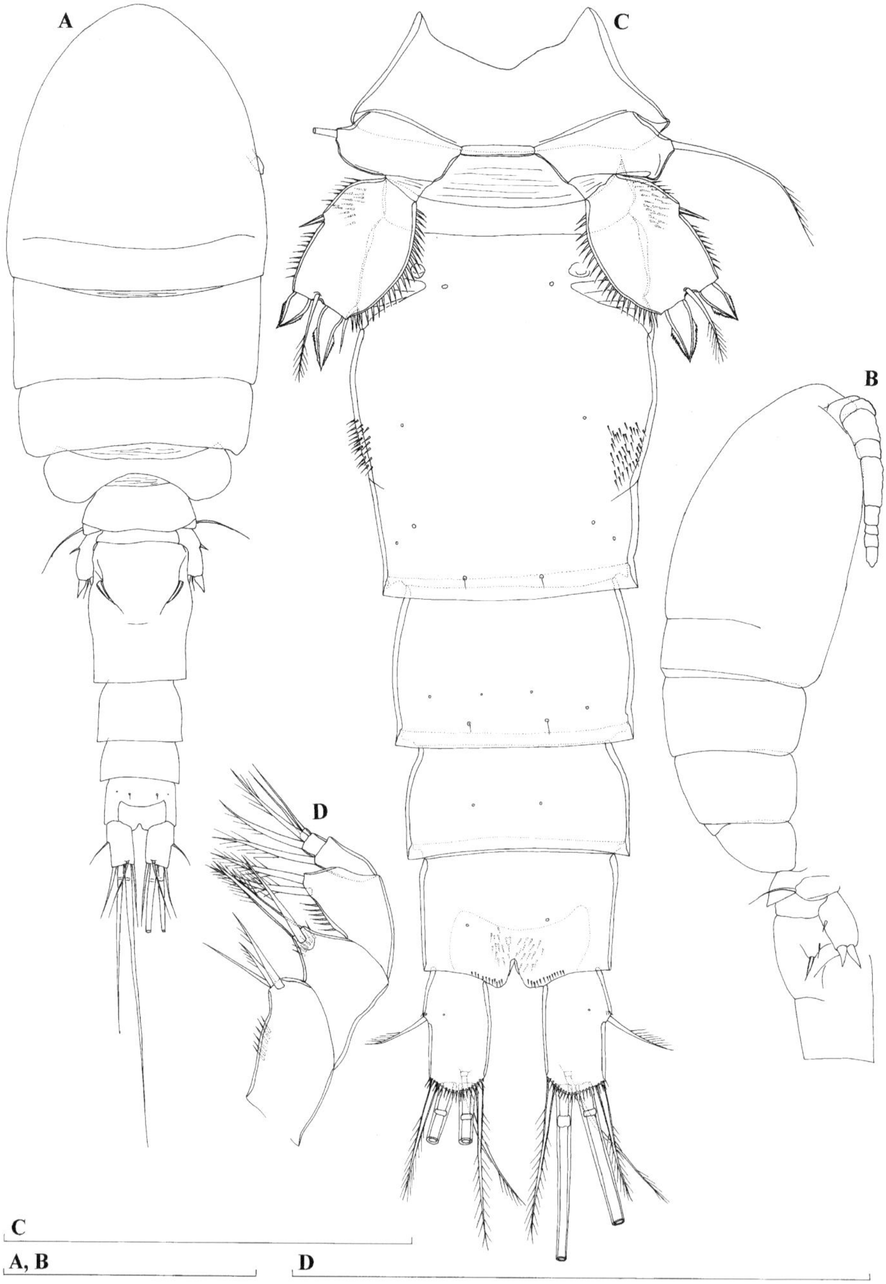

Fig. 26. *Allocyclopina australiensis* n. sp., holotype female. A, habitus, dorsal view; B, part of habitus, lateral view; C, urosome, ventral view; D, maxilliped. Scales = 0.1 mm.

rotized joint between fifth pedigerous somite and genital double somite well developed, almost as a pseudosomite (fig. 26C).

Genital double somite (fig. 26C) almost cylindrical, with anterior part only 1.1 times as wide as posterior and only slightly expanded at middle, 1.2 times as long as wide, ornamented with characteristic pattern of eight pores and two sensilla ventrally, as well as with a patch of spinules laterally at 3/5. No cuticular recesses or windows visible. Hyaline fringe of genital double and two succeeding somites narrow and smooth, both dorsally and ventrally. Copulatory pores and ovipores lying together within lateral genital apertures, which are situated at first third of genital double somite's length and covered by the reduced sixth legs. Copulatory ducts and seminal receptacle not visible. Third urosomite ornamented with four pores and two sensilla ventrally, while fourth urosomite ornamented with only two ventral pores. Anal somite with smooth, very broad and convex anal operculum, which represents 62% of the somite's width and reaches just beyond the middle of the somite's length; ornamented with two large sensilla and two pores dorsally, two pores ventrally, as well as with an interrupted row of minute spinules along posterior margin ventrally. Anal sinus widely open dorsally, ornamented with a large patch of robust spinules. Length ratio of abdominal somites, beginning with genital double somite, 100 : 44 : 28 : 36.

Caudal rami (fig. 26C) cylindrical, 1.7 times as long as wide (ventral view) and slightly shorter than anal somite, parallel, with space between them about one ramus' width; they represent 5% of total body length. Distal margin ventrally convex, ornamented with uninterrupted row of large spinules, with medial cuticular pore not clearly visible. The only additional ornamentation consists of one ventral pore at first third of the ramus' length and several small spinules at base of lateral seta. Dorsal seta 1.4 times as long as ramus, inserted on small protuberance near posterior margin, uniarticulate at base and unipinnate distally. Lateral seta slender and also unipinnate, arising at 2/5, and about 0.6 times as long as ramus. Outermost apical seta strong and bipinnate, as long as ramus, and 0.7 times as long as innermost apical seta, which is also strong and bipinnate. Principal apical setae with breaking planes and bipinnate; inner seta about 1.7 times as long as outer one and 0.3 times as long as body.

Antennula (fig. 27A) short and stout, reaching 3/5 of cephalothorax length, nine-segmented, ornamented with a small patch of spinules on first segment, armed with one slender aesthetasc laterally on sixth segment, one subapically on eighth segment, and one apically on ninth segment, and with setal formula:

3.5.11.6.7.2.2.2.7. No setae with breaking planes and only two setae on ninth segment articulating on basal part; most setae smooth and slender (only ten setae pinnate); two setae on fourth segment very short and spiniform, but smooth. Length ratio of antennular segments, from proximal end and along caudal margins, 1 : 0.4 : 1.7 : 1 : 2.9 : 1.2 : 0.7 : 0.6 : 1.3.

Antenna (fig. 27C) four-segmented, comprising coxobasis and three-segmented endopod, with geniculation between first and second endopodal segments; length ratio of antennal segments, 1 : 1 : 0.7 : 0.7. Coxobasis about 1.6 times as long as wide, ornamented with a short row of spinules along distal margin and armed with a single bipinnate seta at inner distal corner and a similar seta on distal outer corner (representing exopod); seta at inner distal corner about 1.4 times as long as exopodal seta, which is as long as coxobasis. First endopodal segment robust and twice as long as wide, armed with one bipinnate seta at middle, ornamented with a longitudinal row of minute spinules along internal margin proximally. Second endopodal segment about 1.7 times as long as wide, ornamented with a longitudinal row of spinules along outer (caudal) margin, armed with five smooth setae (three lateral, two subapical; one subapical seta geniculate, much longer and stronger than any other seta on that segment). Third endopodal segment 1.9 times as long as wide, ornamented as previous segment, and armed with seven apical setae, four of which very strong and geniculate.

Labrum not mounted properly to allow detailed examination.

Mandibula (fig. 27D) large, comprising coxa with well developed gnathobase and distal palp; palp consisting of basis, two-segmented endopod, and four-segmented exopod. Coxal gnathobase cutting edge divided into three regions: inner group of four complex and strong but short teeth; middle group of four simple teeth, inner two smooth, outer two bipinnate; and outermost unipinnate and short seta. Basis large, reaching somewhat beyond posterior margin of coxal gnathobase, 2.6 times as long as wide, unornamented, and armed with one plumose seta laterally. Endopod small, with first segment attached to basis apically and armed with three pinnate setae subapically; second segment about as long as first one and armed apically with six long and plumose setae. Exopod slender, 0.6 times as long as basis and 1.5 times as long as endopod, unornamented, with all segments of about same length and with setal formula 1.1.1.2.

Maxillula (fig. 28A) composed of syncoxa with well developed arthrite, basis, and one-segmented endopod and exopod. Arthrite of syncoxa with nine elements; three apical elements strong, bipinnate, and geniculate, proximalmost element longest and plumose; other three lateral elements strong

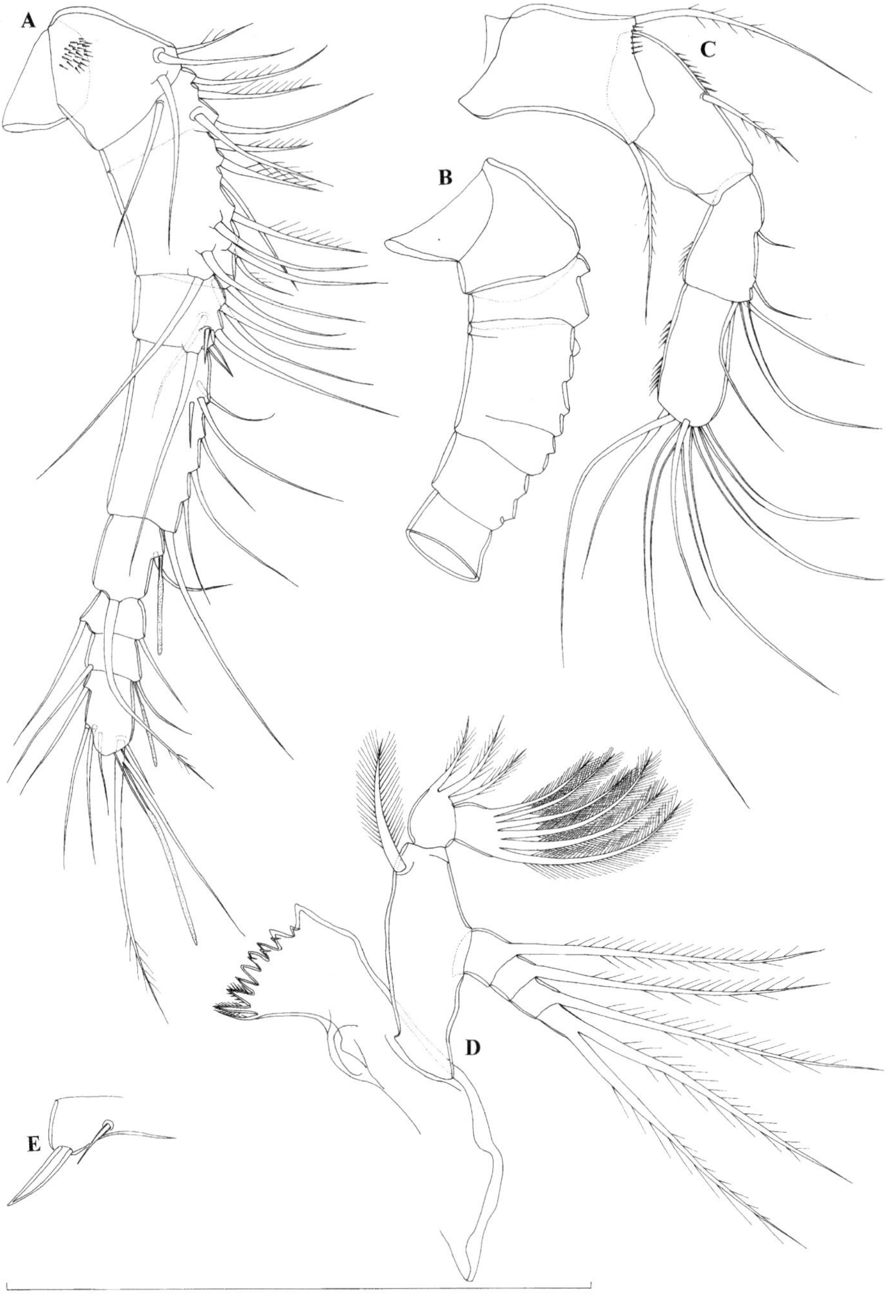

Fig. 27. *Allocyclopina australiensis* n. sp., holotype female. A, antennula; B, proximal part of opposite antennula without armature; C, antenna; D, mandibula; E, sixth leg, lateral view. Scale = 0.1 mm.

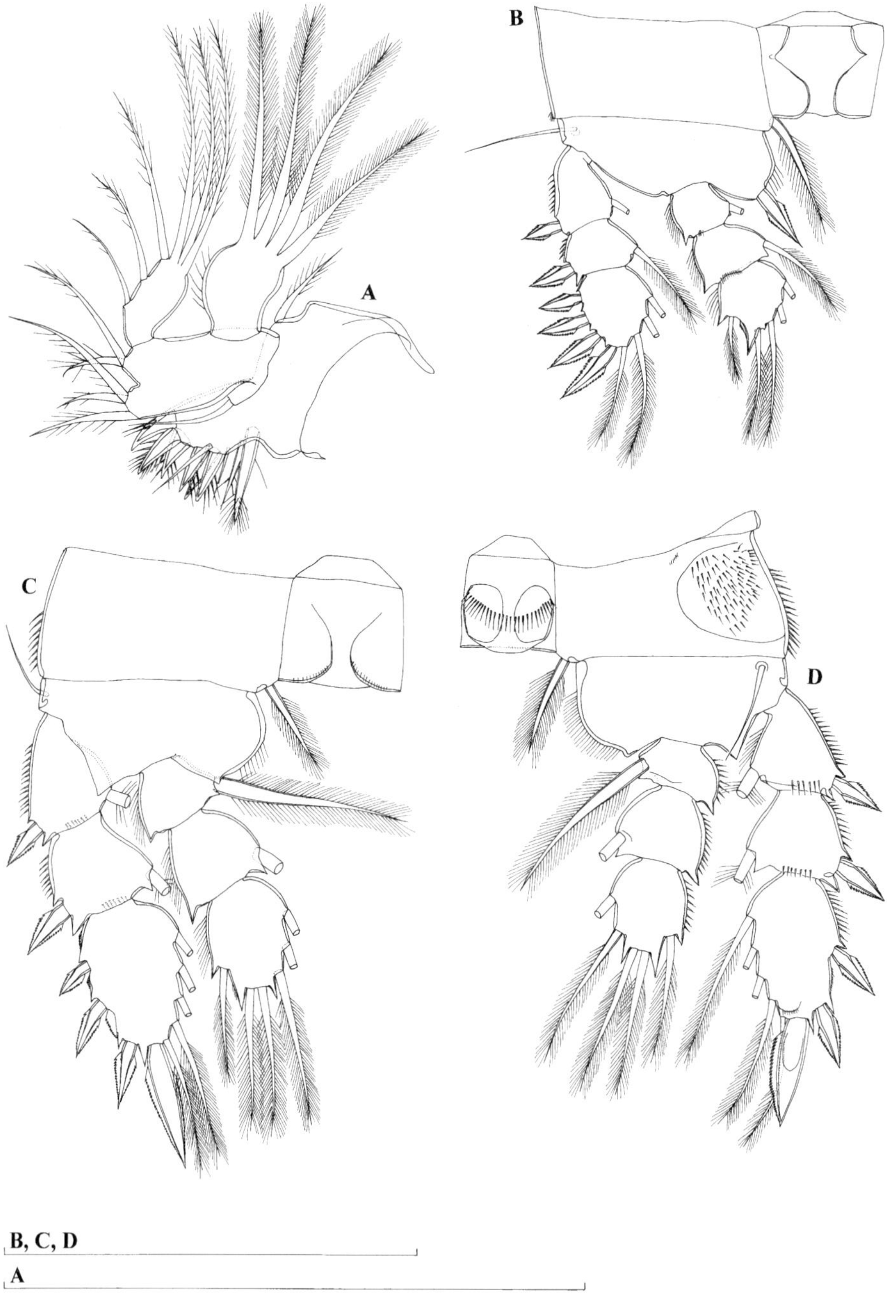

Fig. 28. *Allocyclopina australiensis* n. sp., holotype female. A, maxillula; B, first swimming leg; C, third swimming leg; D, fourth swimming leg. Scales = 0.1 mm.

and bipinnate, and only two lateral elements slender and smooth. Coxa unornamented and armed with one epipodal, bipinnate and slender seta; coxal endites weakly discernible, unarmed. Basis quadriform, longer than arthrite of syncoxa, 2.2 times as long as wide, unornamented, and armed with three setae on proximal and two setae on distal endite; all setae pinnate and relatively slender. Endopod somewhat longer than basis' width and about as long as exopod but more slender, unornamented, and armed with two lateral setae on inner margin and four long setae apically; all endopodal setae slender and pinnate. Exopod 1.2 times as long as wide, with a longitudinal row of hair-like spinules along inner margin, which is also convex, and armed apically with four slender, plumose, and very long setae.

Maxilla (fig. 29A) robust but not very large, five-segmented, comprising praecoxa, coxa, basis, and two-segmented endopod. Praecoxa quadriform, 1.2 times as wide as long, unornamented, with four setae on proximal and only one seta on distal endite. Coxa about as long as praecoxa but not so wide, armed with a single pinnate seta on proximal, elongated endite; distal endite highly mobile, ornamented with arched, transverse row of long spinules and bearing three slender and pinnate setae. Basis much shorter than coxa, expanded into spinulose, geniculate, and unipinnate claw, and armed with one strong and one minute setae. First endopodal segment half as long as basis, armed with four robust and pinnate setae. Second endopodal segment about as long as first but much narrower, armed with three strong and two slender setae.

Maxilliped (fig. 26D) robust but not large, six-segmented, composed of praecoxa, coxa, basis, and three-segmented endopod. Praecoxa 1.6 times as long as greatest width, ornamented with a longitudinal row of long spinules along inner margin proximally, with proximal endite weakly discernible and unarmed, and distal endite armed with three setae. Coxa about as long as praecoxa along outer margin, ornamented with a row of spinules at base of only endite, which is armed with two strong setae. Basis 1.3 times as long as wide, much narrower at basal part, ornamented with several large spinules along inner margin proximally, and armed with two pinnate setae distally on inner margin. Endopod slightly shorter than basis, unornamented, with setal formula 1.2.2. Length ratio of maxillipedal segments, from proximal end, 100 : 115 : 116 : 71 : 21 : 11.

All swimming legs (fig. 28B, C, D) with three-segmented exopod and endopod; their armature formula as follows (legend: inner/outer spine or seta; inner/terminal/outer):

Segments	Exopod 1	2	3	Endopod 1	2	3
First leg	1/1	1/1	3/2/3	1/0	1/0	2/2/1
Second leg	1/1	1/1	4/2/3	1/0	1/0	3/2/1
Third leg	1/1	1/1	4/2/3	1/0	1/0	3/2/1
Fourth leg	1/1	1/1	4/2/2	1/0	1/0	2/2/1

Third exopodal segment spine formula 4.4.4.3 and no spines on endopods. All intercoxal sclerites quadriform, with straight distal margin; that of first leg unornamented, those of second and third legs with minute spinules posteriorly along lateral lobes, while sclerite of fourth leg with continuous, arched row of large spinules at middle. Coxae ornamented with several spinules along outer margin (coxa of fourth leg with additional large patch of spinules at outer half), and armed with plumose seta at inner distal corner. Basis of each leg ornamented with a row of hair-like spinules along inner margin, and armed with slender seta on outer margin; basis of first leg with additional short and stout spine on distal inner corner, which is half as long as endopod; inner distal corners of second, third, and fourth leg pronounced as short spiniform processes; similar, but somewhat larger process also between exopod and endopod. All exopodal and endopodal setae slender and plumose. All exopodal spines stout and with both margins sharply serrated; outer apical spine on fourth leg unusually wide and with smooth outer margin. All endopodal segments with spinules along outer margin and most exopodal segments with spinules along outer and posterior margins. Third endopodal segment of fourth swimming leg about 1.2 times as long as wide; both apical setae of same length, 1.7 times as long as the segment.

Fifth leg (fig. 26C) with very short and broad intercoxal sclerite and two-segmented, comprising coxobasis and one-segmented exopod. Coxobasis almost twice as wide as long, unornamented, and armed with a single unipinnate basal seta on outer margin. Exopod ovoid, 1.9 times as long as basis and 1.4 times as long as its greatest width, much narrower at basal part, armed with three spines and a single seta, and ornamented with strong spinules along both inner and outer margins, as well as with additional patch of spinules proximally on caudal side; seta bipinnate at distal part, positioned centrally, and as the most apical element, 0.6 times as long as exopod, 1.4 times as long as inner apical spine, about 1.8 times as long as outer apical spine, and 2.6 times as long as lateral spine; lateral spine minute and smooth, apical spines very strong and with sharply serrated margins.

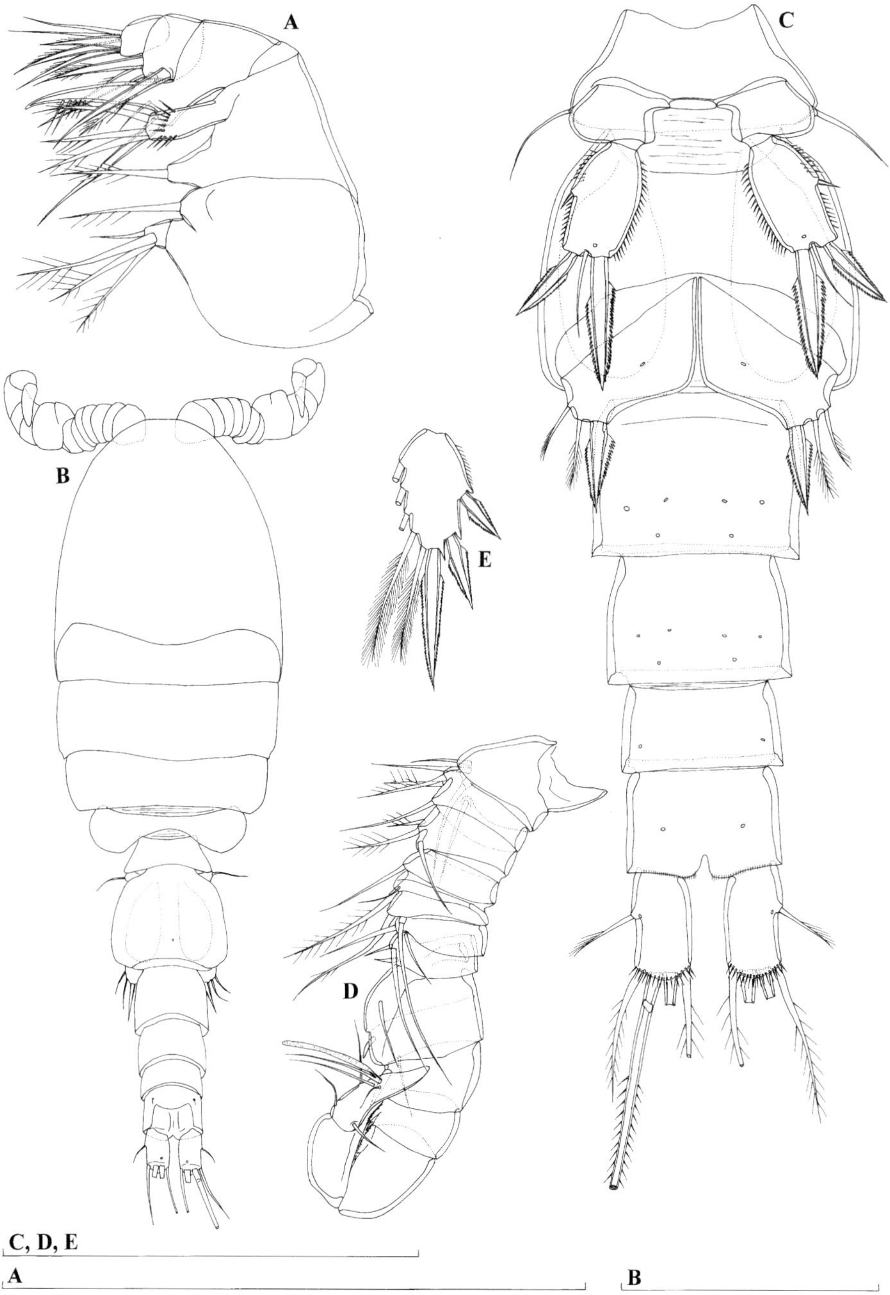

Fig. 29. *Allocyclopina australiensis* n. sp., A, holotype female; B-E, male from Cowell (0.442 mm). A, maxilla; B, habitus, dorsal view; C, urosome, ventral view; D, antennula; E, third exopodal segment of fourth swimming leg. Scales = 0.1 mm.

Sixth leg (fig. 27E) not clearly distinct, cuticular plate, armed with outer smooth and dorsally directed spine, and inner smooth and slender seta; spine 1.7 times as long as seta.

Male from Cowell. Body length, excluding caudal setae, 0.442 mm. Habitus (fig. 29B) similar to that of female, with prosome/urosome ratio 1.3 and greatest width at posterior end of cephalothorax. Body length/width ratio about 3.2; cephalothorax twice as wide as genital somite.

Cephalothorax (fig. 29B) 1.1 times as long as its greatest width, with first pedigerous somite not completely fused dorsally. Arthrodial membrane well developed between third and fourth pedigerous somites, as well as between prosome and urosome. Ornamentation of prosomites, colour, and nauplius eye similar to those of female. Hyaline fringes of all somites smooth.

Genital somite (fig. 29B, C) rather large, 1.3 times as wide as long (dorsal view), with two completely formed and large spermatophores inside, ornamented with only one pore dorsally; third and fourth urosomites ornamented with six ventral pores each, and fifth urosomite with only two ventral pores; anal somite similar to that of female.

Caudal rami (fig. 29C) about 1.8 times as long as wide, parallel, with space between them less than one ramus' width. Armature and ornamentation similar to that in female, but dorsal and innermost apical setae broken.

Antennula (fig. 29D) shorter than cephalothorax, 14-segmented, strongly geniculate between twelfth and thirteenth segments, so not all armature clearly visible.

Antenna, labrum, mandibula, maxillula, maxilla, maxilliped, and swimming legs all very similar to those of female.

Third exopodal segment of fourth swimming leg (fig. 29E) more slender and also with more slender outer apical spine than in female.

Fifth leg (fig. 29C) similar to that of female but much more slender. Exopod 1.8 times as long as wide, additionally ornamented with cuticular pore at base of apical seta. Inner apical spine much longer than in female, slightly longer than exopod, 1.8 times as long as apical seta (which is smooth), 1.5 times as long as outer apical spine, and 5.6 times as long as minute outer spine.

Sixth leg (fig. 29C) distinct, large plate, ornamented with a single pore and armed with two pinnate setae and innermost strong spine; all elements of about same length.

Variability. — Only one female and one male were collected and examined. The holotype female has one antennula with incompletely divided second and third segments (fig. 27A), while the opposite antennula has these segments completely divided and also some additional short subdivision lines on the third segment (fig. 27B).

Etymology. — As a first representative of the genus *Allocyclopina* from Australia, this species is named *australiensis*, by adding the Latin suffix for place "-ensis" to the name of the continent. The specific name consequently is an adjective for place.

Distribution. — *Allocyclopina australiensis* n. sp. is known from two localities on two different peninsulas in South Australia.

Discussion. — Only two species have been described so far in the genus *Allocyclopina* Kiefer, 1954 and, like the new species described here, both are from the marine interstitial of sandy beaches. *Allocyclopina madagassica* Kiefer, 1954 was described after both sexes twice as a new species and a new genus by Kiefer (1954, 1956), from the edge of the Maroantsetra Estuary in Madagascar. Both descriptions are almost identical, although the first one is in German (Kiefer, 1954) and the second one in French (Kiefer, 1956), and the same figures were used (although differently arranged), but the latter paper contains an additional drawing of the cephalothorax. Also, the German version does not contain the armature formula of the swimming legs. *Allocyclopina ambigua* Kiefer, 1960 was described after two females from a beach at St. Pierre, island of Réunion. As Réunion lies only slightly over 500 km east of Madagascar, my new species extends the known range of the genus significantly eastward.

Allocyclopina australiensis n. sp. differs from *A. madagassica* and *A. ambigua* by its shorter caudal rami, the presence of four spines on the third exopodal segment of the first swimming leg (three in the other two species), and by its nine-segmented antennula (10-segmented in the other two species). However, one antennula of *A. australiensis* (fig. 27B) shows a partial division of the third segment, which would make the appendage segmented just as in the other two species if fully expressed. *Allocyclopina madagassica* differs from the new species additionally by having only two armature elements on the first endopodal segment of the mandibula, but the mouth appendages of both *A. madagassica* and *A. ambigua* have been only incompletely described. Nevertheless, the position of my new species in the genus *Allocyclopina* is unquestionable, as it shares many peculiar characters with the other two species, including an almost identical swimming leg armature formula, along with the shape, armature, and ornamentation of the fifth leg. The segmentation and armature formula of the fifth leg of *Allocyclopina* is not unique in the family, and indeed it is shared with the previously mentioned genus *Hemicyclopina* Herbst, 1953, but in *Allocyclopina* the exopod is always swollen, heavily ornamented on both the inner and outer margins, and has a very small lateral spine, which is always the smallest of the three. Also, the male fifth leg is very similar to that of the female in both *A. australiensis* and *A. madagassica* (the male of *A. ambigua* is unknown, as mentioned previously), although with a somewhat more slender exopod. Maybe I should mention here that the male of *A. australiensis* has three armature elements on its sixth leg, while that of *A. madagassica* has only two, which (among other things) makes *A. australiensis* somewhat more distant from the other two species. Also, *A. madagassica* and *A. ambigua* have very similar caudal rami, the same segmentation of the antennula, and the very same swimming legs armature formula, which all suggest that the two species are more closely related than either is to *A. australiensis*. Another important character that all three species share is the absence of a ventrally located, single copulatory pore. This was not specifically stated by Kiefer (1954, 1956, 1960a) for either of the two species described by him, but his drawings of the ventral aspect of the abdomen are pretty clear in this regard.

The copulatory pore and ovipore lie together beneath the reduced sixth legs on each side of the genital double somite. This is the situation found in Poecilostomatoida, in the cyclopoid family Oithonidae Dana, 1853, and in some members of the family Cyclopinidae G. O. Sars, 1913. Besides *Allocyclopina*, this condition is so far confirmed for the genus *Cyclopetta* G. O. Sars, 1913 by Martínez Arbizu (2000a), but may also be the case in the genera *Cyclopuella* Por, 1979, *Paracyclopetta* Wells, 1967, and *Paracyclopina* Herbst, 1955, as far as can be judged from the drawings alone (see Por, 1979; Wells, 1967; Lotufo, 1995). As all of the above-mentioned genera, families, and orders have very little in common, I believe that this character state originated independently a number of times in the Cyclopoida from the ancestral state (a single, ventrally located copulatory pore), while it is the ancestral state for Poecilostomatoida. As can be seen in the genus *Hemicyclops* Boeck, 1873 (figs. 1B, 2B in the present monograph), in poecilostomatoids the spermatophores are also attached laterally, whereas in those Cyclopoida with laterally placed copulatory pores of which we have some knowledge about the placement of the spermatophores (all from the family Oithonidae), they are mostly attached ventrally, close to each other, and have long tubules that reach these pores (see Huys & Boxshall, 1991). The lateral migration of the copulatory pores could be connected with the planktonic way of life in the family Oithonidae (maybe for better protection against algal growth and blockage), but it would be very hard to guess what triggered this transformation in the family Cyclopinidae. In any case, this happened later in the evolution of this family, as all very primitive forms have a single, ventrally located copulatory pore (see, for example, Huys & Boxshall, 1990). Martínez Arbizu (2000a) made remarks on this character in Oithonidae and Cyclopinidae, and even mentioned a possibility that "oithonids evolved as a specialized lineage within the paraphyletic Cyclopinidae". However, he left them as a separate family, which is one of the things that show why his commenced revision of the Cyclopinidae (Martínez Arbizu, 2000a, b, 2001a, b) is unacceptable already at this stage: basically, he split the paraphyletic family Cyclopinidae in order to create more paraphyletic families.

The closest relative of *Allocyclopina* seems to be the monospecific genus *Afrocyclopina* Wells, 1967, described from Inhaca Island, Mozambique, by Wells (1967), but it may prove to be a synonym of the former genus after re-examination. The main characters that distinguish *Afrocyclopina platypus* Wells, 1967 from the three members of *Allocyclopina* are its two-segmented endopod of the first swimming leg and the absence of the lateral spine on the exopod of the fifth leg. However, this spine may easily be broken during dissection, as Wells' (1967) fig. 4L shows a clear notch in the place where this spine is attached in *Allocyclopina*. *Afrocyclopina* also has a very similar shape of caudal rami, as well as a similar armature formula of the swimming legs. I prefer not to synonymize it here, without an examination of the type material.

Genus **Cyclopina** Claus, 1862

[= *Troglocyclopina* Jaume & Boxshall, 1996b;
= *Heptnerina* Ivanenko & Defaye, 2004]

10. **Cyclopina unisetosa** n. sp.

(figs. 30-32)

Material examined. — Type only: Holotype ovigerous female dissected on 2 slides (WAM C37221): Australia, Western Australia, Coral Bay, 3 August 2003, leg. T. Karanovic, 23°08.577′S 113°46.163′E.

Description. — Ovigerous female (holotype). Body length, excluding caudal setae, 0.425 mm. Habitus (fig. 30A) relatively robust, with prosome/urosome ratio 1.4 and greatest width at posterior end of cephalothorax. Body length/width ratio about 3.1; cephalothorax about 2.7 times as wide as genital double somite. Posterior corners of cephalothorax and first two pedigerous somites without pronounced lateral expansions; posterior corners of third and fourth pedigerous somites only slightly expanded laterally. Urosome narrow, and fifth pedigerous somite about as wide as genital double somite. Two egg sacs, each with three large eggs, attached to anterior part of genital double somite and reaching middle of anal somite. Colour of preserved specimen milky white. Nauplius eye not visible. Rostrum broadly rounded and furnished with two large sensilla.

Prosome (fig. 30A) comprising cephalothorax and four free pedigerous somites (first pedigerous somite free), ovoid, almost twice as long as wide (dorsal view); length ratio of prosomal somites, beginning with cephalothorax, 100 : 24 : 28 : 21 : 6. Cephalothorax slightly shorter than its greatest width, representing 31% of total body length; surface of cephalic shield with several small sensilla and cuticular pores. Hyaline fringe of prosomites narrow and smooth. Fifth pedigerous somite (figs. 30B, C, 31A) ornamented with two large sensilla and three pores dorsally and one pore laterally, with smooth fringe dorsally and finely serrated fringe laterally. Arthrodial membrane between third and fourth pedigerous somites visible from dorsal view. Sclerotized joint between fifth pedigerous somite and genital double somite well developed, almost as a pseudosomite (fig. 30B, C).

Genital double somite (figs. 30B, C, 31A) with anterior part inflated laterally, about 1.2 times as long as wide (dorsal view), ornamented with six pores dorsally, four pores laterally, and six pores ventrally; two ventral cuticular pores very close to copulatory pore. Hyaline fringe of genital double somite and two succeeding somites narrow, not frilled, and completely smooth. No cuticular recesses, windows, or sensilla visible. The single copulatory pore relatively large and ovoid, situated midventrally at 1/4 of the double somite's length; copulatory duct short and narrow, not rigidly sclerotized, directed anteriorly and dorsally. Seminal receptacles fused, small and tubular, hardly wider than copulatory duct, and without any visible border with receptacle ducts; seminal receptacles and receptacle ducts form simple, arched tube, which forms a T-shaped structure with the copulatory duct in ventral view. Paired ovipores located dorsolaterally at 1/3 of double somite's length, covered by reduced sixth legs and connected to receptacle via long

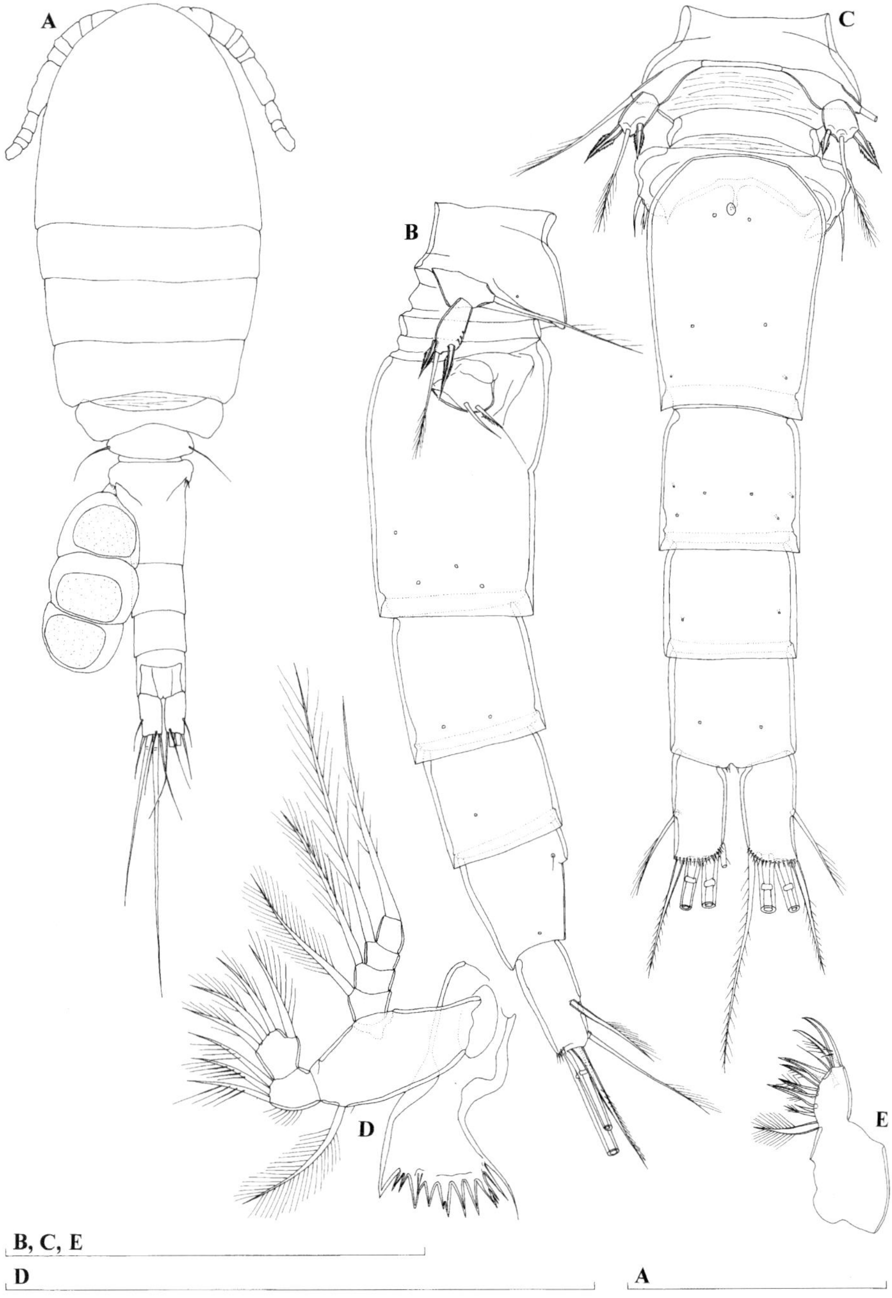

Fig. 30. *Cyclopina unisetosa* n. sp., holotype female. A, habitus, dorsal view; B, urosome, lateral view; C, urosome, ventral view; D, mandibula; E, part of maxillula. Scales = 0.1 mm.

and narrow receptacle ducts. Third urosomite ornamented with three dorsal, two lateral, and six ventral pores, while fourth urosomite has one dorsal, one lateral, and two ventral pores as only ornamentation. Anal somite with smooth, very broad, and slightly convex anal operculum, which represents 73% of the somite's width and reaches to only 2/5 of the somite's length; ornamented with two large dorsal sensilla, one pore laterally on each side, and two pores ventrally; posterior margin smooth and unornamented. Anal sinus widely open dorsally, ornamented with two short longitudinal rows of minute spinules. Length ratio of abdominal somites, beginning with genital double somite, 100 : 56 : 45 : 45.

Caudal rami (figs. 30B, C, 31A) cylindrical, twice as long as wide (ventral view) and about as long as anal somite, parallel, with very small space between them; they represent 6% of total body length. Distal margin ventrally convex and ornamented with continuous row of spinules (as only ramal ornamentation), with medial cuticular pore not clearly visible; posterior outer corners of caudal ramus reaching beyond posterior inner corners. Dorsal seta relatively long, 1.3 times as long as ramus, inserted on small protuberance reaching posterior margin and close to inner dorsal corner, uniarticulate at base, and bipinnate distally. Lateral seta also slender, about as long as ramus, arising at 3/5 of the ramus' length and somewhat dorsolaterally. Outermost apical seta strong and bipinnate, slightly longer than ramus, and half as long as innermost apical seta, which is also strongly bipinnate but more slender. Principal apical setae with breaking planes and bipinnate; inner seta about 1.5 times as long as outer one and 0.35 times as long as body.

Antennula (fig. 31B) short and stout, reaching 3/4 of cephalothorax length, 10-segmented, unornamented, with one short and slender aesthetasc on fifth and seventh segments laterally, one on ninth segment subapically, one longer aesthhetasc on tenth segment apically, and with a setal formula: 3.5.7.4.5.7.3.2.1.7. No setae with breaking planes and only one subapical seta on sixth segment, one on seventh segment and two apical setae on tenth segment articulating on basal part; most setae smooth (only three pinnate) and slender; one seta on fifth segment very short and spiniform; one apical seta on tenth segment fused basally to aesthetasc. Length ratio of antennular segments, from proximal end and along caudal margins, 1 : 0.8 : 1.2 : 0.5 : 0.9 : 3.6 : 1.1 : 0.6 : 0.8 : 1.1.

Antenna (fig. 31C, D) four-segmented, comprising coxobasis and three-segmented endopod, with geniculation between first and second endopodal segments; length ratio of antennal segments, 1 : 0.8 : 0.5 : 0.8. Coxobasis

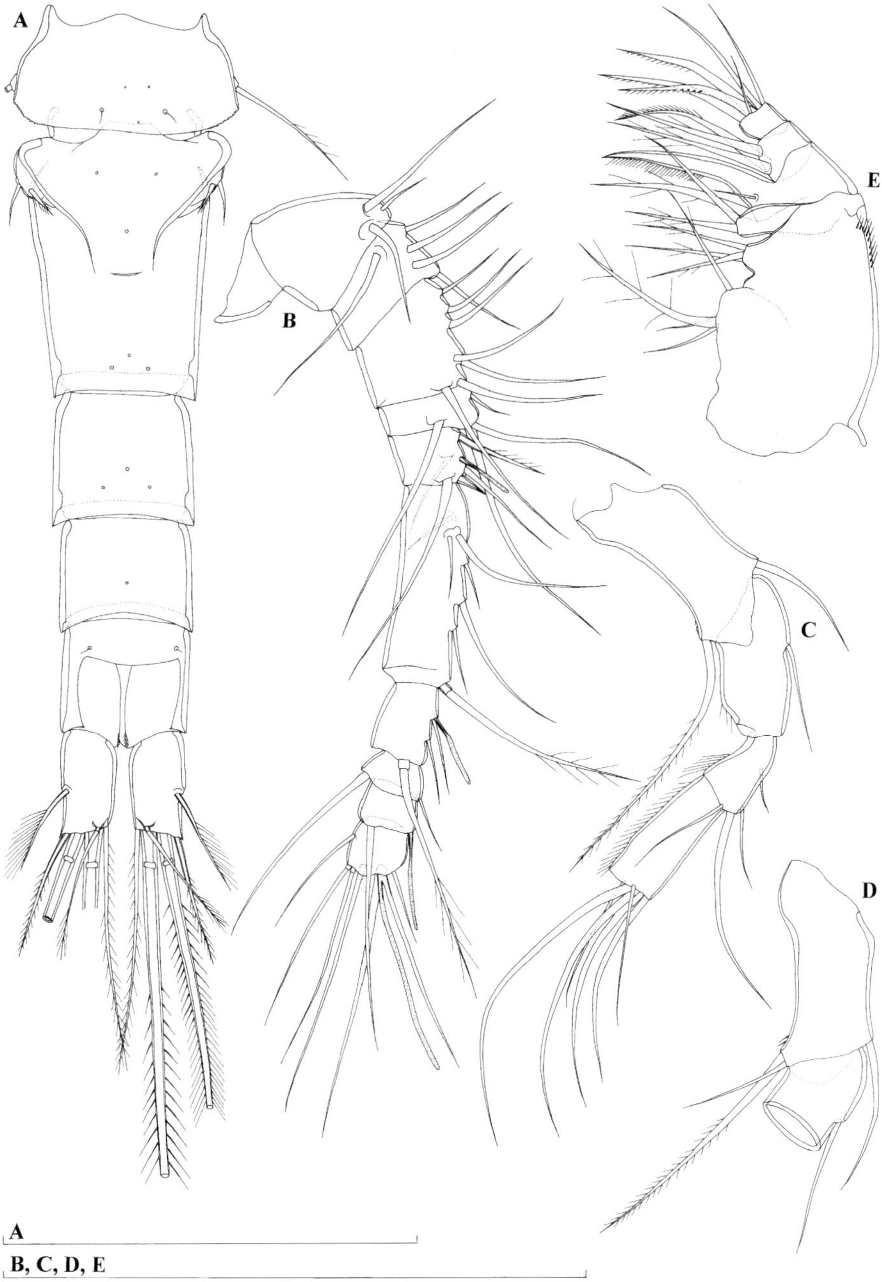

Fig. 31. *Cyclopina unisetosa* n. sp., holotype female. A, urosome, dorsal view; B, antennula; C, antenna; D, proximal part of opposite antenna; E, maxilla. Scales = 0.1 mm.

about 2.2 times as long as wide, ornamented with two or three small spinules on outer (caudal) margin distally, armed with a single smooth seta at inner distal corner, and a much stronger and bipinnate seta on outer distal corner (representing exopod); one antenna with additional smooth and short exopodal seta; pinnate exopodal seta 1.4 times as long as coxobasis and twice as long as smooth seta on inner distal corner. First endopodal segment twice as long as wide, armed with one smooth seta at middle, ornamented with longitudinal row of hair-like spinules along external margin. Second endopodal segment about 1.6 times as long as wide, ornamented as previous segment, armed with four smooth setae (two lateral, two subapical; one subapical seta geniculate, much longer and stronger than any other seta on that segment). Third endopodal segment 2.4 times as long as wide, armed with six smooth apical setae, two of which strong and geniculate, and ornamented with hair-like spinules along outer margin.

Labrum not mounted satisfactorily to allow detailed examination.

Mandibula (fig. 30D) large, comprising coxa with well developed gnathobase and distal palp; palp consisting of basis, two-segmented endopod, and four-segmented exopod. Coxal gnathobase cutting edge wide, armed with nine sharp teeth (four of which complex), one long spinule, and outermost smooth seta. Basis large but not reaching beyond posterior margin of coxal gnathobase, 2.2 times as long as wide, unornamented, and armed with one plumose seta laterally. Endopod small, with first segment attached apically to basis and armed with three pinnate setae subapically; second segment about as long as first one and armed apically with four short and pinnate setae. Exopod slender, 0.6 times as long as basis and 1.3 times as long as endopod, unornamented, with all segments of about the same length, and with setal formula 1.1.1.1.

Maxillula (figs. 30E, 32A) composed of syncoxa with well developed arthrite, basis, and one-segmented endopod and exopod. Arthrite of syncoxa with nine unequal lateral elements and one smooth seta on posterior surface; four apical elements strong and geniculate, proximalmost element longest and plumose. Coxa unornamented and armed with two epipodal setae (one of which moved to base of basis); coxal endites not discernible, unarmed. Basis quadriform, much longer than arthrite of syncoxa, 2.1 times as long as wide, unornamented, and armed with three elements (two spines and one seta) on proximal, and two setae on distal endite. Endopod short and wide, shorter than basis' width, and exopod unornamented and armed with two lateral setae on inner margin and four long setae apically; all endopodal setae slender and all except one plumose. Exopod 1.3 times as long as wide, with longitudinal row of hair-like spinules along inner margin, which is also convex, and armed apically with four slender, plumose, and long setae.

Maxilla (fig. 31E) robust but not very large, five-segmented, comprising praecoxa, coxa, basis, and two-segmented endopod (although praecoxa and coxa partly fused). Praecoxa unornamented, armed with three setae on proximal, and only one seta on distal endite. Coxa ornamented with longitudinal row of spinules along outer margin, and armed with three slender and pinnate setae on proximal endite; distal endite highly mobile, unornamented, bearing two slender setae. Basis much shorter than coxa, expanded into spinulose, geniculate, and unipinnate claw, and armed with one strong and one minute setae. First endopodal segment half as long as basis, armed with three robust and pinnate setae. Second endopodal segment about as long as first but much narrower, armed with three strong and three slender setae.

Maxilliped (fig. 32B) slender, seven-segmented, composed of praecoxa, coxa, basis, and four-segmented endopod. Praecoxa about as long as its greatest width, unornamented, with proximal endite weakly discernible and armed with a single seta, and distal endite armed with three setae. Coxa about as long as praecoxa along outer margin, ornamented with a row of hair-like spinules along outer margin, armed with two strong setae on only endite. Basis the longest segment, 1.7 times as long as wide, much narrower at basal part, ornamented with several hair-like spinules along inner margin proximally, and armed with two pinnate setae distally on inner margin. Endopod slightly shorter than basis, unornamented, with setal formula 0.0.0.3. Length ratio of maxillipedal segments, from proximal end, 100 : 100 : 119 : 33 : 33 : 26 : 15.

All swimming legs (fig. 32C, D, E, F) with three-segmented exopod and endopod; their armature formula as follows (legend: inner/outer spine or seta; inner/terminal/outer):

	Exopod			Endopod		
Segments	1	2	3	1	2	3
First leg	0/1	1/1	3/2/3	1/0	1/0	3/2/1
Second leg	1/1	1/1	4/2/3	1/0	2/0	3/2/1
Third leg	1/1	1/1	4/2/3	1/0	2/0	3/2/1
Fourth leg	1/1	1/1	4/2/2	1/0	2/0	2/2/1

Third exopodal segment spine formula 4.4.4.3 and no spines on endopods. All intercoxal sclerites quadriform, with slightly convex distal margin; those

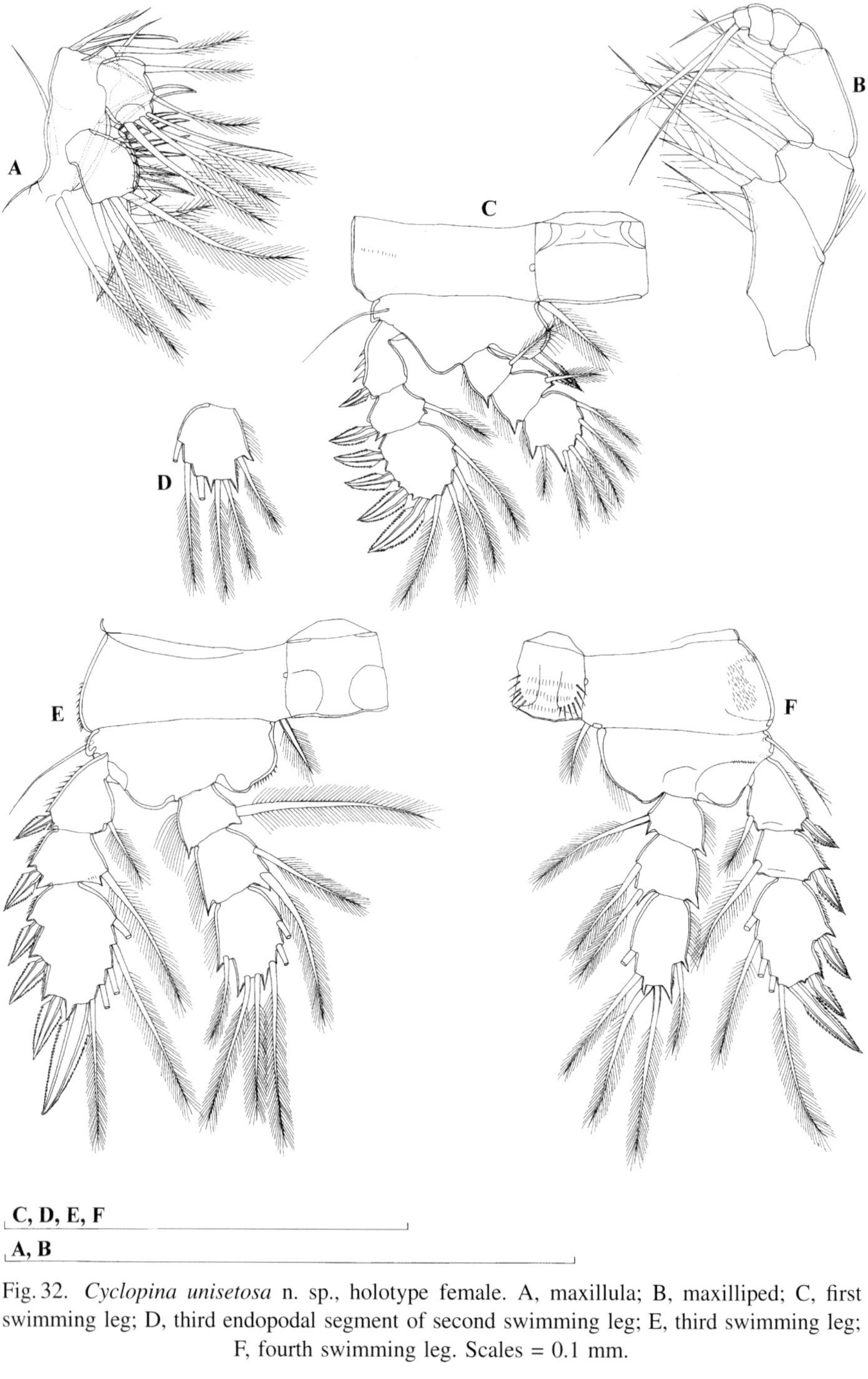

Fig. 32. *Cyclopina unisetosa* n. sp., holotype female. A, maxillula; B, maxilliped; C, first swimming leg; D, third endopodal segment of second swimming leg; E, third swimming leg; F, fourth swimming leg. Scales = 0.1 mm.

of first, second, and third legs unornamented, while sclerite of fourth leg with interrupted, arched row of large spinules and three uninterrupted rows of small spinules at middle. Coxae ornamented with several spinules along outer margin (coxa of fourth leg with additional large patch of spinules on outer half), armed with plumose seta at inner distal corner. Basis of each leg ornamented with a row of spinules along inner margin, and armed with a slender seta on outer margin; basis of first leg with additional short and stout spine on distal inner corner, half as long as endopod; inner distal corners of second, third, and fourth leg only slightly and bluntly pronounced. All exopodal and endopodal setae slender and plumose. All exopodal spines stout and with both margins sharply serrated. All endopodal segments with hair-like spinules along outer margin and most exopodal segments with short spinules along outer margin. Third endopodal segment of fourth swimming leg about 1.5 times as long as wide; apical setae of similar length, twice as long as segment.

Fifth leg (fig. 30B, C) small, with very short and broad intercoxal sclerite, and two-segmented, comprising coxobasis and a one-segmented exopod. Coxobasis about 1.7 times as wide as long, unornamented, and armed with a single unipinnate basal seta on outer margin. Exopod 1.4 times as long as basis and twice as long as its greatest width, somewhat narrower at basal part, armed with two spines and a single seta apically, and ornamented with several spinules along outer margin distally; seta bipinnate at distal part, positioned centrally, and as the most apical element; 1.8 times as long as exopod, 2.6 times as long as inner apical spine, and about 1.9 times as long as outer apical spine; both spines strong and with sharply serrated margins.

Sixth leg (fig. 30B) not clearly distinct, ovoid cuticular plate, armed with two short and slender setae.

Male. Unknown.

Variability. — A single female was collected and examined. The only asymmetry observed is on the antenna, with one appendage having only one exopodal seta (fig. 31C) and the opposite appendage bearing an additional small and smooth exopodal seta (fig. 31D).

Etymology. — The specific name refers to the presence of only one seta on the ultimate exopodal segment of the mandibula. It is to be treated as an adjective, agreeing in gender with the feminine generic name.

Distribution. — *Cyclopina unisetosa* n. sp. is known from the type locality only.

Discussion. — *Cyclopina* Claus, 1862 is the type and oldest cyclopinid genus. About 60 species and subspecies have been described, but during the last 140 years many of these have been either transferred to newly established genera, or synonymized. Nevertheless, with more than 30 valid species, *Cyclopina* is still the largest cyclopinid genus, containing around 30% of the family's diversity (Boxshall & Halsey, 2004). Despite its size and the systematic confusion surrounding the family Cyclopinidae, *Cyclopina* is relatively well defined by its characteristic fifth legs, short antennulae (10-13 segmented), characteristic genital field, short caudal rami armed with only six setae, conservative armature formula of the swimming legs, and characteristic mouth appendages. In fact, the genus is so conservative that species delineation is often difficult, and this is compounded by the many descriptions that lack detail and by significant variability of some of its members; this is one of the copepod genera where alternative taxonomic methods could make a real difference. The genus *Troglocyclopina* Jaume & Boxshall, 1996, described by Jaume & Boxshall (1996b) to accommodate a single anchialine species from the Balearic Islands, in my opinion is a synonym of *Cyclopina*. The authors did not compare their new genus with *Cyclopina*, but rather relied on the key provided by Huys & Boxshall (1990) for establishing closer relationships with other genera. The only character that distinguishes *Cyclopina balearica* (Jaume & Boxshall, 1996) n. comb. from other congeners is a discrete coxal endite on the maxillula (almost certainly a plesiomorphic feature), but its armature differs little. The same could be said for the genus *Heptnerina* Ivanenko & Defaye, 2004, which was erected to accommodate a single species from the Mid-Atlantic Ridge by Ivanenko & Defaye (2004). This genus was properly compared with *Cyclopina*, but it was diagnosed by an atavistic and very unstable character (i.e., not present in all specimens). Although the presence of a very reduced endopod on the female fifth leg in some specimens of *Cyclopina confusa* (Ivanenko & Defaye, 2004) n. comb. is a very interesting feature, this kind of atavism could not be used as a sole generic character. For example, Petkovski & Karanovic (1997) reported an atavistic male fifth leg for a new harpacticoid species from Ohrid Lake, which in all other aspects is a sister species to another, more widely distributed congener.

The six new species described in this monograph constitute the first record of the genus *Cyclopina* from Australia. Although each one of them is easily distinguishable from the other five (despite a significant variability of some of them) and from other members of the genus, this is not always the case when trying to distinguish previously described representatives. However, I was not able to positively identify any of the six Australian representatives as a previously described species: they could always be distinguished by at least one or two characters. While it is possible that some species of *Cyclopina* will turn out to belong to a widely distributed and highly variable species, the six species described here are obviously very close, but are very easy to distinguish when studied in detail, and all show a very restricted distribution.

Cyclopina unisetosa n. sp., as its name suggests, is the only species in the genus with only one seta on the ultimate exopodal segment of the mandibula (fig. 30D). All other species (including all other Australian ones) have two setae on this segment and the segmentation and armature of the mandibular exopod is unusually conservative in this genus. The only other case of an armature formula of the mandibular exopod being different from the normal 1.1.1.2 is that reported by Sars (1913) for *Cyclopina gracilis* Claus, 1863, with three setae on the ultimate article, but this was probably erroneous. *Cyclopina unisetosa* is most closely related to *C. kasignete* n. sp., which is the only other Australian representative with a 10-segmented female antennula. However, the latter species has a much longer copulatory duct, smaller eggs, and more of those in its egg sac, a different abdominal pore pattern, as well as slightly longer caudal rami, and a more robust inner spine on the fifth leg. Unfortunately, male characters could not be compared, as *C. unisetosa* is known only from females. Another unusual character of *C. unisetosa* is the absence of the inner seta from the first exopodal segment of the first leg, which was so far reported only for some specimens of *C. steueri* Früchtl, 1923 by Pleşa (1963). This character is also present in two other Australian representatives: *C. amita* n. sp. and *C. soror* n. sp. It should be mentioned here, that all six Australian species lack the apical tuft of hairs on the tip of the outer apical (or only apical in *C. unisetosa*) seta of the mandibular exopod, which is otherwise very common in the genus *Cyclopina*. This may suggest their closer relationship, as it may be easy to interpret this character as a synapomorphy. However, this would be hard to prove on morphological grounds only, as a complete morphology is known for only a handful of *Cyclopina* species so far.

11. **Cyclopina amita** n. sp.
(figs. 33-36)

Material examined. — Types: Holotype female dissected on 1 slide (WAM C37222); paratype female in alcohol (WAM C37223): Australia, Western Australia, Dampier, 4 August 2003, leg. T. Karanovic, 20°40.051′S 116°42.244′E.

Other material: Australia, Western Australia, Karratha, near jetty, 4 August 2003, leg. T. Karanovic, 20°43.298′S 116°53.458′E: 4 males + 2 females + 2 copepodids (1 male (WAM C37224) and 1 female (WAM C37225) dissected on 1 slide each, 1 female on 1 slide in toto (TMAG G5776), others in alcohol (WAM C37226).

Description. — Female (holotype). Body length, excluding caudal setae, 0.438 mm. Habitus (fig. 33A) relatively robust, with prosome/urosome ratio 1.5 and greatest width at posterior end of cephalothorax. Body length/width ratio about 3.0; cephalothorax 2.9 times as wide as genital double somite. Posterolateral corners of cephalothorax forming carapace-like extensions and concealing lateral margins of first pedigerous somite. First three pedigerous somites without pronounced lateral expansions; posterior corners of fourth pedigerous somite somewhat expanded laterally. Urosome narrow, and fifth pedigerous somite only slightly narrower than genital double somite. Colour of preserved specimen milky white. Nauplius eye not visible. Rostrum broadly rounded and furnished with two large sensilla.

Prosome (fig. 33A) comprising cephalothorax and four free pedigerous somites (first pedigerous somite free, although partly concealed by cephalothorax), ovoid, almost twice as long as wide (dorsal view); length ratio of prosomal somites, beginning with cephalothorax, 100 : 21 : 25 : 18 : 7. Cephalothorax slightly longer than its greatest width, representing 35% of total body length; surface of cephalic shield with several small sensilla and cuticular pores. Hyaline fringe of prosomites narrow and smooth. Fifth pedigerous somite (fig. 33C) ornamented with two large sensilla dorsally and one pore laterally, and with smooth fringe dorsally and laterally. Arthrodial membrane between third and fourth pedigerous somites visible from dorsal view. Sclerotized joint between fifth pedigerous somite and genital double somite well developed, almost as a pseudosomite.

Genital double somite (fig. 33B, C) with anterior part inflated laterally, about 1.2 times as long as wide (dorsal view), ornamented with three pores dorsally, two pores laterally, and two pores ventrally. Hyaline fringe of genital double somite and two succeeding somites narrow, not frilled, and completely smooth. No cuticular recesses, windows, or sensilla visible. The single copulatory pore very small and ovoid, situated just anteriorly of midlength; copulatory duct long and narrow, well sclerotized, directed anteriorly, with posterior part wider than anterior. Seminal receptacles fused, well sclerotized, small and tubular, as wide as anterior part of copulatory duct; receptacle ducts weakly chitinized and not clearly visible. Paired ovipores located dorsolaterally at 1/3 of the double somite's length, covered by the reduced sixth legs.

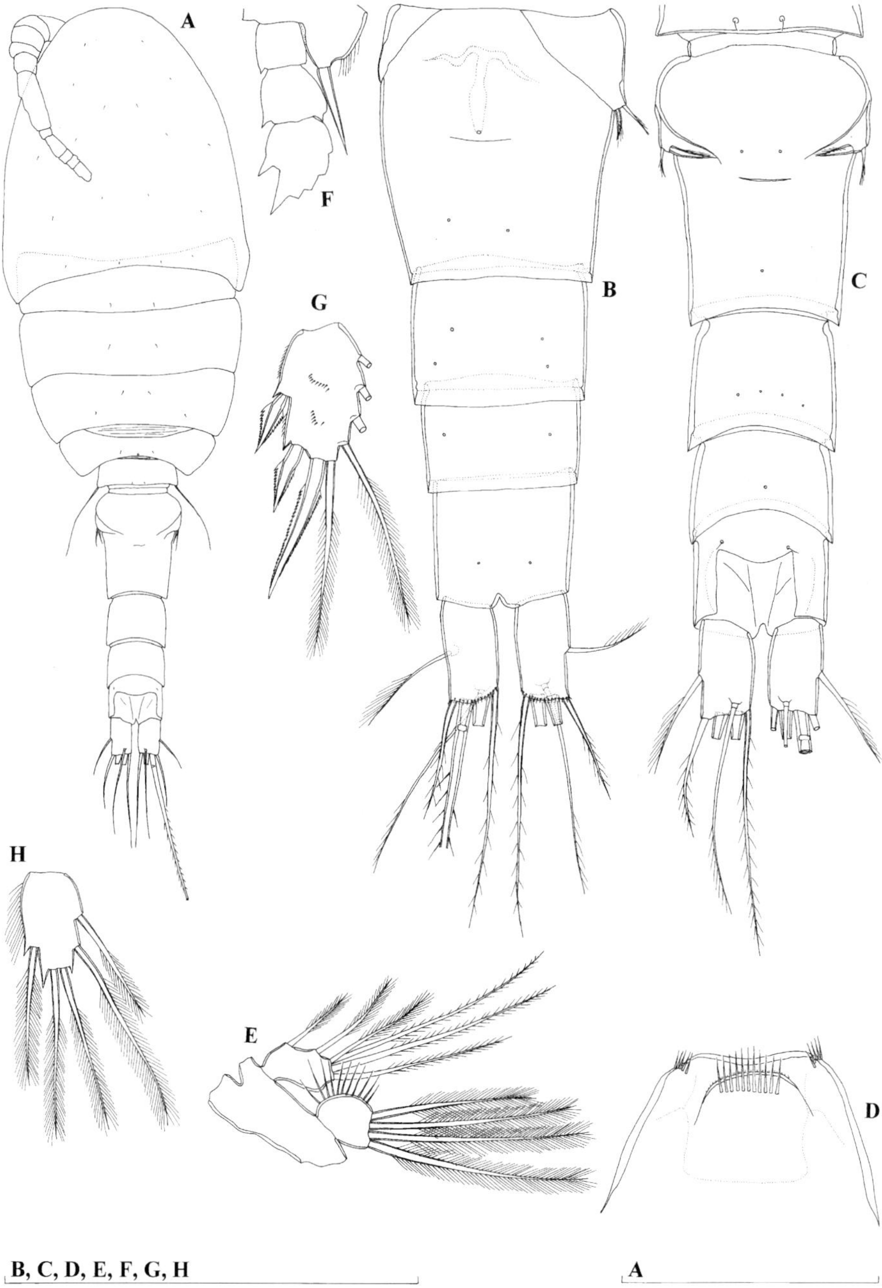

Fig. 33. *Cyclopina amita* n. sp., holotype female. A, habitus, dorsal view; B, abdomen, ventral view; C, urosome, dorsal view; D, labrum; E, palp of maxillula without basal armature; F, part of basis and endopod of first swimming leg without armature; G, third exopodal segment of fourth swimming leg; H, third endopodal segment of fourth swimming leg. Scales = 0.1 mm.

Third urosomite ornamented with four dorsal, two lateral, and four ventral pores, while fourth urosomite with one dorsal, one lateral, and two ventral pores as only ornamentation. Anal somite with smooth, broad, and convex anal operculum, which represents 55% of the somite's width and reaches to only 2/5 of the somite's length; ornamented with two large dorsal sensilla, one pore laterally on each side, and two pores ventrally; posterior margin smooth and unornamented. Anal sinus widely open dorsally, smooth. Length ratio of abdominal somites, beginning with genital double somite, 100 : 52 : 37 : 47.

Caudal rami (fig. 33B, C) cylindrical, 1.8 times as long as wide (ventral view) and nearly 0.9 times as long as anal somite; parallel, with space between them less than half of one ramus' width; they represent 5% of total body length. Distal margin ventrally convex and ornamented with a continuous row of spinules as the only ramal ornamentation, with medial cuticular pore not clearly visible; posterior outer corners of caudal ramus reaching beyond posterior inner corners. Dorsal seta relatively long, 2.1 times as long as ramus, inserted on small protuberance close to posterior margin, uniarticulate at base and unipinnate distally. Lateral seta also slender, slightly longer than ramus, arising at middle of ramus' length. Outermost apical seta strong and bipinnate, 1.2 times as long as ramus and half as long as innermost apical seta, which is also bipinnate and strong. Principal apical setae with breaking planes and bipinnate, but all broken in the present collection.

Antennula (fig. 34A) short and stout, reaching 3/4 of cephalothorax in length, 11-segmented, unornamented, armed with one short and slender aesthetasc on fifth segment laterally, one on seventh and tenth segments subapically, one long aesthetasc on eleventh segment apically, and with setal formula: 3.5.8.4.5.7.1.2.2.1.7. No setae with breaking planes and only one subapical seta on sixth segment, one on eighth, and three apical setae on eleventh segment articulating on basal part; most setae smooth (only eight pinnate) and slender; one seta on fifth segment stout and spiniform, but not very short; one apical seta on tenth segment fused to aesthetasc basally. Length ratio of antennular segments, from proximal end and along caudal margins, 1 : 0.9 : 1.2 : 0.5 : 0.6 : 3.4 : 0.5 : 1.1 : 0.6 : 0.8 : 1.1.

Antenna (fig. 34B, C) four-segmented, comprising coxobasis and three-segmented endopod, with geniculation between first and second endopodal segments; length ratio of antennal segments, 1 : 0.8 : 0.5 : 0.7. Coxobasis about 2.9 times as long as wide, unornamented, armed with one or two setae at inner distal corner, and with or without exopodal seta. First endopodal

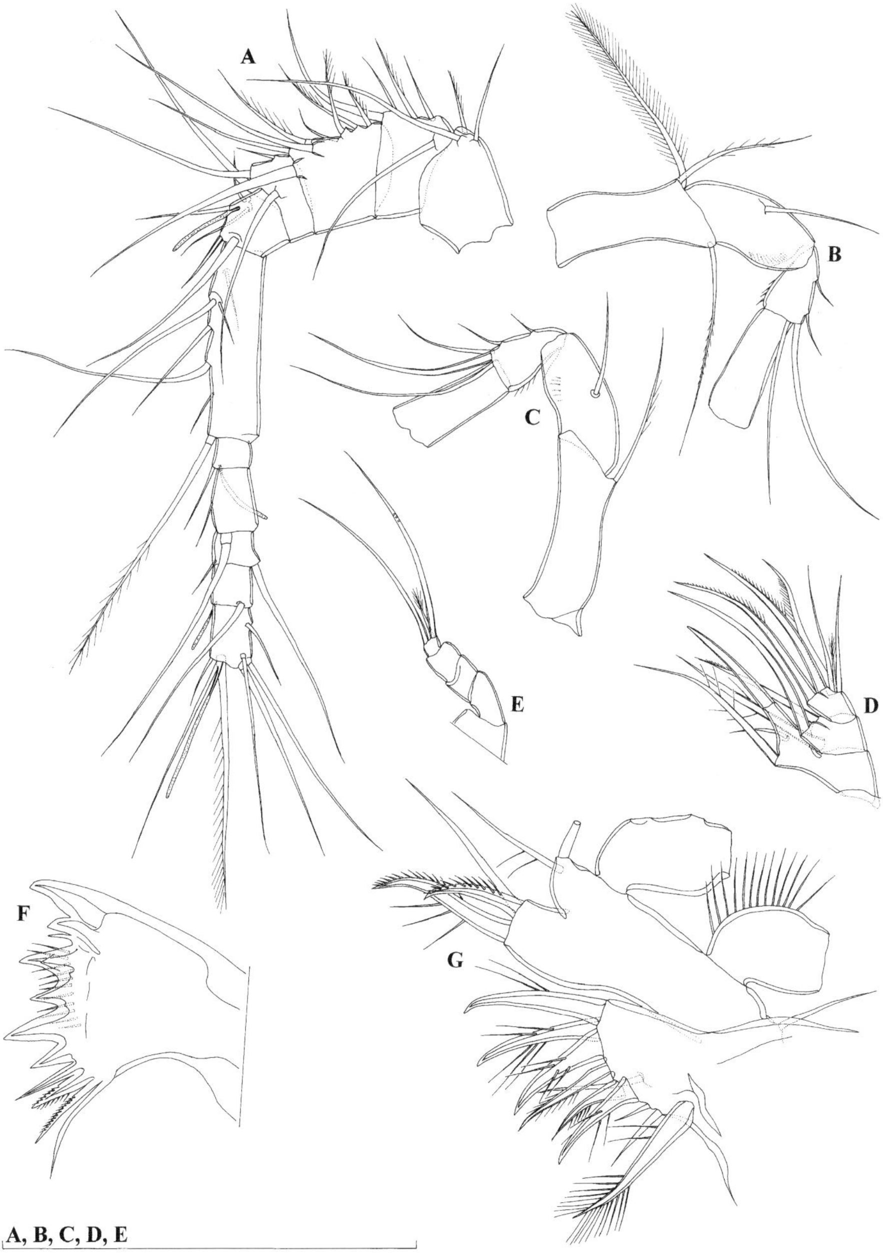

Fig. 34. *Cyclopina amita* n. sp., holotype female. A, antennula; B, antenna; C, opposite antenna; D, distal part of maxilla; E, distal part of maxilliped; F, cutting edge of mandibula; G, maxillula without endopodal and exopodal armature. Scales = 0.1 mm.

segment 2.2 times as long as wide, armed with one smooth seta at middle, ornamented with longitudinal row of spinules close to outer (caudal) margin. Second endopodal segment about 1.6 times as long as wide, ornamented with a few spinules along outer margin, and armed with four or five smooth setae (two or three lateral and two subapical; one subapical seta geniculate and much stronger than any other seta on that segment). Third endopodal segment 2.2 times as long as wide, armed with six smooth apical setae, two of which strong and geniculate, unornamented.

Labrum (fig. 33D) trapezoidal, not very large, with smooth and slightly concave distal margin, ornamented with a few large spinules at distal outer corners and with a short, transverse row of 11 long spinules medially, closer to distal margin; tips of medial spinules reaching beyond distal margin.

Mandibula (fig. 34F) large, comprising coxa with well developed gnathobase and distal palp; palp consisting of basis, two-segmented endopod, and four-segmented exopod. Cutting edge of coxal gnathobase divided into three regions: inner group of three complex, strong, and smooth teeth, and 11 long spinules; middle group of three simple and sharp teeth, of which only the outermost is bipinnate; and outermost group of two short setae, inner one unipinnate, outer one smooth. Basis large but not reaching beyond posterior margin of coxal gnathobase, 2.3 times as long as wide, unornamented, and armed with one plumose seta laterally. Endopod small, with first segment attached apically to basis and armed with three pinnate setae subapically; second segment longer than first and armed apically with four short setae. Exopod slender, 0.7 times as long as basis and 1.3 times as long as endopod, unornamented, with all segments of about the same length, and with setal formula 1.1.1.2.

Maxillula (figs. 33E, 34G) composed of syncoxa with well developed arthrite, basis, and one-segmented endopod and exopod. Arthrite of syncoxa with nine unequal lateral elements and one smooth seta on posterior surface; four apical elements strong and geniculate, proximalmost element longest and plumose. Coxa unornamented and armed with one epipodal seta; coxal endites not discernible, unarmed. Basis quadriform, much longer than arthrite of syncoxa, 2.4 times as long as wide, unornamented, and armed with three elements (two spines and one seta) on proximal, and two setae on distal endite. Endopod relatively slender, 1.2 times as long as basis' width and 1.7 times as long as wide, unornamented, and armed with two lateral setae on inner margin and four long setae apically; all endopodal setae slender and pinnate or plumose. Exopod 1.4 times as long as wide and somewhat shorter than endopod, with longitudinal row of long spinules along inner margin, which is also convex, and armed apically with four slender, plumose, and long setae.

Maxilla (fig. 34D) robust but not very large, five-segmented, comprising praecoxa, coxa, basis, and two-segmented endopod (although praecoxa and coxa partly fused). Praecoxa unornamented, armed with three setae on proximal, and only one seta on distal endite. Coxa ornamented with longitudinal row of spinules along outer margin, armed with three slender and pinnate setae on proximal endite; distal endite highly mobile, unornamented, bearing two slender setae. Basis much shorter than coxa, expanded into spinulose, geniculate, and smooth claw, and armed with one strong and one minute setae. First endopodal segment somewhat shorter than basis, armed with three robust setae. Second endopodal segment about as long as first but much narrower, armed with three strong and three slender setae.

Maxilliped (fig. 34E) slender, seven-segmented, composed of praecoxa, coxa, basis, and four-segmented endopod. Praecoxa about as long as greatest width, unornamented, with proximal endite weakly discernible and armed with a single seta, and distal endite armed with three setae. Coxa about as long as praecoxa along outer margin, ornamented with a row of hair-like spinules along outer margin, armed with two strong setae on only endite. Basis the longest segment, 1.7 times as long as wide, much narrower at basal part, ornamented with several hair-like spinules along inner margin proximally, and armed with two pinnate setae distally on inner margin. Endopod slightly shorter than basis, unornamented, with setal formula 0.0.0.3.

All swimming legs (fig. 33F, G, H) with three-segmented exopod and endopod; their armature formula as follows (legend: inner/outer spine or seta; inner/terminal/outer):

	Exopod			Endopod		
Segments	1	2	3	1	2	3
First leg	0/1	1/1	3/2/3	1/0	1/0	3/2/1
Second leg	1/1	1/1	4/2/3	1/0	2/0	3/2/1
Third leg	1/1	1/1	4/2/3	1/0	2/0	3/2/1
Fourth leg	1/1	1/1	4/2/2	1/0	2/0	2/2/1

Third exopodal segment spine formula 4.4.4.3 and no spines on endopods. All intercoxal sclerites quadriform, with slightly convex distal margin; those of first, second, and third legs unornamented, while sclerite of fourth leg with several parallel rows of spinules at middle. Coxae ornamented with several spinules along outer margin (coxa of fourth leg with additional large patch of spinules at outer half), armed with plumose seta at inner distal corner. Basis of each leg ornamented with a row of spinules along inner margin, armed with a slender seta on outer margin; basis of first leg (fig. 33F) with additional short and smooth spine on distal inner corner, reaching 1/3 of third endopodal segment; inner distal corners of second, third, and fourth leg only slightly and bluntly pronounced. All exopodal and endopodal setae slender and plumose. All exopodal spines stout and with both margins sharply serrated. All endopodal segments with hair-like spinules along outer margin and most exopodal segments with short spinules along outer margin. Third exopodal segment of fourth leg (fig. 33G) additionally with two short and arched rows of spinules on posterior surface. Third endopodal segment of fourth swimming leg (fig. 33H) about 1.5 times as long as wide; apical setae of similar length, 1.8 times as long as the segment.

Fifth leg not mounted satisfactorily to allow drawing, but very similar to that of female from Karratha (fig. 35F), relatively small, with very short and broad intercoxal sclerite and two-segmented, comprising coxobasis and a one-segmented exopod. Coxobasis about 1.4 times as wide as long, armed with a single unipinnate basal seta on outer margin, and ornamented with one pore and a row of short spinules on anterior surface, as well as with long spinules along inner margin and several small spinules on outer distal corner. Exopod 1.4 times as long as basis and more than twice as long as its greatest width, somewhat narrower at basal part, armed with two spines and a single seta apically and ornamented with three transverse rows of long spinules on anterior surface; seta bipinnate at distal part, positioned centrally, and as the most apical element; 1.8 times as long as exopod, 3.0 times as long as inner apical spine, and about 1.3 times as long as outer apical spine; both spines with sharply serrated margins, but outer spine much stronger and longer, 1.3 times as long as the segment, and 2.3 times as long as inner spine.

Sixth leg (fig. 33B, C) not clearly distinct, ovoid cuticular plate, armed with outer short spine, which is directed dorsally, and an inner slender seta; both elements unipinnate.

Male from Karratha. Body length, excluding caudal setae, 0.481 mm. Habitus (fig. 36A) similar to that of female, but first pedigerous somite completely fused to cephalothorax, with prosome/urosome ratio 1.8, and greatest width at posterior end of cephalothorax. Body length/width ratio about 3.0; cephalothorax 2.9 times as wide as genital somite.

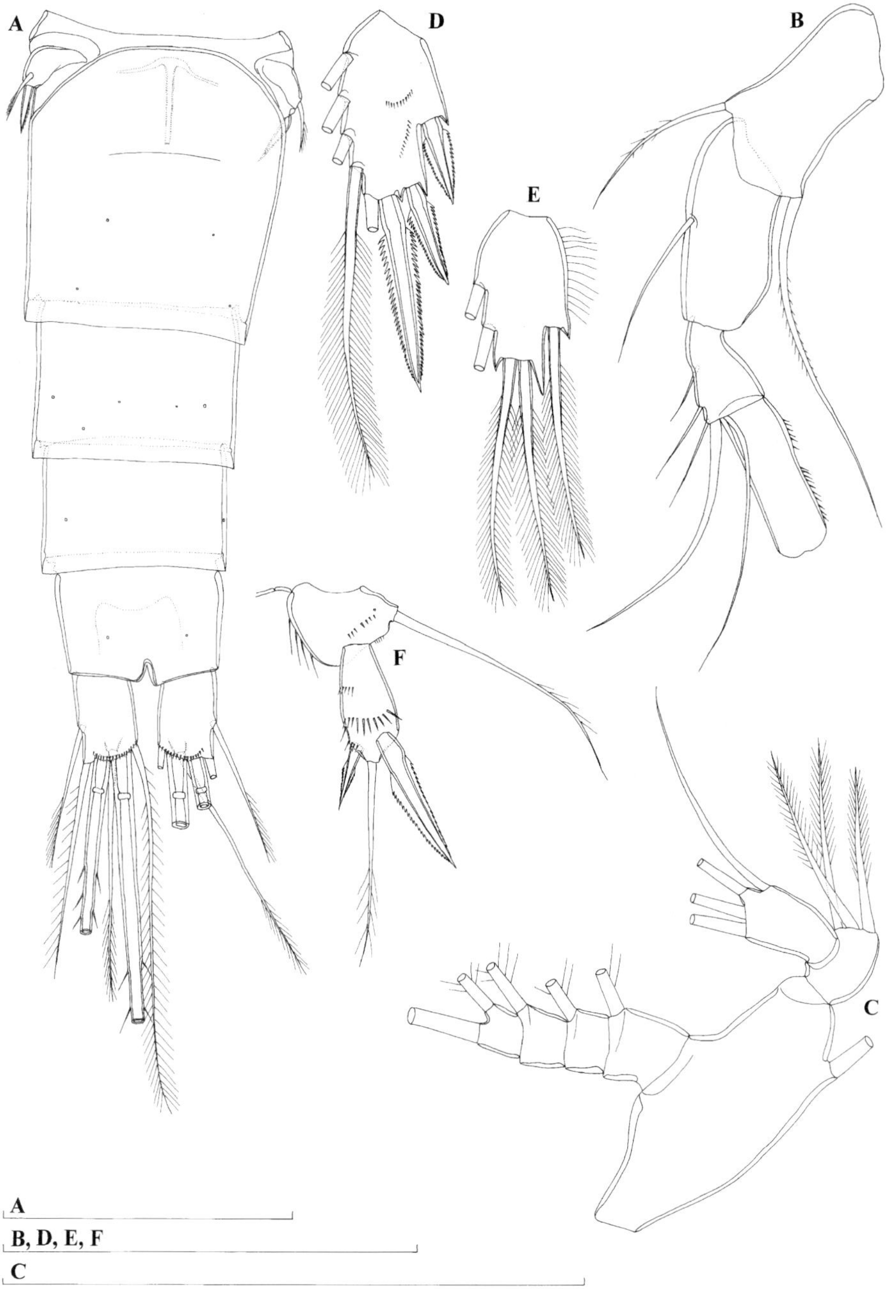

Fig. 35. *Cyclopina amita* n. sp., female from Karratha (0.65 mm). A, abdomen, ventral view; B, antenna; C, palp of mandibula; D, third exopodal segment of fourth swimming leg; E, third endopodal segment of fourth swimming leg; F, fifth leg. Scales = 0.1 mm.

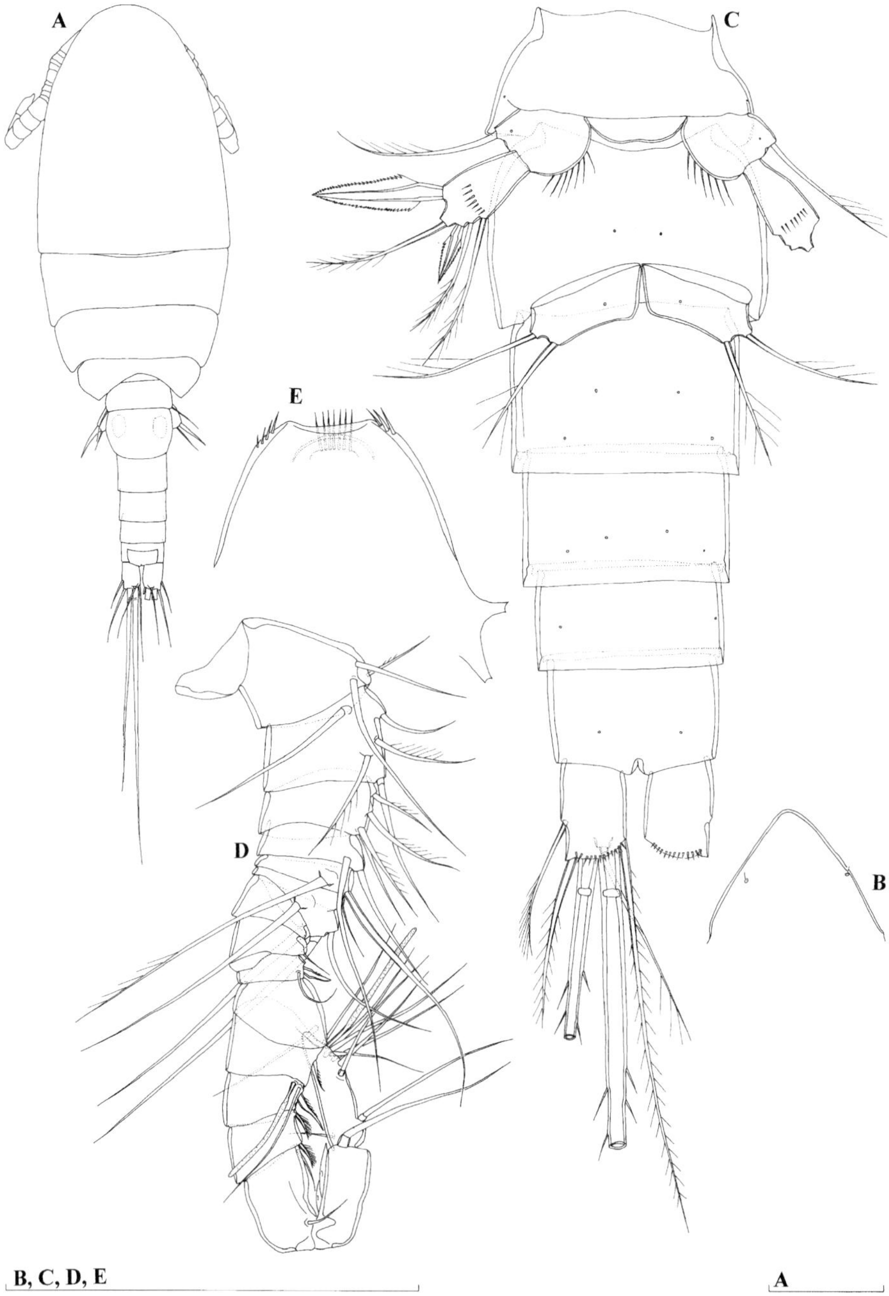

Fig. 36. *Cyclopina amita* n. sp., male from Karratha (0.481 mm). A, habitus, dorsal view; B, rostrum; C, urosome, ventral view; D, antennula; E, labrum. Scales = 0.1 mm.

Cephalothorax (fig. 36A) almost 1.3 times as long as its greatest width, representing 43% of total body length. Arthrodial membrane between prosome and urosome well developed. Ornamentation of prosomites, colour, and nauplius eye similar to those of female. Hyaline fringes of all somites smooth. Rostrum (fig. 36B) similar to that of female.

Genital somite (fig. 36C) rather small, 1.4 times as wide as long (dorsal view), with two not completely formed, large spermatophores inside, ornamented with two pores dorsally, one pore laterally, and two pores ventrally; third and fourth urosomites armed with four pores ventrally each and fifth urosomite with only two ventral pores; anal somite similar to that of female.

Caudal rami (fig. 36C) about 1.4 times as long as wide, parallel, with space between them less than half of one ramus' width. Armature and ornamentation similar to those of female, but all setae comparatively longer. Principal apical setae with breaking planes and bipinnate; inner seta 1.4 times as long as outer one and 0.5 times as long as body.

Antennula (fig. 36D) shorter than cephalothorax, 16-segmented, strongly geniculate between fourteenth and fifteenth segments, with one aesthetasc on twelfth segment and one on sixteenth segment, and with setal formula 3.4.4.2.2.5.0.1.2.1.1.2.2.2.1.10.

Antenna similar to that of female, with one exopodal seta and one seta on inner distal corner of coxobasis.

Labrum (fig. 36E) also similar to that of female, but central row with only seven long spinules.

Mandibula, maxillula, maxilla, maxilliped, and swimming legs very similar to those of female.

Fifth leg (fig. 36C) similar to that of female in segmentation and ornamentation, additionally armed with two slender setae on inner margin of exopod; outer apical exopodal spine as robust as in female.

Sixth leg (fig. 36C) distinct, large plate, ornamented with a single pore and armed with two pinnate and slender setae; outer seta about 1.3 times as long as inner one.

Variability. — Body length of females ranging from 0.430 mm to 0.658 mm (0.544 mm average; n = 4), while in males it ranges from 0.469 mm to 0.512 mm (0.490 mm average; n = 4). Cuticular pore pattern variable among specimens from the same sample, and sometimes even asymmetrical on the same specimen (fig. 35A). Armature of the antenna also variable (fig. 34B, C), although the most usual situation is that with one exopodal and one normal seta on the inner distal corner of the coxobasis, one seta on the first endopodal segment, five setae on the second, and six setae on the third endopodal segment (fig. 35B). One female from Karratha has shorter caudal rami and a narrower copulatory duct (fig. 35A) than the holotype female, but all the other details, including the mandibular palp (fig. 35C), swimming legs (fig. 35D, E), and fifth leg (fig. 35F) are the same.

Etymology. — The name of the species is the Latin noun "amita", meaning "aunt on father's side" and refers to its close relationship with the other five Australian congeners. It thus is a noun (gender feminine) in apposition to the generic name.

Distribution. — *Cyclopina amita* n. sp. was collected from two localities in the Pilbara region in Western Australia.

Discussion. — *Cyclopina amita* n. sp. is the only Australian species with an ornamented anterior surface of the fifth leg. Only two other species have been reported so far with this character and these are both mentioned in the Discussion of the previous new Australian species: they are *C. balearica* (Jaume & Boxshall, 1996) n. comb. and *C. confusa* (Ivanenko & Defaye, 2004) n. comb. They also share the same armature of the male fifth leg, with two spines and three setae on the exopod, but this was also reported for *C. caissara* Lotufo, 1994 and *C. esilis* Brian, 1938 (see Monchenko, 1979b; Lotufo, 1994; Jaume & Boxshall, 1996b). However, as both characters are probably plesiomorphic, they are not useful to establish closer relationships in the genus. As many members of *Cyclopina* are still known only after females, the state of the latter character is known for only 20 (out of the approx. 30) species.

Cyclopina amita n. sp. shares with *C. kasis* n. sp. an exceptionally robust outer spine on the fifth leg, which is also more than twice as long as the inner one. However, the two species are easily distinguishable by the segmentation of their antennulae, as well as by the segmentation of the maxillipedal endopods. *Cyclopina kasis* also has somewhat longer caudal rami than *C. amita* and a different ventral pore pattern on the genital double somite. The outer spine on the female fifth leg being more than twice as long as the inner one is found in only five other *Cyclopina* species: *C. americana* Herbst, 1982; *C. caissara* Lotufo, 1994; *C. esilis* Brian, 1938; *C. heterospina* Shen & Bai, 1956; and *C. kieferi* Schäfer, 1936. However, they all have a different segmentation of the female antennula and could be distinguished from *C. amita* by some other additional characters as well (see Brian, 1938; Herbst, 1953,

1962, 1982; Shen & Bai, 1956; Monchenko, 1979b; Lotufo, 1994; Jaume & Boxshall, 1996a; Gomez & Martínez Arbizu, 2004). The status of *C. heterospina* is questionable, as this species was very poorly described and Shen & Bai (1956) figured a female antennula composed of at least 18 segments. I think this species should be placed as incertae sedis in the family until a full redescription is presented, and I formally exclude it here from the genus *Cyclopina*.

Cyclopina amita has a male sixth leg with only two slender setae, while a majority of the *Cyclopina* species, for which males are known, have an additional small spine, just like *C. kasignete* n. sp. (fig. 41E). Only two *Cyclopina* species share this character with *C. amita*: *C. americana* Herbst, 1982 and *C. mediterranea* Steuer, 1940 (see Herbst, 1964, 1982; Lotufo, 1984). Although this may be a synapomorphy for the latter two species, which share many other characters (like 10-segmented female antennulae, armature of the male fifth leg, etc.), I think it originated independently in *C. amita* and could be interpreted as a convergence. The 11-segmented female antennula of *C. amita* is also a pretty unusual feature and is only present in two other Australian representatives: *C. adelphe* n. sp. and *C. soror* n. sp. A great majority of the *Cyclopina* species have a 10-segmented female antennula, while only three species (*C. brevifurca* G. O. Sars, 1913; *C. caissara* Lotufo, 1994; and *C. laurentica* Nicholls, 1939) have been reported with a 12-segmented one, and only one species (*C. pacifica* Smirnov, 1929) with a 13-segmented appendage. *Cyclopina adelphe* differs from *C. amita* by many characters (see below), and *C. soror* has a completely different shape of the female fifth leg and a different armature formula of the maxillipedal endopod. However, *C. soror* seems to be the closest known living relative of *C. amita*, as far as can be judged by their morphology alone. Their common ancestor was probably widely distributed along the Australian coast during warmer periods in Australian geological history (possibly in the Tertiary), fragmenting its distribution afterwards. *Cyclopina amita* survived in the tropical Pilbara region of Western Australia, while *C. soror* became trapped in relatively sheltered waters of South Australia, which are known to harbour many tropical relicts (Weston & Crisp, 1994; Bridgewater, 1999; Sanmartín & Ronquist, 2004).

12. **Cyclopina adelphe** n. sp.
(figs. 37, 38)

Material examined. — Type only: Holotype female dissected on 1 slide (WAM C37227): Australia, Western Australia, Dampier, 4 August 2003, leg. T. Karanovic, 20°40.051′S 116°42.244′E.

Description. — Female (holotype). Body length, excluding caudal setae, 0.496 mm. Habitus (fig. 37A) robust, with prosome/urosome ratio 1.6 and greatest width at posterior end of cephalothorax. Body length/width ratio about 2.5; cephalothorax 2.7 times as wide as genital double somite. Posterolateral corners of cephalothorax forming short carapace-like extensions and concealing anterior parts of lateral margins of first pedigerous somite. First three pedigerous somites without pronounced lateral expansions; posterior corners of fourth pedigerous somite somewhat expanded laterally. Fourth pedigerous somite also partly concealed by third pedigerous somite. Urosome narrow and fifth pedigerous somite narrower than genital double somite. Specimen transparent, collected long dead (with almost all soft tissues disintegrated) and damaged, with all but one caudal setae and most antennal setae broken. Integument not strongly chitinized. Nauplius eye not visible. Rostrum broadly rounded and furnished with two large sensilla.

Prosome (fig. 37A) comprising cephalothorax and four free pedigerous somites (first pedigerous somite free, although partly concealed by cephalothorax), ovoid, about 1.6 times as long as wide (dorsal view); length ratio of prosomal somites, beginning with cephalothorax, 100 : 23 : 21 : 15 : 8. Cephalothorax slightly shorter than its greatest width, representing 37% of total body length; surface of cephalic shield with several small sensilla and cuticular pores. Hyaline fringe of prosomites narrow and smooth. Fifth pedigerous somite ornamented with only two large dorsal sensilla, and with smooth fringe dorsally and laterally. Arthrodial membrane between prosome and urosome visible from dorsal view and no sclerotized joint between fifth pedigerous somite and genital double somite.

Genital double somite (fig. 37B) with anterior part laterally inflated, slightly shorter than wide (dorsal view), ornamented with four pores dorsally, two pores laterally, and four pores ventrally. Hyaline fringe of genital double somite and two succeeding somites narrow, not frilled, and completely smooth. No cuticular recesses, windows, or sensilla visible. The single copulatory pore very small and ovoid, situated at 2/5 of the double somite's length, and medially; copulatory duct long and very wide, ovoid, well sclerotized, directed anteriorly. Seminal receptacles fused, well sclerotized, small and tubular, much narrower than copulatory duct; receptacle ducts weakly chitinized and not clearly visible. Paired ovipores located dorsolaterally at 2/5 of the double somite's length, covered by the reduced sixth legs. Third and fourth urosomites ornamented with two dorsal and two ventral pores each. Anal somite (fig. 38A) with smooth, very broad, and slightly convex

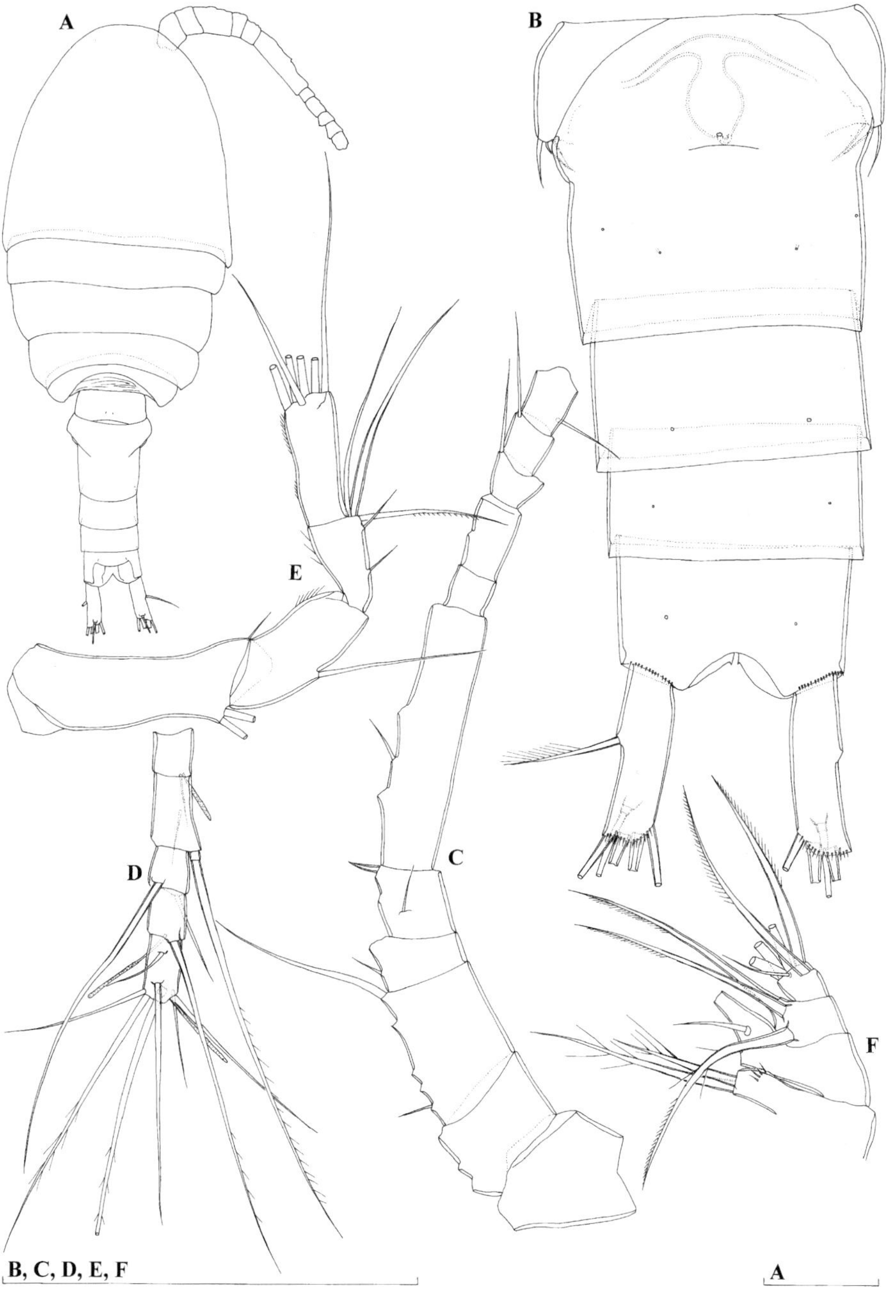

Fig. 37. *Cyclopina adelphe* n. sp., holotype female. A, habitus, dorsal view; B, abdomen, ventral view; C, antennula with incomplete armature; D, distal part of antennula; E, antenna; F, distal part of maxilla. Scales = 0.1 mm.

anal operculum, which represents 64% of the somite's width and reaches only to 2/5 of the somite's length; ornamented with two large dorsal sensilla, one pore laterally on each side, and two pores ventrally; posterior margin ornamented with a row of small spinules ventrally. Anal sinus widely open dorsally, smooth. Length ratio of abdominal somites, beginning with genital double somite, 100 : 56 : 43 : 47.

Caudal rami (figs. 37B, 38A) slender and cylindrical, almost three times as long as wide (ventral view) and nearly 1.2 times as long as anal somite; only slightly divergent, with enormous space between them; they represent 7% of total body length. Distal margin ventrally convex and ornamented with continuous row of spinules, with medial cuticular pore not clearly visible; posterior outer corners of caudal ramus reaching beyond posterior inner corners. One dorsal cuticular pore as the only other ramal ornamentation. All setae, except a single lateral seta, broken. Dorsal seta relatively strong, inserted on small protuberance close to, but not reaching posterior margin, uniarticulate at base. Lateral seta slender, 0.7 times as long as ramus, arising just before middle of the ramus' length. Principal apical setae with breaking planes.

Antennula (fig. 37C, D) relatively long, but not reaching posterior margin of cephalothorax, 11-segmented, unornamented, with one short and slender aesthetasc on fifth segment laterally, one on seventh and tenth segments subapically, and one long aesthetasc on eleventh segment apically. Almost all setae broken on one antennula (fig. 37C), while the opposite appendage has a complete armature on the last five segments only (fig. 37D). Armature very similar to that of previous species. One seta on fifth segment short and spiniform. Length ratio of antennular segments, from proximal end and along caudal margins, 1 : 1 : 1.4 : 0.4 : 0.9 : 3.5 : 0.5 : 1 : 0.5 : 0.6 : 0.9.

Antenna (fig. 37E) four-segmented, comprising coxobasis and three-segmented endopod, with geniculation between first and second endopodal segments; length ratio of antennal segments, 1 : 0.8 : 0.4 : 0.6. Coxobasis about 2.9 times as long as wide, unornamented, armed with two setae at inner distal corner (both broken), and with a single, minute exopodal seta. First endopodal segment 2.4 times as long as wide, armed with one smooth seta at middle, ornamented with longitudinal row of spinules along outer (caudal) margin. Second endopodal segment about 1.8 times as long as wide, ornamented with a few spinules along outer margin, armed with five setae (three lateral and two subapical; one subapical seta geniculate and much stronger than any other seta on that segment). Third endopodal segment 2.4 times as

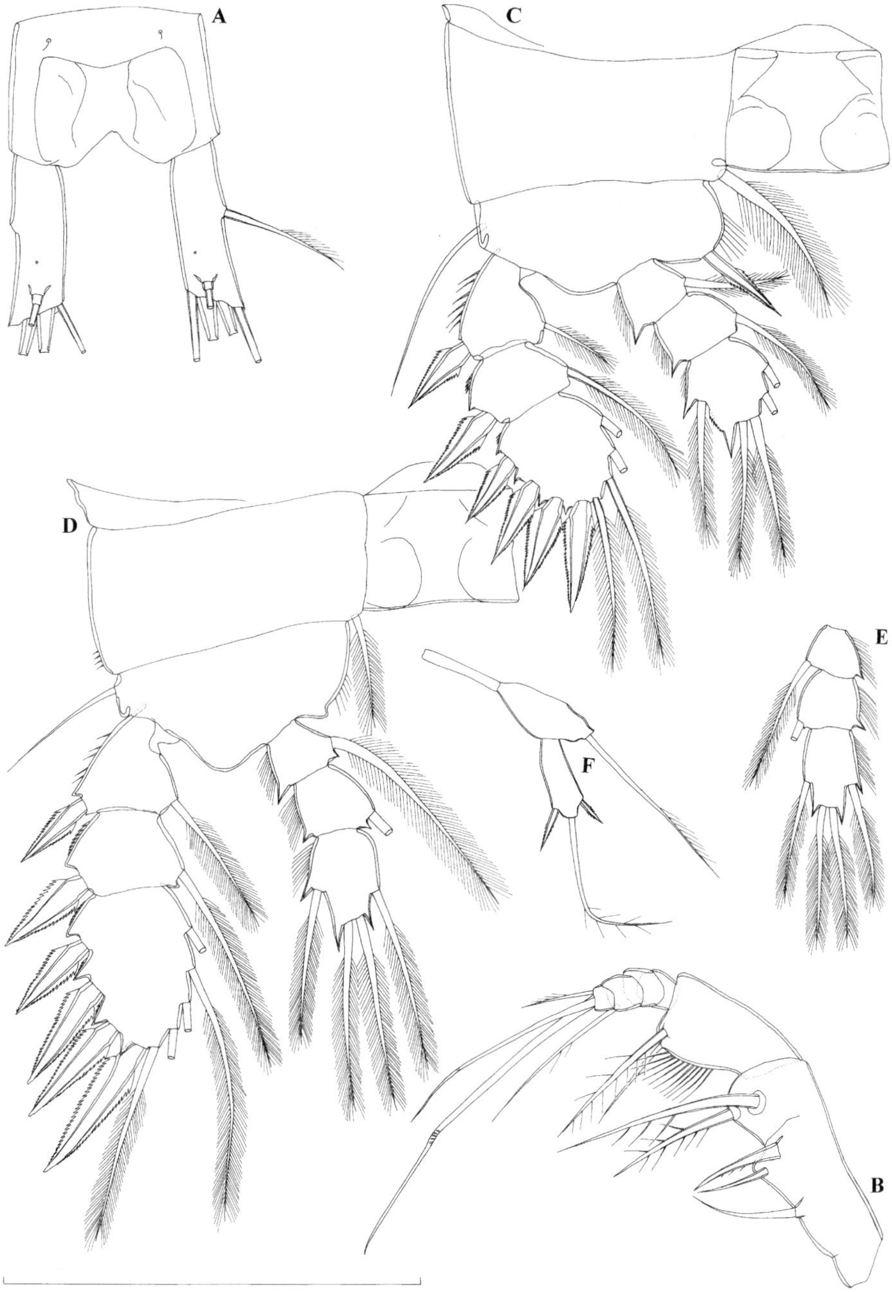

Fig. 38. *Cyclopina adelphe* n. sp., holotype female. A, anal somite and caudal rami, dorsal view; B, maxilliped; C, first swimming leg; D, second swimming leg; E, endopod of fourth swimming leg; F, fifth leg. Scale = 0.1 mm.

long as wide, armed with six smooth apical setae, two of which strong and geniculate, ornamented as previous two segments.

Labrum not mounted satisfactorily to allow detailed examination.

Mandibula large, comprising coxa with well developed gnathobase and distal palp; palp consisting of basis, two-segmented endopod, and four-segmented exopod. Cutting edge of coxal gnathobase divided into three regions: inner group of three complex, strong, and smooth teeth, and several long spinules; middle group of three simple and sharp teeth; and an outermost group of two short setae. Basis large but not reaching beyond posterior margin of coxal gnathobase, 2.3 times as long as wide, unornamented, and armed with one seta laterally. Endopod small, with first segment attached apically to basis and armed with three pinnate setae subapically; second segment longer than first and armed apically with four short setae. Exopod slender, 0.6 times as long as basis and 1.2 times as long as endopod, unornamented, with all segments of about the same length and with setal formula 1.1.1.2.

Maxillula composed of syncoxa with well developed arthrite, basis, and one-segmented endopod and exopod. Arthrite of syncoxa with nine unequal lateral elements and one smooth seta on posterior surface; four apical elements strong and geniculate, proximalmost element longest and plumose. Coxa unornamented and armed with one epipodal seta; coxal endites not discernible, unarmed. Basis quadriform, much longer than arthrite of syncoxa, 2.4 times as long as wide, unornamented, and armed with three elements (two spines and one seta) on proximal, and two setae on distal endite. Endopod relatively slender, 1.2 times as long as the basis' width and 1.7 times as long as wide, unornamented, and armed with two lateral setae on inner margin and four long setae apically; all endopodal setae slender and plumose. Exopod 1.4 times as long as wide and somewhat shorter than endopod, with longitudinal row of long spinules along inner margin, which is also convex, and armed apically with four slender, plumose, and long setae.

Maxilla (fig. 37F) robust but not very large, five-segmented, comprising praecoxa, coxa, basis, and two-segmented endopod (although praecoxa and coxa partly fused). Praecoxa unornamented, armed with three setae on proximal, and only one seta on distal endite. Coxa ornamented with longitudinal row of spinules along outer margin, armed with three slender and pinnate setae on proximal endite; distal endite highly mobile, ornamented with short transverse row of four spinules, bearing two slender and pinnate setae. Basis much shorter than coxa, expanded into spinulose, geniculate, and smooth claw (which is distinct at base) and armed with one strong and one minute setae. First endopodal segment shorter than basis, armed with three robust and unipinnate setae. Second endopodal segment about as long as first but much narrower, armed with six setae.

Maxilliped (fig. 38B) slender, prehensile, seven-segmented, composed of praecoxa, coxa, basis, and four-segmented endopod, although praecoxa and coxa partly fused. Praecoxa unornamented, with proximal endite weakly discernible and armed with a single seta, and distal endite armed with two setae. Coxa half as long as praecoxa, also unornamented, armed with two strong setae on its only endite. Basis about as long as praecoxa, 1.9 times as long as wide, narrower at basal part, ornamented with very long spinules along inner margin proximally, armed with two pinnate setae distally on inner margin. Endopod much shorter than basis, unornamented, with setal formula 0.0.1.3.

All swimming legs (fig. 38C, D, E) with three-segmented exopod and endopod; their armature formula as follows (legend: inner/outer spine or seta; inner/terminal/outer):

	Exopod			Endopod		
Segments	1	2	3	1	2	3
First leg	1/1	1/1	3/2/3	1/0	1/0	2/2/1
Second leg	1/1	1/1	4/2/3	1/0	1/0	1/2/1
Third leg	1/1	1/1	4/2/3	1/0	1/0	1/2/1
Fourth leg	1/1	1/1	4/2/2	1/0	1/0	1/2/1

Third exopodal segment spine formula 4.4.4.3 and no spines on endopods. All intercoxal sclerites quadriform, with almost straight distal margin; those of first, second, and third legs unornamented, while sclerite of fourth leg with several parallel rows of spinules at middle. Coxae ornamented with several spinules along outer margin (coxa of fourth leg with additional large patch of spinules at outer half), armed with plumose seta at inner distal corner. Basis of each leg ornamented with a row of spinules along inner margin, armed with a slender seta on outer margin; basis of first leg (fig. 38C) with additional stout and serrated spine on distal inner corner, which reaches distal margin of second endopodal segment; inner distal corners of second, third, and fourth leg only slightly and bluntly pronounced. All exopodal and endopodal setae slender and plumose. All exopodal spines stout and with both margins sharply serrated. All endopodal segments with hair-like spinules along outer margin and most exopodal segments with short spinules along outer margin. Endopods of second, third, and fourth leg considerably shorter than corresponding exopods; second (fig. 38D) and third leg without any visible difference. Third endopodal segment of fourth swimming leg (fig. 38E) about 1.4 times as long as wide; apical setae of similar length, 1.8 times as long as segment.

Fifth leg (fig. 38F) very small, unornamented, with very short and broad intercoxal sclerite, and two-segmented, comprising coxobasis and a one-segmented exopod. Coxobasis about 2.4 times as wide as long and armed with a single unipinnate basal seta on outer margin. Exopod 1.9 times as long as basis and about 2.5 times as long as its greatest width, somewhat narrower at basal part, armed with two spines and one seta apically; seta bipinnate in distal part, positioned centrally, and as the most apical element, 2.2 times as long as exopod, 4.0 times as long as inner apical spine, and about 5.7 times as long as outer apical spine; both spines pinnate and not very robust.

Sixth leg (fig. 37B) not clearly distinct, ovoid cuticular plate, armed with outer short spine, which is directed dorsally, and an inner slender seta; both elements smooth.

Male. Unknown.

Variability. — Only one female was collected and examined, and no asymmetries were observed.

Etymology. — The name of this species is the Greek noun "adelphe", meaning "sister". It is to be treated as a noun (gender feminine) in apposition to the generic name.

Distribution. — *Cyclopina adelphe* n. sp. was collected from a beach in Dampier (Pilbara region, Western Australia), where it lives sympatrically with its previously described congener.

Discussion. — *Cyclopina adelphe* n. sp. could be easily distinguished from all its congeners by two clear autapomorphies: its very widely spaced caudal rami, and the reduced endopodal armature of its swimming legs. The second endopodal segment of the second, third, and fourth legs of *C. adelphe* bears only one seta, while all other species have two setae on this segment. The only exception is *C. parapsammophila* Monchenko, 1980, described from the Black Sea by Monchenko (1980), which has one seta on the second leg, but two setae on the third and fourth. The armature formula of the third endopodal segment of *C. adelphe* is 5.4.4.4, whereas in almost all other *Cyclopina* species (including the other five Australian ones) it is 6.6.6.5. The only exceptions are *C. hadzii* Petkovski, 1955 and some specimens of

C. steueri Früchtl, 1923, which have the formula 5.6.6.5 (see Petkovski, 1955a; Pleşa, 1963). Another unusual feature of *C. adelphe* is its very wide copulatory duct (fig. 37B), which was recorded so far only in *C. caissara* Lotufo, 1994, a species described from Brazil by Lotufo (1994) and later also reported from Mexico by Gomez & Martínez Arbizu (2004).

Although *C. adelphe* occupies an isolated position in the genus *Cyclopina*, among its Australian congeners it seems to be most closely related to *C. soror* n. sp. Both species have an 11-segmented antennula, a feature they share only with *C. amita* n. sp., as mentioned before. *Cyclopina adelphe* and *C. soror* also have very similar fifth legs, with the outer exopodal spine shorter than the inner. This condition is very rare in the genus *Cyclopina*, and it was so far reported only for the Brazilian *C. dorae* Lotufo, 1994 (see Lotufo, 1994). Thus, I believe that *C. adelphe* and *C. soror* are more closely related than their different habitus and swimming legs armature formula would suggest at a first glance.

13. **Cyclopina kasis** n. sp.
(fig. 39)

Material examined. — Type only: Holotype female dissected on 1 slide (WAM C37228): Australia, Queensland, Tin Can Bay, 14 August 2003, leg. T. Karanovic, 25°16.728′S 152°50.400′E.

Description. — Female (holotype). Body length, excluding caudal setae, 0.500 mm. Habitus (fig. 39A) relatively robust, with prosome/urosome ratio 1.6 and greatest width at posterior end of cephalothorax. Body length/width ratio about 2.8; cephalothorax 3.6 times as wide as genital double somite. Posterolateral corners of cephalothorax forming carapace-like extensions and concealing most of first pedigerous somite. First three pedigerous somites without pronounced lateral expansions; posterior corners of fourth pedigerous somite somewhat expanded laterally. Urosome very narrow, and fifth pedigerous somite only slightly wider than genital double somite. Colour of preserved specimen milky white. Nauplius eye not visible. Rostrum broadly rounded and furnished with two large sensilla.

Prosome (fig. 39A) comprising cephalothorax and four free pedigerous somites (first pedigerous somite free, although mostly concealed by cephalothorax), ovoid, about 1.8 times as long as wide (dorsal view); length ratio of prosomal somites, beginning with cephalothorax, 100 : 22 : 19 : 25 : 10. Cephalothorax slightly longer than its greatest width, representing 38%

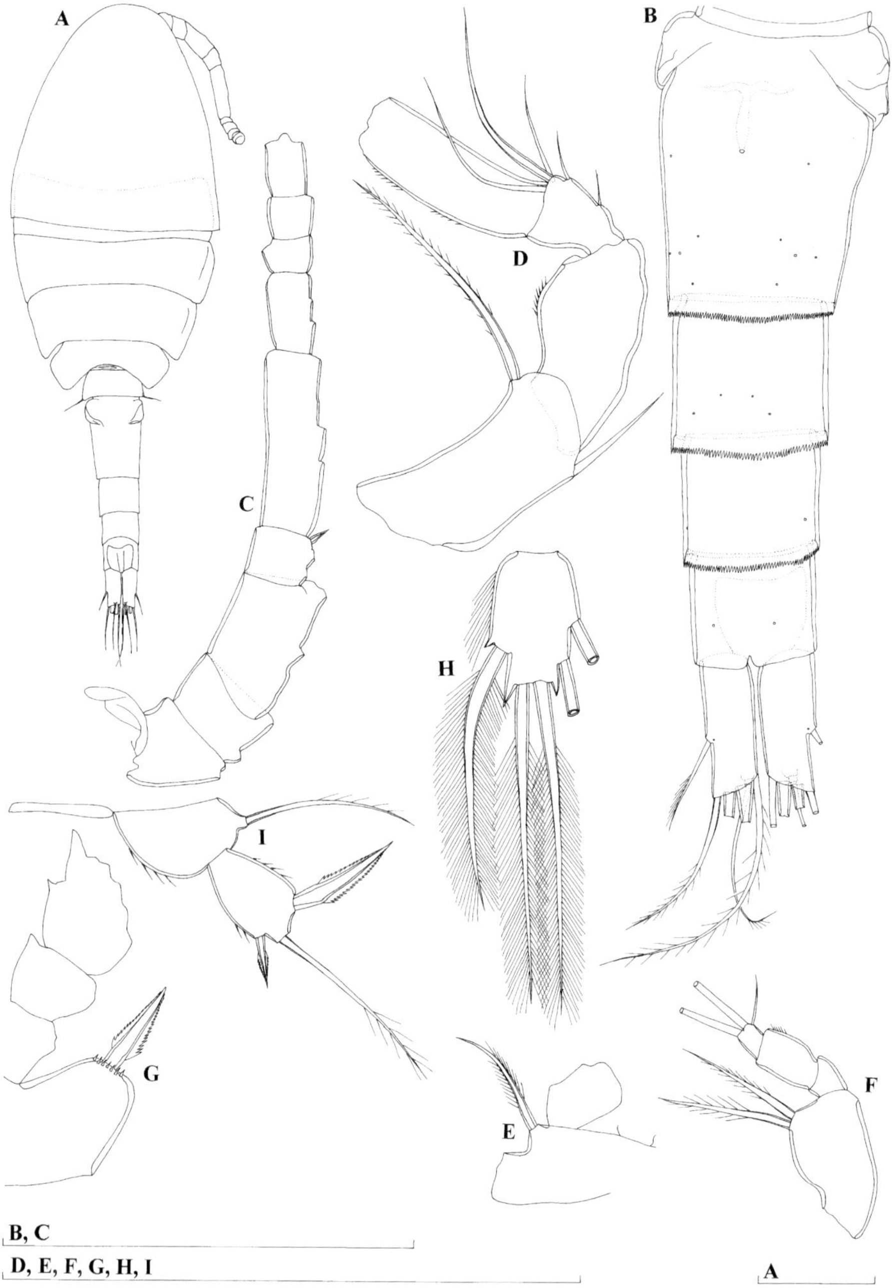

Fig. 39. *Cyclopina kasis* n. sp., holotype female. A, habitus, dorsal view; B, abdomen, ventral view; C, antennula without armature; D, antenna; E, basis of maxillula; F, distal part of maxilliped; G, part of basis and endopod of first swimming leg without armature; H, third endopodal segment of fourth swimming leg; I, fifth leg. Scales = 0.1 mm.

of total body length; surface of cephalic shield with several small sensilla and cuticular pores. Hyaline fringe of prosomites narrow and smooth. Fifth pedigerous somite ornamented with two large sensilla dorsally and one pore laterally, and with smooth fringe dorsally and laterally. Arthrodial membrane between prosome and urosome visible from dorsal view. Sclerotized joint between fifth pedigerous somite and genital double somite well developed, almost as a pseudosomite (fig. 39B).

Genital double somite (fig. 39B) with anterior part inflated laterally, about 1.4 times as long as wide (dorsal view), ornamented with six pores dorsally, two pores laterally, and ten pores ventrally. Hyaline fringe of genital double somite and two succeeding somites narrow and finely serrated both ventrally and dorsally. No cuticular recesses, windows, or sensilla visible. The single copulatory pore very small and ovoid, situated at 2/5 of the double somite's length, and medially; copulatory duct long and narrow, only slightly inflated at middle, well sclerotized, directed anteriorly. Seminal receptacles fused, well sclerotized, small and tubular, as wide as anterior part of copulatory duct, forming a pretty regular T-structure with it; receptacle ducts weakly chitinized and not clearly visible. Paired ovipores located dorsolaterally at 1/3 of the double somite's length, covered by the reduced sixth legs. Third urosomite ornamented with four dorsal, two lateral, and four ventral pores, while fourth urosomite with one dorsal, one lateral, and two ventral pores as only ornamentation. Anal somite with smooth, broad, and convex anal operculum, which represents 62% of the somite's width and reaches to only 2/5 of the somite's length; ornamented with two large dorsal sensilla, one pore laterally on each side, and two pores ventrally; posterior margin smooth and unornamented. Anal sinus widely open dorsally, smooth. Length ratio of abdominal somites, beginning with genital double somite, 100 : 55 : 47 : 39.

Caudal rami (fig. 39B) cylindrical, 2.4 times as long as wide (ventral view) and nearly 1.2 times as long as anal somite; parallel, with almost no space between them; they represent 6% of total body length. Distal margin ventrally convex and smooth, with medial cuticular pore not clearly visible; posterior outer corners of caudal rami reaching beyond their posterior inner corners. One pore ventrally and one pore dorsally as only ramal ornamentation. Dorsal seta relatively long and very slender, 1.3 times as long as ramus, inserted on a small protuberance reaching posterior margin, uniarticulate at base and unipinnate distally. Lateral seta also slender and unipinnate, 0.7 times as long as ramus, arising at 3/5 of the ramus' length. Outermost apical seta strong and bipinnate, 1.2 times as long as ramus and 0.6 times as long as innermost apical seta, which is also bipinnate and strong. Principal apical setae with breaking planes, and all broken.

Antennula (fig. 39C) short, reaching 3/4 of cephalothorax in length, nine-segmented (with fused third and fourth, and seventh and eighth ancestral segments, from supposedly 11-segmented ancestral stage), unornamented, with one short and slender aesthetasc on fourth and sixth segments laterally, one on eighth segment subapically, one long aesthetasc on ninth segment apically, and as setal formula: 3.5.12.5.7.3.2.1.7. Although the segmentation differs from that of other species, the armature is basically the same. Length ratio of antennular segments, from proximal end and along caudal margins, 1 : 1.8 : 2.9 : 1.5 : 5.6 : 2.4 : 1 : 1.3 : 2.

Antenna (fig. 39D) four-segmented, comprising coxobasis and three-segmented endopod, with geniculation between first and second endopodal segments; length ratio of antennal segments, 1 : 1 : 0.4 : 0.8. Coxobasis about 1.9 times as long as wide, unornamented, armed with one smooth seta at inner distal corner, and with one bipinnate exopodal seta on outer distal corner; exopodal seta 1.2 times as long as coxobasis and 2.2 times as long as seta at inner distal corner. First endopodal segment twice as long as wide, unarmed, ornamented with longitudinal row of short spinules along outer (caudal) margin. Second endopodal segment about 1.3 times as long as wide, unornamented, and armed with five smooth setae (three lateral and two subapical; one subapical seta geniculate and much stronger than any other seta on that segment). Third endopodal segment 2.8 times as long as wide, armed with six smooth apical setae, two of which strong and geniculate, ornamented with longitudinal row of small spinules along outer margin.

Labrum trapezoidal, not very large, with smooth and slightly concave distal margin, ornamented with a few large spinules at distal outer corners and with a short transverse row of long spinules medially, closer to the distal margin; tips of medial spinules reaching beyond distal margin.

Mandibula large, comprising coxa with well developed gnathobase and distal palp; palp consisting of basis, two-segmented endopod, and four-segmented exopod. Cutting edge of coxal gnathobase divided into three regions: an inner group of three complex, strong, and smooth teeth, and several long spinules; a middle group of three simple and sharp teeth; and an outermost group of two short setae. Basis large but not reaching beyond posterior margin of coxal gnathobase, more than twice as long as wide, unornamented, and armed with one plumose seta laterally. Endopod small, with first segment attached apically to basis and armed with two pinnate setae subapically; second segment longer than first and armed apically with four short setae. Exopod slender, half as long as basis and slightly longer than endopod, unornamented, with all segments of about the same length and with setal formula 1.1.1.2.

Maxillula (fig. 39E) composed of syncoxa with well developed arthrite, basis, and one-segmented endopod and exopod. Arthrite of syncoxa with nine unequal lateral elements and one smooth seta on posterior surface; four apical elements strong and geniculate, proximalmost element longest and plumose. Coxa unornamented and armed with one epipodal seta; coxal endites not discernible, unarmed. Basis quadriform, much longer than arthrite of syncoxa, 2.4 times as long as wide, unornamented, and armed with three elements (two spines and one seta) on proximal, and only one strong seta on distal endite. Endopod relatively slender, 1.2 times as long as the basis' width and 1.7 times as long as wide; unornamented and armed with two lateral setae on inner margin and four long setae apically; all endopodal setae slender and pinnate. Exopod 1.4 times as long as wide and somewhat shorter than endopod, with longitudinal row of long spinules along inner margin, which is also convex, and armed apically with four slender, plumose, and long setae.

Maxilla robust but not very large, five-segmented, comprising praecoxa, coxa, basis, and two-segmented endopod (although praecoxa and coxa partly fused). Praecoxa unornamented, armed with three setae on proximal, and only one seta on distal endite. Coxa ornamented with longitudinal row of spinules along outer margin, armed with three slender and pinnate setae on proximal endite; distal endite highly mobile, unornamented, bearing two slender setae. Basis much shorter than coxa, expanded into spinulose, geniculate, and smooth claw, and armed with one strong and one minute setae. First endopodal segment somewhat shorter than basis, armed with three robust setae. Second endopodal segment about as long as first but much narrower, armed with three strong and three slender setae.

Maxilliped (fig. 39F) slender, six-segmented, composed of praecoxa, coxa, basis, and three-segmented endopod. Praecoxa about as long as its greatest width, unornamented, with proximal endite weakly discernible and armed with a single seta, and distal endite armed with three setae. Coxa about as long as praecoxa along outer margin, ornamented with a row of hair-like spinules along outer margin, and armed with two strong setae on its only endite. Basis the longest segment, 1.7 times as long as wide, much narrower at basal part, unornamented, and armed with two pinnate setae distally on inner margin. Endopod 0.8 times as long as basis, ornamented with several spinules on outer margin of second segment, which is also the largest, and with setal formula 0.0.3.

All swimming legs (fig. 39G, H) with three-segmented exopod and endopod; their armature formula as follows (legend: inner/outer spine or seta; inner/terminal/outer):

	Exopod			Endopod		
Segments	1	2	3	1	2	3
First leg	1/1	1/1	3/2/3	1/0	1/0	3/2/1
Second leg	1/1	1/1	4/2/3	1/0	2/0	3/2/1
Third leg	1/1	1/1	4/2/3	1/0	2/0	3/2/1
Fourth leg	1/1	1/1	4/2/2	1/0	2/0	2/2/1

Third exopodal segment spine formula 4.4.4.3 and no spines on endopods. All intercoxal sclerites quadriform, with almost straight distal margin; those of first, second, and third legs unornamented, while sclerite of fourth leg with several parallel rows of spinules at middle. Coxae ornamented with several spinules along outer margin (coxa of fourth leg with additional large patch of spinules at outer half), armed with plumose seta at inner distal corner. Basis of each leg ornamented with a row of spinules along inner margin, armed with a slender seta on outer margin; basis of first leg (fig. 39B) with additional short and sharply serrated spine on distal inner corner, which reaches posterior margin of second endopodal segment; inner distal corners of second, third, and fourth leg only slightly and bluntly pronounced. All exopodal and endopodal setae slender and plumose. All exopodal spines stout and with both margins sharply serrated. All endopodal segments with hair-like spinules along outer margin and most exopodal segments with short spinules along outer margin. Third endopodal segment of fourth swimming leg (fig. 39H) about 1.5 times as long as wide; apical setae of similar length, 2.5 times as long as the segment.

Fifth leg (fig. 39I) relatively small, with very short and broad intercoxal sclerite, and two-segmented, comprising coxobasis and a one-segmented exopod. Coxobasis about 1.9 times as wide as long, armed with a single unipinnate basal seta on outer margin, and ornamented with several spinules along convex inner margin. Exopod 1.5 times as long as basis and 1.6 times as long as its greatest width, much narrower at basal part, armed with two spines and a single seta apically, and ornamented with several spinules along inner and outer margins; seta bipinnate in distal part, positioned centrally, and as the most apical element; 1.9 times as long as exopod, 3.9 times as long as inner apical spine, and about 1.7 times as long as outer apical spine; both spines with sharply serrated margins, but outer spine much stronger and longer, somewhat longer than the segment and 2.3 times as long as the inner spine.

Sixth leg not clearly distinct, ovoid cuticular plate, armed with short outer spine, which is directed dorsally, and an inner slender seta; both elements smooth.

Male. Unknown.

Variability. — Only one female was collected and no asymmetries were observed.

Etymology. — The name of this new species is the Greek noun "kasis", which means "sister" and is of feminine gender. It is to be treated as a noun in apposition to the generic name.

Distribution. — This species is known from the type locality only.

Discussion. — *Cyclopina kasis* n. sp. differs from all its congeners by its 9-segmented female antennula, which is the lowest number recorded in the genus so far. However, the armature of this appendage is quite the same as in the other five Australian representatives, and it clearly shows that the 9-segmented state originated by fusion of the third and fourth, and the seventh and eighth segments of an 11-segmented antennula. Although this will slightly widen the generic diagnosis (to include members with 9-13-segmented antennulae), I do not think this is a character that would allow the erection of a new genus. Especially so, because *C. kasis* shows similarity in most other important characters with the other Australian representatives, and is particularly close to *C. amita* n. sp. The two species share a very similar shape of the fifth leg, as discussed previously (see the discussion section for *C. amita*), which is not very common in the genus.

Cyclopina kasis differs from its five Australian congeners also by the reduced segmentation of the maxillipedal endopod, which is 3-segmented in *C. kasis* and 4-segmented in all other species. A 3-segmented maxillipedal endopod is so far observed in three other *Cyclopina* species: *C. balearica* (Jaume & Boxshall, 1996) n. comb.; *C. brevifurca* G. O. Sars, 1913; and *C. tuberculata* Herbst, 1962 (see Sars, 1913; Herbst, 1962; Jaume & Boxshall, 1996b). Sars (1913) also illustrated a 3-segmented maxillipedal endopod for *C. gracilis* Claus, 1863, but that is likely to be either a mistake, or a result of intraspecific variability. Monchenko (1976) did illustrate a 3-segmented maxillipedal endopod for the Black Sea population of *C. steueri* Früchtl, 1923, although he was not entirely sure of the specific status of that population (he identified it as *C.* cf. *steueri*). However, of the five above-mentioned species, only *C. balearica* and *C. steueri* have the middle endopodal segment unarmed, as is the situation in *C. kasis*. As the three species show very few similarities, I think the reduction of the maxillipedal endopod happened independently, and not just in *C. balearica*, *C. steueri*, and *C. kasis*, but in all five species. Thus, I consider it to be a convergence in the genus *Cyclopina*.

14. **Cyclopina kasignete** n. sp. (figs. 40, 41)

Material examined. — Types: Holotype ovigerous female dissected on 1 slide (WAM C37229); allotype male dissected on 1 slide (WAM C37230); paratypes, 1 male + 6 females (2 females on 1 slide in toto (TMAG G5777), others in alcohol (WAM C37231)): Australia, South Australia, Robe, 27 December 2002, leg. T. Karanovic, 39°09.725′S 139°45.153′E.

Other material: Australia, South Australia, Eyre Peninsula, Venus Bay, 25 December 2002, leg. T. Karanovic, 33°13.896′S 134°40.347′E: 3 females (1 ovigerous) in alcohol (WAM C37232).

Three females (2 ovigerous) + 5 copepodids (1 female on 1 slide in toto (TMAG G5778), others in alcohol (WAM C37233)): Australia, South Australia, Streaky Bay, 25 December 2002, leg. T. Karanovic, 32°47′S 134°13′E.

Description. — Ovigerous female (holotype). Body length, excluding caudal setae, 0.537 mm. Habitus (fig. 40A) relatively slender, with prosome/urosome ratio 1.3 and greatest width at posterior end of cephalothorax. Body length/width ratio about 3.0; cephalothorax 2.8 times as wide as genital double somite. Posterolateral corners of cephalothorax forming carapace-like extensions and concealing lateral margins of first pedigerous somite (fig. 40B). First three pedigerous somites without pronounced lateral expansions; posterior corners of fourth pedigerous somite somewhat expanded laterally. Urosome narrow, and fifth pedigerous somite narrower than genital double somite. Colour of preserved specimen milky white. Nauplius eye not visible. Rostrum broadly rounded and furnished with two large sensilla.

Prosome (fig. 40A, B) comprising cephalothorax and four free pedigerous somites (although first pedigerous somite laterally fused to cephalothorax), ovoid, almost twice as long as wide (dorsal view); length ratio of prosomal somites, beginning with cephalothorax, 100 : 26 : 27 : 27 : 10. Cephalothorax slightly shorter than its greatest width, representing 29% of total body length; surface of cephalic shield with several small sensilla and cuticular pores. Hyaline fringe of prosomites narrow and smooth. Fifth pedigerous

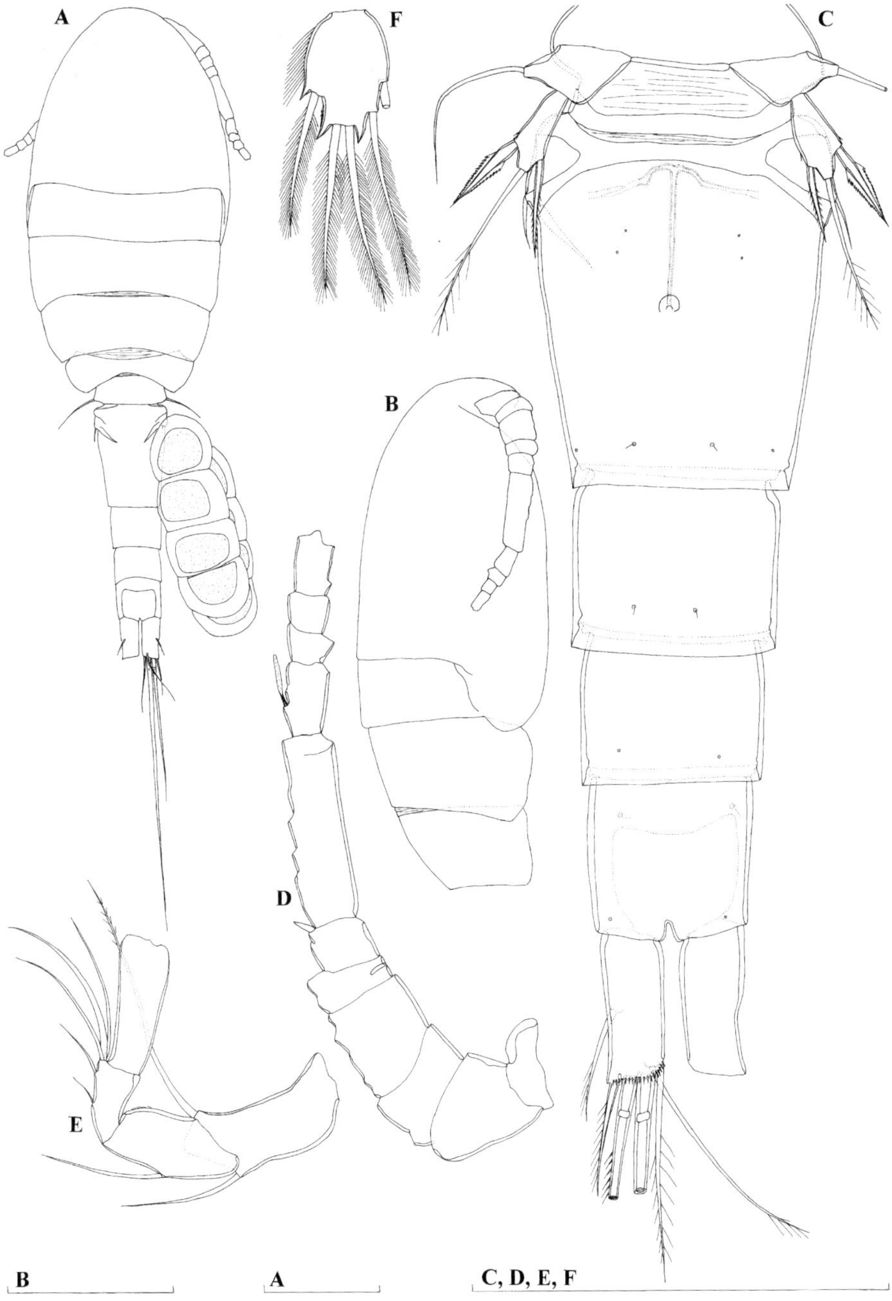

Fig. 40. *Cyclopina kasignete* n. sp., holotype ovigerous female. A, habitus, dorsal view; B, prosome, lateral view; C, urosome, ventral view; D, antennula without armature; E, antenna; F, third endopodal segment of fourth swimming leg. Scales = 0.1 mm.

somite ornamented with two large sensilla dorsally and one pore laterally, and with smooth fringe dorsally and laterally. Arthrodial membrane between second and third, and between third and fourth pedigerous somites, as well as between prosome and urosome, visible from dorsal view. Sclerotized joint between fifth pedigerous somite and genital double somite well developed, almost as a pseudosomite.

Genital double somite (fig. 40C) with anterior part slightly inflated laterally, about 1.4 times as long as wide (dorsal view), ornamented with six pores dorsally, two pores laterally, and two sensilla and six pores ventrally. Hyaline fringe of genital double somite and two succeeding somites narrow, not frilled, and completely smooth. No cuticular recesses or windows visible. The single copulatory pore relatively large and ovoid, situated medially at midlength; copulatory duct long and very narrow, well sclerotized, straight, and directed anteriorly. Seminal receptacles fused, well sclerotized, small and tubular, as wide as copulatory duct; receptacle ducts weakly chitinized and not clearly visible. Paired ovipores located dorsolaterally at 1/3 of the double somite's length, covered by the reduced sixth legs. Two egg sacs, each with eight eggs, attached to anterior part of genital double somite, and reaching midlength of caudal rami. Third urosomite ornamented with two pores dorsally, one pore laterally, and two sensilla ventrally, while fourth urosomite with one dorsal, one lateral, and two ventral pores as only ornamentation. Anal somite with smooth, very broad, and slightly convex anal operculum, which represents 70% of the somite's width and reaches to only 2/5 of the somite's length; ornamented with two large dorsal sensilla, one pore laterally on each side, and two pores ventrally; posterior margin smooth and unornamented. Anal sinus widely open dorsally, smooth. Length ratio of abdominal somites, beginning with genital double somite, 100 : 51 : 41 : 47.

Caudal rami (fig. 40C) cylindrical, 2.3 times as long as wide (ventral view), and nearly 0.9 times as long as anal somite; parallel, with space between them less than half one ramus' width; they represent 6% of total body length. Distal margin ventrally convex and ornamented with continuous row of spinules as the only ramal ornamentation, with medial cuticular pore not clearly visible; posterior outer corners of caudal rami reaching beyond posterior inner corners. Dorsal seta relatively long, 1.6 times as long as ramus, inserted on a small protuberance close to posterior margin and dorsal inner corner, uniarticulate at base and unipinnate distally. Lateral seta also slender, 0.8 times as long as ramus, arising at midlength of ramus and slightly dorsolaterally. Outermost apical seta strong and unipinnate, 0.8 times as long as ramus, and half as long as innermost apical seta, which is bipinnate and not so strong. Principal apical setae with breaking planes and bipinnate; inner seta about 1.9 times as long as outer one and 0.4 times as long as body.

Antennula (fig. 40D) short and stout, reaching 3/4 along length of cephalothorax, 10-segmented, unornamented, with one short and slender aesthetasc on fifth and seventh segments laterally, one on ninth segment subapically, one longer aesthetasc on tenth segment apically, and setal formula: 3.5.7.4.5.7.3.2.1.7. No setae with breaking planes and only one subapical seta on sixth segment, one on seventh, and two apical setae on tenth segment articulating on basal part; most setae smooth (only three pinnate) and slender; one seta on fifth segment very short and spiniform; one apical seta on tenth segment fused basally to aesthetasc. Length ratio of antennular segments, from proximal end and along caudal margins, 1 : 1 : 1.5 : 0.6 : 1 : 4.4 : 1.7 : 0.7 : 0.9 : 1.5.

Antenna (fig. 40E) four-segmented, comprising coxobasis and three-segmented endopod, with geniculation between first and second endopodal segments; length ratio of antennal segments, 1 : 1 : 0.5 : 0.9. Coxobasis about 2.3 times as long as wide, unornamented, and armed with one smooth seta at inner distal corner and with one very long exopodal seta at outer distal corner; exopodal seta almost 1.8 times as long as coxobasis and twice as long as seta at inner distal corner. First endopodal segment 2.1 times as long as wide, armed with one smooth seta at middle, ornamented with longitudinal row of minute spinules along outer (caudal) margin. Second endopodal segment about 1.8 times as long as wide, unornamented, and armed with five smooth setae (two or three lateral and two subapical; one subapical seta geniculate and much stronger than any other seta on that segment). Third endopodal segment 2.5 times as long as wide, unornamented, and armed with six smooth apical setae, two of which strong and geniculate.

Labrum trapezoidal, not very large, with smooth and slightly concave distal margin, ornamented with a few large spinules at distal outer corners and with a short, transverse row of long spinules medially, closer to the distal margin; tips of medial spinules reaching beyond distal margin.

Mandibula (fig. 41A) large, comprising coxa with well developed gnathobase and distal palp; palp consisting of basis, two-segmented endopod, and four-segmented exopod. Cutting edge of coxal gnathobase divided into three regions: inner group of three complex, strong and smooth teeth and several long spinules; middle group of three simple and sharp teeth; and outermost group of two short setae. Basis large but not reaching beyond posterior margin

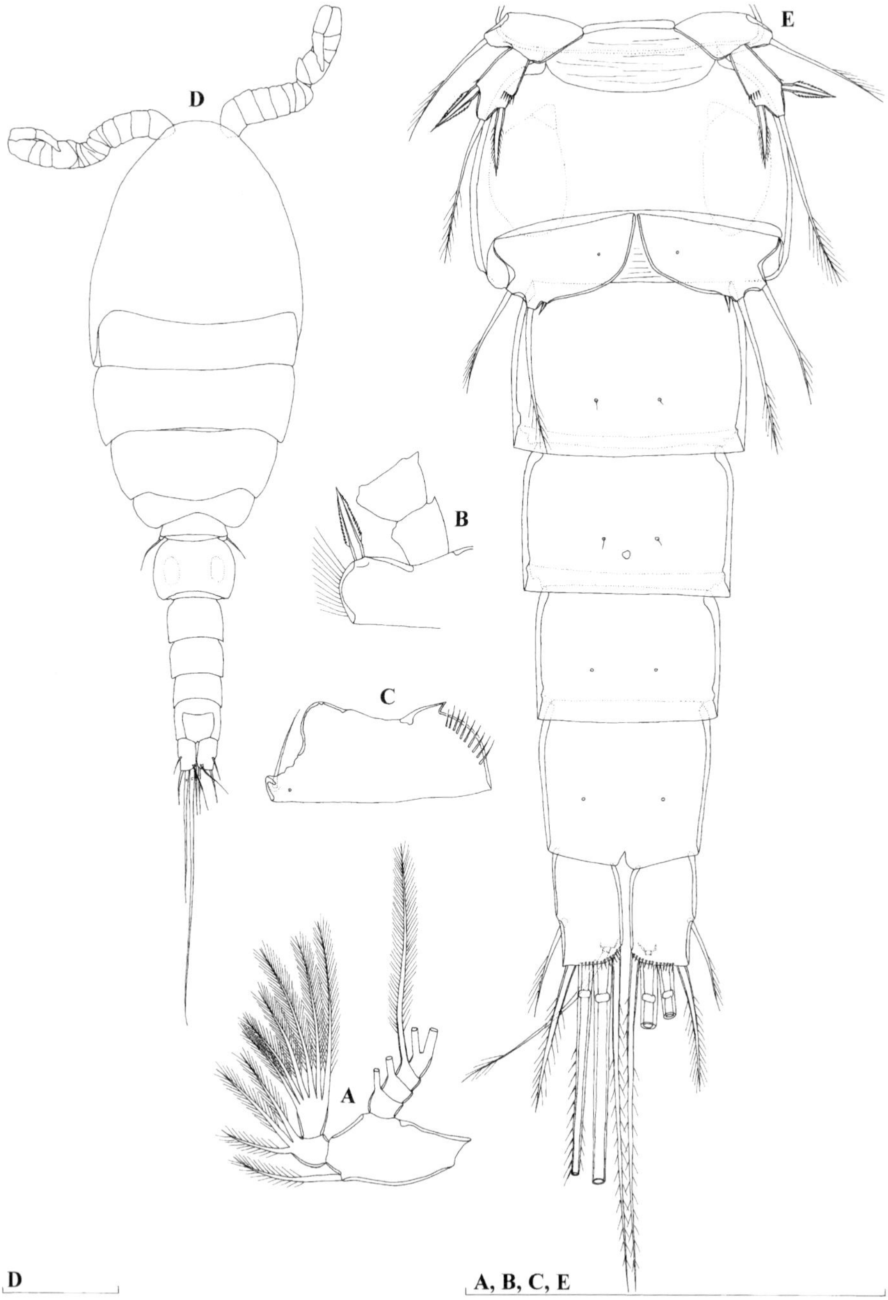

Fig. 41. *Cyclopina kasignete* n. sp., A-C, holotype ovigerous female; D-E, allotype male. A, palp of mandibula; B, part of basis and endopod of first swimming leg without armature; C, basis of fourth swimming leg; D, habitus, dorsal view; E, urosome, ventral view. Scales = 0.1 mm.

of coxal gnathobase, twice as long as wide, unornamented, and armed with one pinnate seta laterally. Endopod small, with first segment attached apically to basis and armed with three pinnate setae subapically; second segment longer than first and armed apically with six plumose setae. Exopod slender, 0.6 times as long as basis and 1.2 times as long as endopod, unornamented, with all segments of about the same length and with setal formula 1.1.1.2.

Maxillula composed of syncoxa with well developed arthrite, basis, and one-segmented endopod and exopod. Arthrite of syncoxa with nine unequal lateral elements and one smooth seta on posterior surface; four apical elements strong and geniculate, proximalmost element the longest and plumose. Coxa unornamented and armed with one epipodal seta; coxal endites not discernible, unarmed. Basis quadriform, much longer than arthrite of syncoxa, 2.4 times as long as wide, unornamented, and armed with three elements (two spines and one seta) on proximal, and two setae on distal endite. Endopod relatively slender, 1.5 times as long as wide, unornamented, and armed with two lateral setae on inner margin and four long setae apically; all endopodal setae slender and plumose. Exopod 1.3 times as long as wide and somewhat shorter than endopod, with longitudinal row of long spinules along inner margin, which is also convex, and armed apically with four slender, plumose, and long setae.

Maxilla robust but not very large, five-segmented, comprising praecoxa, coxa, basis, and two-segmented endopod (although praecoxa and coxa partly fused). Praecoxa unornamented, armed with three setae on proximal, and only one seta on distal endite. Coxa ornamented with longitudinal row of spinules along outer margin, armed with three slender and pinnate setae on proximal endite; distal endite highly mobile, unornamented, bearing two slender setae. Basis much shorter than coxa, expanded into spinulose, geniculate, and smooth claw, and armed with one strong and one minute setae. First endopodal segment somewhat shorter than basis, armed with three robust setae. Second endopodal segment about as long as first but much narrower, armed with three strong and three slender setae.

Maxilliped slender, seven-segmented, composed of praecoxa, coxa, basis, and four-segmented endopod. Praecoxa about as long as its greatest width, unornamented, with proximal endite weakly discernible and armed with a single seta, and distal endite armed with three setae. Coxa about as long as praecoxa along outer margin, ornamented with a row of hair-like spinules along outer margin, armed with two strong setae on only endite. Basis the longest segment, 1.8 times as long as wide, much narrower at basal part, ornamented with several hair-like spinules along inner margin proximally, and armed with two pinnate setae distally on inner margin. Endopod slightly shorter than basis, unornamented, with setal formula 0.0.0.3.

All swimming legs (figs. 40F, 41B, C) with three-segmented exopod and endopod; their armature formula as follows (legend: inner/outer spine or seta; inner/terminal/outer):

	Exopod			Endopod		
Segments	1	2	3	1	2	3
First leg	1/1	1/1	3/2/3	1/0	1/0	3/2/1
Second leg	1/1	1/1	4/2/3	1/0	2/0	3/2/1
Third leg	1/1	1/1	4/2/3	1/0	2/0	3/2/1
Fourth leg	1/1	1/1	4/2/2	1/0	2/0	2/2/1

Third exopodal segment spine formula 4.4.4.3 and no spines on endopods. All intercoxal sclerites quadriform, with slightly convex distal margin; those of first, second, and third legs unornamented, while sclerite of fourth leg with several parallel rows of spinules at middle. Coxae ornamented with several spinules along outer margin (coxa of fourth leg with additional large patch of spinules at outer half), armed with plumose seta at inner distal corner. Basis of each leg ornamented with a row of spinules along inner margin, armed with a slender seta on outer margin; basis of first leg (fig. 41B) with additional short and sharply serrated spine on distal inner corner, which reaches distal margin of second endopodal segment; inner distal corners of second, third, and fourth leg (fig. 41C) pronounced as sharp, spiniform processes. All exopodal and endopodal setae slender and plumose. All exopodal spines stout, and with both margins sharply serrated. All endopodal segments with hair-like spinules along outer margin and most exopodal segments with short spinules along outer margin. Third endopodal segment of fourth swimming leg (fig. 40F) about 1.4 times as long as wide; apical setae of similar length, 1.6 times as long as the segment.

Fifth leg (fig. 40C) relatively small, with very short and broad intercoxal sclerite, and two-segmented, comprising coxobasis and a one-segmented exopod. Coxobasis about twice as wide as long, armed with a single smooth basal seta on outer margin and unornamented. Exopod 1.7 times as long as basis and more than twice as long as its greatest width, somewhat narrower at basal part, armed with two spines and a single seta apically, and ornamented with one or two spinules on inner and outer margin; seta bipinnate at distal part, positioned centrally and as the most apical element, 1.8 times as long as exopod, twice as long as inner apical spine (which is bipinnate), and about 1.8 times as long as outer apical spine (which has both margins sharply serrated and is more robust than the inner spine).

Sixth leg (fig. 40C) not clearly distinct, ovoid cuticular plate, armed with robust outer spine, which is directed dorsally, and an inner slender seta; both elements smooth.

Male (allotype). Body length, excluding caudal setae, 0.535 mm. Habitus (fig. 41D) similar to that of female, with first pedigerous somite fused to cephalothorax only laterally, with prosome/urosome ratio 1.5, and greatest width near posterior end of cephalothorax. Body length/width ratio about 3.0; cephalothorax 2.6 times as wide as genital somite.

Cephalothorax (fig. 41D) slightly shorter than its greatest width, representing 31% of total body length. Arthrodial membranes not visible from dorsal view. Ornamentation of prosomites, colour, and nauplius eye similar to those of female. Hyaline fringes of all somites smooth.

Genital somite (fig. 41D, E) relatively large, 1.7 times as wide as long (dorsal view), with two not completely formed, large spermatophores inside, ornamented with two pores dorsally and one pore ventrally/laterally; third and fourth urosomites ventrally armed with two sensilla each, and fourth urosomite with large pore in-between; fifth urosomite ornamented with only two ventral pores; anal somite similar to that of female, although ventral pores situated more anteriorly.

Caudal rami (fig. 41E) about 1.6 times as long as wide, parallel, with almost no space between them. Armature and ornamentation similar to those of female, but all setae comparatively longer. Principal apical setae with breaking planes and bipinnate; inner seta 1.9 times as long as outer and 0.4 times as long as body.

Antennula about as long as cephalothorax, 16-segmented, strongly geniculate between fourteenth and fifteenth segments, with one aesthetasc on twelfth segment and one on sixteenth segment, and setal formula 3.4.4.2.2.5.0.1.2.1.1.2.2.2.1.10.

Antenna similar to that of female, with one exopodal seta and one seta on inner distal corner of coxobasis.

Labrum, mandibula, maxillula, maxilla, maxilliped, and swimming legs very similar to those of female.

Fifth leg (fig. 41E) also very similar to that of female, but without additional setae on inner margin of exopod; outer apical exopodal spine slightly less robust than in female.

Sixth leg (fig. 41E) distinct, large plate, ornamented with a single pore, and armed with two pinnate and slender setae, and one very short and smooth spine as innermost element; middle seta about 1.3 times as long as outermost.

Variability. — Body length of females ranging from 0.527 mm to 0.600 mm (0.560 mm average; n = 10), while in males it ranges from 0.438 mm to 0.535 mm (0.487 mm average; n = 2). The cuticular pore pattern is variable among specimens from the same sample, and sometimes even asymmetrical. No other type of variability was observed.

Etymology. — The name of the species is the Greek noun "kasignete", which is of feminine gender and means "sister". It refers to a close relationship of this species to the other five Australian congeners, and is to be treated as a noun in apposition to the generic name.

Distribution. — *Cyclopina kasignete* n. sp. was collected from three different localities on the exposed part of the eastern South Australian coast, but not in the sheltered waters of Spencer Gulf, where it is replaced by another congener, *C. soror* n. sp.

Discussion. — *Cyclopina kasignete* n. sp. does not have a single autapomorphy that would easily distinguish it from previously described species, but shares with the other five Australian representatives a lack of the apical tuft of hairs on top of the apical seta on the mandibular exopod. It differs from all those five species by the armature of its mandibular exopod, which bears six setae on the second segment and only four setae in the other species, but this is a plesiomorphic character state and quite common in the genus, so it cannot be used for establishing closer phylogenetic relationships (except for the assumption that the other five Australian species share a common ancestor, of which their allegedly synapomorphic character state would bear testimony).

As mentioned before, of all Australian congeners, *C. kasignete* seems to be most closely related to *C. unisetosa* n. sp. (see the discussion section for that species), but the two species could be distinguished by a number of small differences. All other Australian species differ even more, as they all have a different antennular segmentation. *Cyclopina kasignete* is superficially similar to *C. steueri* Früchtl, 1923, which is a widely distributed species and also known from Indonesia, i.e., the closest locality of any *Cyclopina* species to Australia (see Monchenko, 1976). However, *C. steueri* has a different armature of the male fifth leg, which bears two spines and two setae on the exopod (the most common situation in the genus), while the male fifth leg of *C. kasignete* is the same as that of the female. The latter situation is known in only five other species described so far: *C. caiala* Lotufo & Da Rocha, 1991; *C. ensifera* Grandori, 1926; *C. gracilis* Claus, 1863; *C. parapsammophila* Monchenko, 1980; and *C. psammophila* Steuer, 1940 (see Sars, 1913; Grandori, 1926; Steuer, 1940; Monchenko, 1980; Lotufo & Da Rocha, 1991). A male fifth leg with only one seta on the exopod was also reported for *C. kieferi* Schäfer, 1936 by Herbst (1962), although not by Petkovski (1955a), but it is obvious that they have studied two different species. Herbst (1962) indicated his uncertainty about his identification by inserting a question mark in the figure legend for this species (his p. 192), although not in the taxonomic subtitle. Unfortunately, as males are known only for one half of the *Cyclopina* species described, it is very hard to predict the significance of this character. However, male characters may help much in distinguishing closely related species, as becomes obvious when comparing the fifth and sixth legs of *C. amita* n. sp. and *C. kasignete*, the only two Australian species with known males.

15. **Cyclopina soror** n. sp.
(fig. 42)

Material examined. — Type: Holotype female dissected on 1 slide (WAM C37234): Australia, South Australia, Eyre Peninsula, Cowell, 25 December 2002, leg. T. Karanovic, 33°41.180′S 136°55.786′E.

Other material: Australia, South Australia, Eyre Peninsula, Tumby Bay, 25 December 2002, leg. T. Karanovic, 34°22.417′S 136°06.262′E: 1 damaged juvenile male in alcohol (WAM C37235).

Description. — Female (holotype). Body length, excluding caudal setae, 0.619 mm. Habitus (fig. 42A) relatively robust, with prosome/urosome ratio 1.5 and greatest width at posterior end of cephalothorax. Body length/width ratio about 3.0; cephalothorax 2.9 times as wide as genital double somite. Posterolateral corners of cephalothorax forming carapace-like extensions and concealing lateral margins of first pedigerous somite. First three pedigerous somites without pronounced lateral expansions; posterior corners of fourth pedigerous somite somewhat expanded laterally. Urosome narrow and fifth pedigerous somite about as wide as genital double somite. Colour of preserved specimen milky white. Nauplius eye not visible. Rostrum broadly rounded and furnished with two large sensilla.

Prosome (fig. 33A) comprising cephalothorax and four free pedigerous somites (although first pedigerous somite laterally fused to cephalothorax),

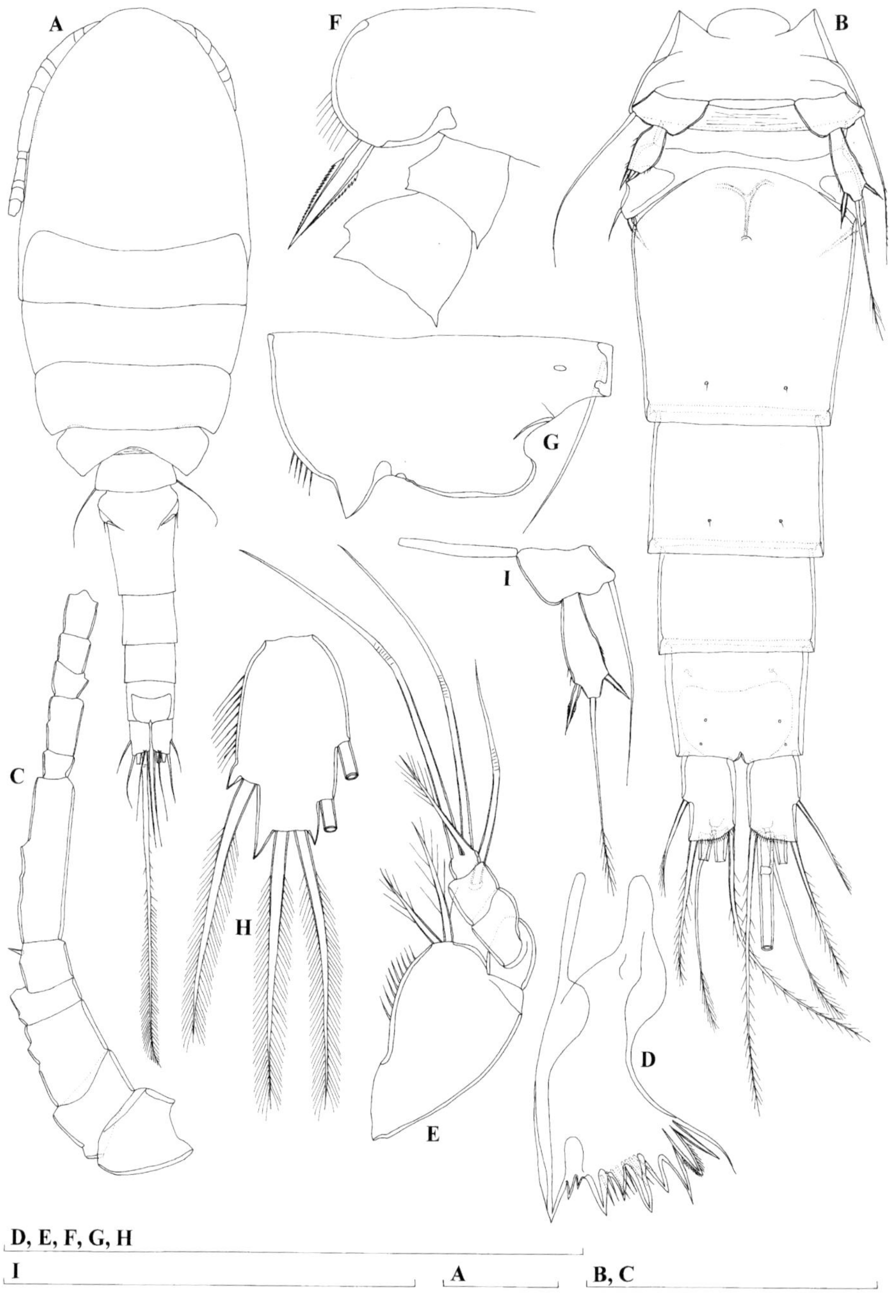

Fig. 42. *Cyclopina soror* n. sp., holotype female. A, habitus, dorsal view; B, urosome, ventral view; C, antennula without armature; D, coxa of mandibula; E, distal part of maxilliped; F, part of basis and endopod of first swimming leg without armature; G, basis of fourth swimming leg; H, third endopodal segment of fourth swimming leg; I, fifth leg. Scales = 0.1 mm.

ovoid, almost twice as long as wide (dorsal view); length ratio of prosomal somites, beginning with cephalothorax, 100 : 23 : 23 : 26 : 7. Cephalothorax about as long as wide, representing 33% of total body length; surface of cephalic shield with several small sensilla and cuticular pores. Hyaline fringe of prosomites narrow and smooth. Fifth pedigerous somite ornamented with two large sensilla dorsally and one pore laterally, and with smooth fringe dorsally and laterally. Arthrodial membrane between prosome and urosome visible from dorsal view. Sclerotized joint between fifth pedigerous somite and genital double somite well developed, almost as a pseudosomite.

Genital double somite (fig. 42B) with anterior part inflated laterally, about 1.2 times as long as wide (dorsal view), ornamented with four pores dorsally, two pores laterally, and two sensilla ventrally. Hyaline fringe of genital double somite and two succeeding somites narrow, not frilled, and completely smooth. No cuticular recesses or windows visible. The single copulatory pore small and ovoid, situated just after 1/3, and medially; copulatory duct short and narrow, cylindrical, well sclerotized, directed anteriorly. Seminal receptacles fused, well sclerotized, small and tubular, as wide as copulatory duct; receptacle ducts weakly chitinized and not clearly visible. Paired ovipores located dorsolaterally at 1/3 of the double somite's length, covered by the reduced sixth legs. Third urosomite ornamented with four pores dorsally and two sensilla ventrally, fourth urosomite without any ornamentation. Anal somite with smooth, broad, and convex anal operculum, which represents 77% of the somite's width and reaches to only 2/5 of the somite's length; ornamented with two large dorsal sensilla, one pore laterally on each side, and four pores ventrally; posterior margin smooth and unornamented. Anal sinus widely open dorsally, smooth. Length ratio of abdominal somites, beginning with genital double somite, 100 : 52 : 37 : 47.

Caudal rami (fig. 42B) cylindrical, 1.5 times as long as wide (ventral view) and only 0.7 times as long as anal somite, parallel, with space between them less than half of one ramus' width; they represent 4% of total body length. Distal margin ventrally convex and ornamented with continuous row of spinules as the only ramal ornamentation, with medial cuticular pore not clearly visible; posterior outer corners of caudal rami reaching beyond posterior inner corners. Dorsal seta relatively long, 2.5 times as long as ramus, inserted on a small protuberance close to the posterior margin, uniarticulate at base and unipinnate distally. Lateral seta 1.2 times as long as ramus, arising at middle of ramus' length. Outermost apical seta relatively strong and bipinnate, 1.8 times as long as ramus, and half as long as innermost apical seta, which is also bipinnate and strong. Principal apical setae with breaking planes and bipinnate, all except one inner seta broken.

Antennula (fig. 42C) relatively long and slender, almost reaching posterior margin of cephalothorax, 11-segmented, unornamented, with one short and slender aesthetasc on fifth segment laterally, one on seventh and tenth segments subapically, one long aesthhetasc on eleventh segment apically, and setal formula: 3.5.8.4.5.7.1.2.2.1.7. No setae with breaking planes and only one subapical seta on sixth segment, one on eighth, and three apical setae on eleventh segment articulating on basal part; most setae smooth (only eight pinnate) and slender; one seta on fifth segment stout, very short, and spiniform; one apical seta on tenth segment fused basally to aesthetasc. Length ratio of antennular segments, from proximal end and along caudal margins, 1 : 1.2 : 1.4 : 0.6 : 1 : 3.9 : 0.5 : 1.4 : 0.6 : 0.8 : 1.1.

Antenna four-segmented, comprising coxobasis and three-segmented endopod, with geniculation between first and second endopodal segments; length ratio of antennal segments, 1 : 1 : 0.5 : 0.9. Coxobasis about 2.3 times as long as wide, unornamented, and armed with one smooth seta at inner distal corner and with one very long exopodal seta at outer distal corner. First endopodal segment twice as long as wide, armed with one smooth seta at middle, ornamented with longitudinal row of minute spinules along outer (caudal) margin. Second endopodal segment about 1.8 times as long as wide, unornamented, armed with five smooth setae (two or three lateral and two subapical; one subapical seta geniculate and much stronger than any other seta on that segment). Third endopodal segment 2.5 times as long as wide, unornamented, and armed with six smooth apical setae, two of which strong and geniculate.

Labrum trapezoidal, not very large, with smooth and slightly concave distal margin, ornamented with a few large spinules at distal outer corners and with a short, transverse row of long spinules medially, closer to distal margin; tips of medial spinules reaching beyond distal margin.

Mandibula (fig. 42D) large, comprising coxa with well developed gnathobase and distal palp; palp consisting of basis, two-segmented endopod, and four-segmented exopod. Cutting edge of coxal gnathobase divided into three regions: inner group of three complex, strong, and smooth teeth and eight long spinules; middle group of three simple and sharp teeth, of which only the outermost bipinnate; and outermost group of two short and smooth setae. Basis large but not reaching beyond posterior margin of coxal gnathobase, 2.3 times as long as wide, unornamented, and armed with one plumose seta laterally. Endopod small, with first segment attached apically to basis and armed with three pinnate setae subapically; second segment longer than first and armed apically with four short setae. Exopod slender, 0.7 times as long as basis and 1.2 times as long as endopod, unornamented, with all segments of about the same length, and with setal formula 1.1.1.2.

Maxillula composed of syncoxa with well developed arthrite, basis, and one-segmented endopod and exopod. Arthrite of syncoxa with nine unequal lateral elements and one smooth seta on posterior surface; four apical elements strong and geniculate, proximalmost element longest and plumose. Coxa unornamented and armed with one epipodal seta; coxal endites not discernible, unarmed. Basis quadriform, much longer than arthrite of syncoxa, 2.4 times as long as wide, unornamented, and armed with three elements (two spines and one seta) on proximal, and two setae on distal endite. Endopod relatively slender, 1.6 times as long as wide, unornamented, and armed with two lateral setae on inner margin and four long setae apically; all endopodal setae slender and plumose. Exopod 1.4 times as long as wide and somewhat shorter than endopod, with longitudinal row of long spinules along inner margin, which is also convex, and armed apically with four slender, plumose, and long setae.

Maxilla five-segmented, comprising praecoxa, coxa, basis, and two-segmented endopod (although praecoxa and coxa partly fused). Praecoxa unornamented, armed with three setae on proximal, and only one seta on distal endite. Coxa ornamented with longitudinal row of spinules along outer margin, and armed with three slender and pinnate setae on proximal endite; distal endite highly mobile, unornamented, bearing two slender setae. Basis much shorter than coxa, expanded into spinulose, geniculate, and smooth claw, and armed with one strong and one minute setae. First endopodal segment somewhat shorter than basis, armed with three robust setae. Second endopodal segment about as long as first but much narrower, armed with three strong and three slender setae.

Maxilliped (fig. 42E) slender, seven-segmented, composed of praecoxa, coxa, basis, and four-segmented endopod. Praecoxa about as long as greatest width, unornamented, with proximal endite weakly discernible and armed with a single seta, and distal endite armed with three setae. Coxa about as long as praecoxa along outer margin, ornamented with a row of hair-like spinules along outer margin, and armed with two strong setae on its only endite. Basis the longest segment, 1.7 times as long as wide, much narrower at basal part, ornamented with several strong spinules along inner margin proximally, and armed with two pinnate setae distally on inner margin. Endopod slightly shorter than basis, unornamented, with setal formula 0.0.1.3.

All swimming legs (fig. 42F, G, H) with three-segmented exopod and endopod; their armature formula as follows (legend: inner/outer spine or seta; inner/terminal/outer):

	Exopod			Endopod		
Segments	1	2	3	1	2	3
First leg	0/1	1/1	3/2/3	1/0	1/0	3/2/1
Second leg	1/1	1/1	4/2/3	1/0	2/0	3/2/1
Third leg	1/1	1/1	4/2/3	1/0	2/0	3/2/1
Fourth leg	1/1	1/1	4/2/2	1/0	2/0	2/2/1

Third exopodal segment spine formula 4.4.4.3 and no spines on endopods. All intercoxal sclerites quadriform, with slightly convex distal margin; those of first, second, and third legs unornamented, while sclerite of fourth leg with several parallel rows of spinules at middle. Coxae ornamented with several spinules along outer margin (coxa of fourth leg with an additional large patch of spinules at outer half), armed with plumose seta at inner distal corner. Basis of each leg ornamented with a row of spinules along inner margin, armed with a slender seta on outer margin; basis of first leg (fig. 42F) with additional short and sharply serrated spine on distal inner corner, which reaches distal margin of second endopodal segment; inner distal corners of second, third, and fourth leg sharply pronounced. All exopodal and endopodal setae slender and plumose. All exopodal spines stout and with both margins sharply serrated. All endopodal segments with hair-like spinules along outer margin and most exopodal segments with short spinules along outer margin. Third endopodal segment of fourth swimming leg (fig. 42H) about 1.7 times as long as wide; apical setae of similar length, 1.5 times as long as segment.

Fifth leg (fig. 42B, I), relatively small, with very short and broad intercoxal sclerite and two-segmented, comprising coxobasis and one-segmented exopod. Coxobasis about 1.9 times as wide as long, armed with a single unipinnate basal seta on outer margin, and unornamented. Exopod of right fifth leg (fig. 42B) abnormal. Exopod of left leg (fig. 42I) about 1.9 times as long as basis and 2.9 times as long as its greatest width, somewhat narrower at basal part, armed with two spines and a single seta apically, and ornamented with a few minute spinules on outer and inner margins; seta bipinnate at distal part, positioned centrally and as the most apical element, 1.9 times as long as exopod, 3.8 times as long as inner apical spine, and about 6.0 times as long as outer apical spine; both spines not very robust, but inner spine somewhat stronger, serrated, and 1.5 times as long as outer spine, which is bipinnate.

Sixth leg (fig. 42B) not clearly distinct, ovoid cuticular plate, armed with short outer spine, which is directed dorsally, and an inner slender seta; both elements smooth.

Male. Unknown.

Variability. — Only one adult female was collected and examined. The only asymmetrical feature of this specimen is its abnormal right fifth leg (fig. 42B), which could be the result of damage during development.

Etymology. — This species is named after the Latin noun "soror", meaning "sister" and referring to a close relationship of the new species to its Australian congeners. The specific name is to be treated as a noun (gender feminine) in apposition to the generic name.

Distribution. — *Cyclopina soror* n. sp. inhabits sheltered waters of Spencer Gulf in South Australia, where it was found at two localities.

Discussion. — *Cyclopina soror* n. sp. has an 11-segmented female antennula. As mentioned before, this is a pretty unusual feature in the genus and only two other species share it, both of them newly described in this monograph: *C. amita* n. sp. and *C. adelphe* n. sp. *Cyclopina soror* is most closely related to *C. amita* and their similarities, differences, and a possible origin from the same ancestor were mentioned in the discussion section for the latter species. I think *C. soror* is a tropical relict in South Australia and that is why it can be found only in the sheltered waters of Spencer Gulf and not along the coast of South Australia that is exposed to the Southern Ocean. In the latter environment it is replaced by *C. kasignete* n. sp., a species not so closely related to *C. soror* as is the Western Australian *C. amita*, which lives in the tropical Pilbara region.

All my attempts to complete a world key to females of the genus *Cyclopina* Claus, 1862 have failed, just because many species were originally incompletely described and have never been redescribed since. This will certainly pose even more problems in the future, as more new species would be described from different parts of the world. A revision of the genus *Cyclopina* is a priority for starting a proper revision of the family Cyclopinidae G. O. Sars, 1913. With the inclusion of six new Australian species and two species I have transferred to *Cyclopina* from two other genera (see the discussion section for *C. unisetosa* n. sp.), the genus now counts more than 40 members and its revision will not be an easy task. Here I give a key for the easy identification of the known Australian species, which is based entirely on the female morphology.

KEY TO AUSTRALIAN SPECIES OF THE GENUS *CYCLOPINA*

1. Antennula 10- or 11-segmented 2
– Antennula 9-segmented .... *C. kasis* n. sp.
2. Antennula 11-segmented 4
– Antennula 10-segmented 3
3. Copulatory duct short .... *C. unisetosa* n. sp.
– Copulatory duct long .... *C. kasignete* n. sp.
4. Third endopodal segment of second, third, and fourth legs with five setae 5
– Same segment with four setae .... *C. adelphe* n. sp.
5. Outer spine on fifth leg more than twice as long as inner one .... *C. amita* n. sp.
– Outer spine shorter than inner one .... *C. soror* n. sp.

Genus **Heterocyclopina** Pleşa, 1968

Diagnosis (revised). — Moderately large Cyclopinidae, with slender habitus and first pedigerous somite fused to cephalothorax. Urosome five-segmented in female and six-segmented in male. Abdominal somites with serrated hyaline fringes. Female genital double somite with the single copulatory pore situated ventrally, and the paired ovipores located laterally and covered by the reduced sixth legs. Anal somite longer than wide, with very short and broad anal operculum lying very close to the somite's anterior margin, and which is usually covered by the serrated margin of the preanal somite (although not extended posteriorly as a pseudo-operculum). Caudal rami much longer than anal somite and cylindrical, from 3.5 to 12 times as long as wide, armed with one dorsal, two lateral, and four apical elements; proximal lateral element inserted at 1/3 and distal one at 2/3 of the ramus' length; dorsal seta significantly longer than innermost and outermost apical setae. Female antennula 18- or 19-segmented, with large first and second segments. Male antennula geniculate and 16- or 17-segmented. Antenna four-segmented, with or without exopodal setae. Mandibula composed of coxa, basis, two-segmented endopod, and four-segmented exopod; basis armed with a single seta, endopodal armature formula 2.5 or 2.6, and exopodal armature formula 1.1.1.2 or 0.1.1.1.2. Maxillula composed of syncoxa, basis, and one-segmented endopod and exopod; basis armed with three setae on proximal endite and one or two setae on distal endite; endopod armed with six, and exopod with four setae. Maxilla composed of praecoxa, coxa (usually partly fused to praecoxa), basis, and three-segmented endopod. Maxilliped slender and prehensile, composed of syncoxa, basis, and five-segmented endopod (although last two segments may be difficult to discern); first endopodal segment small, second very long, third long but shorter than second, and last two segments minute; five endopodal setae very strong and prehensile. All swimming legs with three-segmented exopods and endopods; third exopodal segments' spine formula 3.4.4.3 and setal formula 5.5.5.5; first exopodal segments of first, second, and third leg without inner seta, that of fourth leg with or without a seta; second endopodal segment of first leg with one seta, while those of other legs with one or two setae; third endopodal segment of first and fourth legs with five setae, those of second and third legs with five or six setae. Female fifth leg with small intercoxal sclerite and located ventrally, composed of coxobasis and one-segmented exopod; coxobasis slightly wider than long, armed with a single basal seta on outer margin; exopod quadriform, only slightly longer than coxobasis, from 1.5 to 1.8 times as long as wide, and armed with two spines and a single seta apically; inner spine longer than exopod or outer spine, but shorter than middle seta. Male fifth leg two- or three-segmented, with armature formula 1.5 or 1.1.4. Female sixth leg armed with inner strong and dorsally directed spine, and outer slender seta. Male sixth leg armed with three elements.

Type species. — *Heterocyclopina vietnamensis* Pleşa, 1968.

Other species. — *Paracyclopina feiticeira* Lotufo, 1995; *Paracyclopina uguaipuku* Lotufo, 1995; *Heterocyclopina plesai* n. sp.

16. **Heterocyclopina plesai** n. sp.
(figs. 43-45)

Material examined. — Types only: Holotype female dissected on 2 slides (WAM C37236); paratypes, 5 females (2 on 1 slide in toto (TMAG G5779), other 3 in alcohol (WAM C37237)): Australia, Queensland, Yeppoon, 13 August 2003, leg. T. Karanovic, 23°08.582′S 150°45.659′E.

Description. — Female (holotype). Body length, excluding caudal setae, 0.577 mm. Habitus (fig. 43A, B) very slender from dorsal view, with prosome/urosome ratio 1.2 and greatest width at posterior end of cephalothorax. Body length/width ratio about 4.2; cephalothorax about 2.4 times as wide as genital double somite. Posterior corners of cephalothorax and first two free pedigerous somites without pronounced lateral expansions; fourth pedigerous somite slightly expanded laterally (dorsal view). Urosome almost cylindrical and fifth pedigerous somite about as wide as genital double somite. Colour of preserved specimen yellowish. Nauplius eye not visible. Rostrum (fig. 43B, C) pointed, directed ventrally, furnished with two large sensilla dorsally and one pore laterally on each side.

Prosome (fig. 34A, B) comprising cephalothorax and three free pedigerous somites (first pedigerous somite completely fused to cephalothorax), ovoid, and only slightly constricted between second and third pedigerous somites, about 2.4 times as long as wide (dorsal view); length ratio of prosomal somites, beginning with cephalothorax, 100 : 20 : 19 : 6. Arthrodial membrane between prosome and urosome visible from dorsal view. Cephalothorax large, 1.6 times as long as its greatest width, representing 37% of total body length; surface of cephalic shield with several small sensilla and pores. Hyaline fringe of prosomites narrow and smooth. Fifth pedigerous somite (fig. 44A, B, C) ornamented with two large sensilla and three pores dorsally, one pore ventrally, and one pore laterally on each side, and with a smooth fringe. No sclerotized joint between fifth pedigerous somite and genital double somite.

Genital double somite (fig. 44A, B, C) with laterally inflated anterior part, 1.3 times as long as wide (dorsal view), ornamented with four pores ventrally, two pores laterally, and four sensilla and one pore dorsally. No cuticular recesses or windows visible. Hyaline fringe of genital double somite smooth dorsally and sharply serrated ventrally. Copulatory pore small and ovoid, situated medially at midlength; copulatory duct long and very narrow, rigidly sclerotized, directed largely anteriorly, attached to very wide bridge between fused seminal receptacles. Seminal receptacles large, expanded posteriorly beyond copulatory pore, and forming a single curved and complex structure; they represent 64% of the double somite's width and 49% of its length. Paired ovipores located laterally at middle, covered by the reduced sixth legs, and connected to receptacles via very short and relatively narrow receptacle ducts. Third urosomite ornamented with two pores ventrally, two pores dorsally, and two pores laterally on each side, with a smooth hyaline fringe dorsally and sharply serrated ventrally. Fourth urosomite without any ornamentation and with a frilled and deeply serrated fringe both ventrally and dorsally. Anal somite with smooth, very broad, and slightly convex anal operculum, which represents 81% of the somite's width and is situated very close to the somite's anterior margin, largely covered by the serrated fringe of the fourth urosomite; ornamented with two sensilla and two pores dorsally,

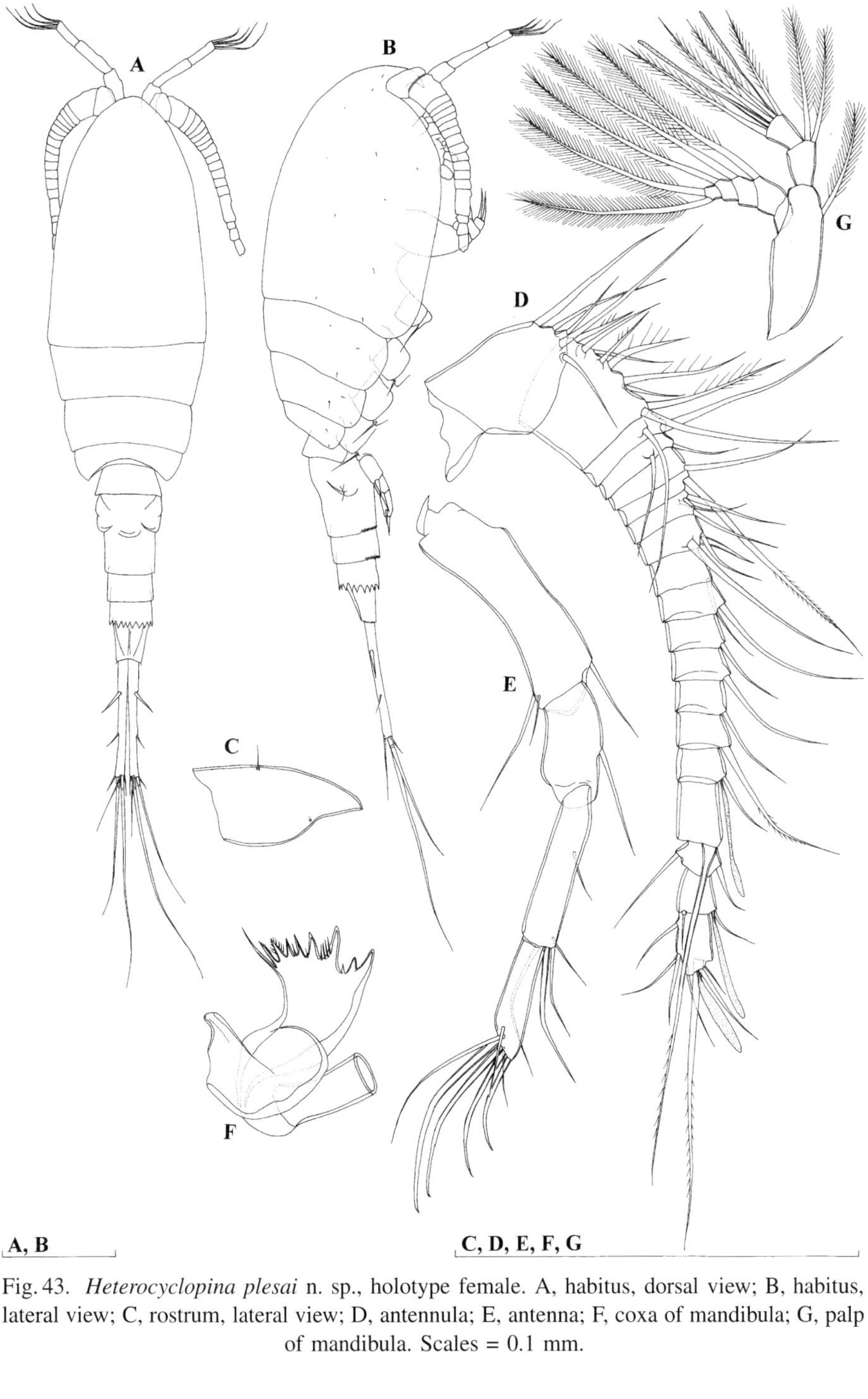

Fig. 43. *Heterocyclopina plesai* n. sp., holotype female. A, habitus, dorsal view; B, habitus, lateral view; C, rostrum, lateral view; D, antennula; E, antenna; F, coxa of mandibula; G, palp of mandibula. Scales = 0.1 mm.

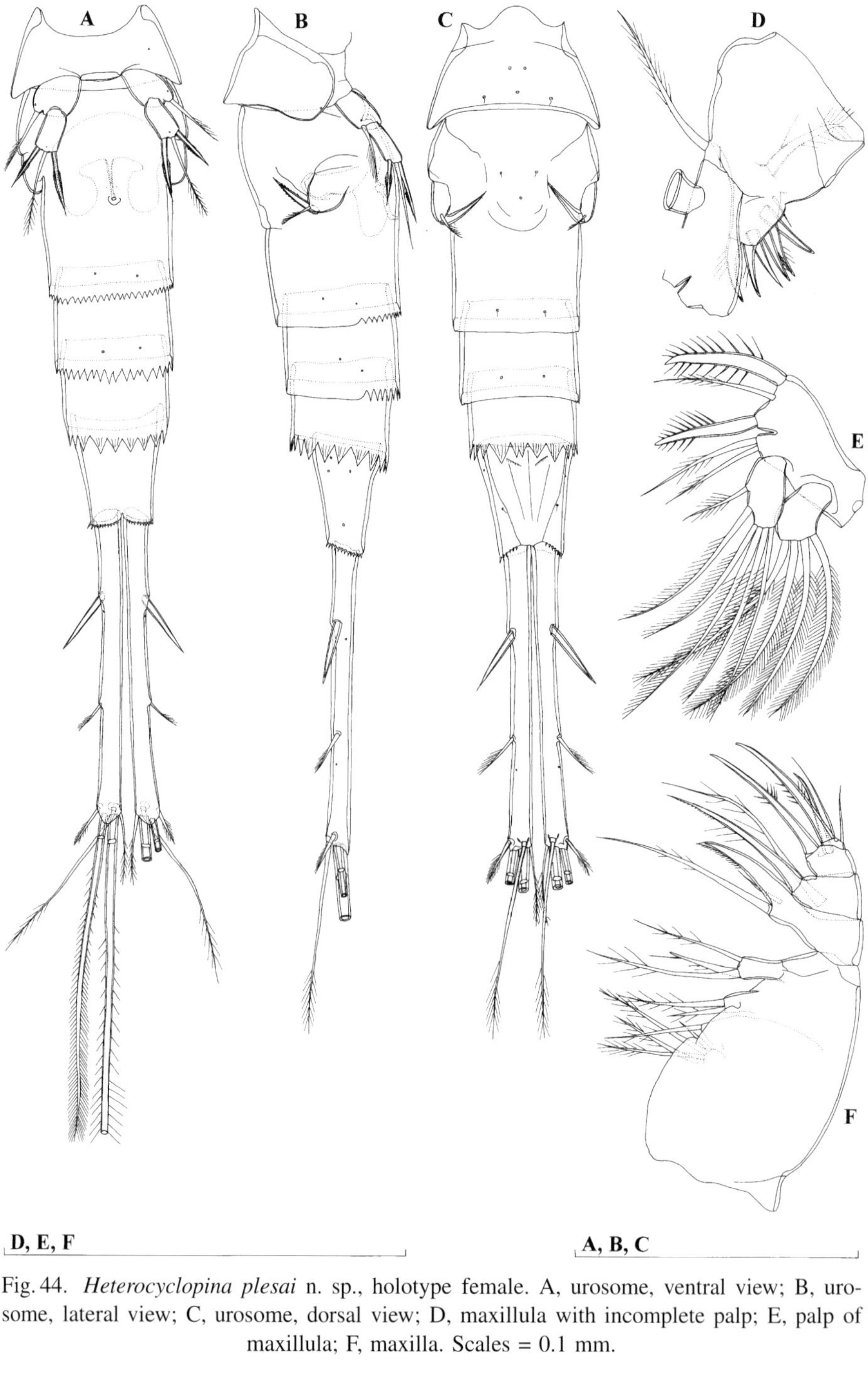

Fig. 44. *Heterocyclopina plesai* n. sp., holotype female. A, urosome, ventral view; B, urosome, lateral view; C, urosome, dorsal view; D, maxillula with incomplete palp; E, palp of maxillula; F, maxilla. Scales = 0.1 mm.

two pores laterally, and an uninterrupted row of small spinules along the posterior margin. Anal sinus widely open dorsally, ornamented with only two short diagonal rows of minute spinules. Length ratio of abdominal somites, beginning with genital double somite, 100 : 55 : 53 : 49.

Caudal rami (fig. 44A, B, C) very long and cylindrical, 12 times as long as wide (ventral view), 2.9 times as long as anal somite and even slightly longer than last three urosomites combined; parallel, with almost no space between them; they represent 17% of total body length. Distal margin ventrally convex and smooth, with medial cuticular pore not clearly visible. The only ornamentation consists of one dorsal and two lateral pores. The armature consists of one dorsal, two lateral, and four apical elements. Dorsal seta slender and relatively long, 0.6 times as long as ramus, inserted on a small protuberance almost reaching posterior margin, uniarticulate at base and bipinnate distally. Proximal lateral element a smooth spine, inserted at first third and slightly dorsolaterally, almost 3.0 times as long as the ramus' width. Distal lateral element a slender and unipinnate seta, arising at 2/3 and about 0.7 times as long as proximal lateral element. Outermost apical seta slender and bipinnate, twice as long as the ramus' width and 0.6 times as long as innermost apical seta, which is also slender and bipinnate. Principal apical setae with breaking planes and bipinnate; inner seta about 1.6 times as long as outer one and 0.3 times as long as body.

Antennula (fig. 43D) short and stout, reaching 2/3 of cephalothorax length, 19-segmented, unornamented, with one short aesthetasc on fifteenth and eighteenth segments subapically, and one on nineteenth segment apically, and with setal formula: 5.9.2.2.1.1.3.3.1.1.1.1.1.1.1.1.2.2.7. No setae with breaking planes and only one seta on eighteenth segment and one apical seta on nineteenth segment articulating on basal part; most setae smooth and slender (only nine setae pinnate); subapical seta on sixteenth segment the longest and strongest; one seta on eighth segment spiniform but not very short. Length ratio of antennular segments, from proximal end and along caudal margins, 1 : 1.6 : 0.5 : 0.3 : 0.3 : 0.4 : 0.5 : 0.5 : 0.7 : 0.6 : 0.6 : 0.6 : 0.7 : 0.8 : 0.7 : 1.2 : 0.5 : 1 : 1.2.

Antenna (fig. 43E) large and slender, unornamented, directed anteriorly, four-segmented, comprising coxobasis and three-segmented endopod, with only small geniculation between first and second endopodal segment; length ratio of antennal segments, 1 : 0.5 : 0.8 : 0.6. Coxobasis about 2.8 times as long as wide, armed with a single smooth and short seta at inner distal corner and two smooth and unequal exopodal setae at outer distal corner. First endopodal segment twice as long as wide, armed with one smooth and short seta at middle. Second endopodal segment about 4.2 times as long as wide, armed with five smooth setae (three lateral, two subapical; one subapical seta geniculate, much stronger than any other seta on that segment). Third endopodal segment 3.5 times as long as wide, armed with seven apical setae, four of which very strong and geniculate.

Labrum small and trapezoidal, with smooth and straight distal margin and ornamented with a few rows of long spinules on each side at distal lateral corner.

Mandibula (fig. 43F, G) comprising coxa with well developed gnathobase and distal palp; palp consisting of basis, two-segmented endopod, and five-segmented exopod. Cutting edge of coxal gnathobase divided into three regions: inner group of two complex and strong teeth, and four spinules at base of outer tooth; middle group of four simple teeth and two spinules (only outermost spine pinnate); and outermost group of two setae (outermost seta smooth and twice as long as inner unipinnate one). Basis large, 3.0 times as long as wide, unornamented, and armed with one plumose seta laterally. Endopod small and unornamented, with first segment attached apically to basis and armed with two plumose setae subapically; second segment about as long as first and armed apically with four pinnate setae and two slender aesthetascs. Exopod slender, 0.7 times as long as basis and 1.6 times as long as endopod, unornamented, and with setal formula 0.1.1.1.2.

Maxillula (fig. 44D, E) composed of syncoxa with well developed arthrite, basis, and one-segmented endopod and exopod. Arthrite of syncoxa with eight elements; four apical elements strong and spiniform, proximalmost element the longest and plumose. Coxa unornamented and armed with one epipodal pinnate seta; coxal endites weakly discernible, unarmed. Basis quadriform, much longer than arthrite of syncoxa, almost 1.8 times as long as wide, unornamented, and armed with three setae on proximal and two setae on distal endite; two setae on proximal endite very strong and spiniform; all setae on basis pinnate. Endopod as long as basis' width, and about as long as exopod but more slender, unornamented, and armed with two lateral setae on inner margin and four long setae apically; all endopodal setae slender and pinnate. Exopod 1.5 times as long as wide, also unornamented, with convex inner margin, armed apically with four slender, plumose, and very long setae.

Maxilla (fig. 44F) robust and large, six-segmented, comprising praecoxa, coxa, basis, and three-segmented endopod (although praecoxa and coxa partly fused). Praecoxa inflated, ovoid, and about as wide as long, unornamented, with four setae on proximal, and only one seta on distal endite. Coxa half as large as praecoxa, also unornamented, armed with two pinnate setae on elongated proximal endite; distal endite highly mobile and bearing three slender and pinnate setae. Basis much shorter than coxa, expanded into spinulose, geniculate, and unipinnate claw, and armed with one minute and one large setae. First endopodal segment about as long as basis, armed with two robust and geniculate setae. Second endopodal segment about half as long as first endopodal segment, armed with two strong setae. Third endopodal segment minute, armed with one strong and two slender setae.

Maxilliped (fig. 45A) large and slender, prehensile, seven-segmented, composed of syncoxa, basis, and five-segmented endopod. Syncoxa 1.8 times as long as greatest width, unornamented, with proximal praecoxal endite weakly discernible and armed with a single short seta, and distal praecoxal endite armed with two pinnate setae; only coxal endite well-developed, armed with two strong and pinnate setae. Basis 0.7 times as long as syncoxa, twice as long as wide, much narrower at basal part, ornamented with several hair-like spinules along inner margin proximally, and armed with two pinnate setae distally on inner margin. Endopod almost twice as long as basis, unornamented, with setal formula 1.2.2.1.4; two apical setae on fifth endopodal segment, as well as all setae on fourth and third segments very strong and geniculate. Length ratio of maxillipedal segments, from proximal end, 100 : 69 : 29 : 52 : 31 : 8 : 14.

All swimming legs (fig. 45B, C, D, E) with three-segmented exopod and endopod of about the same length; their armature formula as follows (legend: inner/outer spine or seta; inner/terminal/outer):

	Exopod			Endopod		
Segments	1	2	3	1	2	3
First leg	0/1	1/1	3/2/3	1/0	1/0	2/2/1
Second leg	0/1	1/1	4/2/3	1/0	1/0	3/2/0
Third leg	0/1	1/1	4/2/3	1/0	2/0	3/2/1
Fourth leg	0/1	1/1	4/2/2	1/0	2/0	2/2/1

Third exopodal segment spine formula 3.4.4.3 and no spines on endopods. All intercoxal sclerites relatively small, quadriform, unornamented, with slightly concave or almost straight distal margin. Coxae also unornamented, armed with a plumose seta at inner distal corner. Basis of each leg unornamented and armed with a slender seta on outer margin; basis of first leg with additional short and robust spine on distal inner corner, which reaches 1/3 of

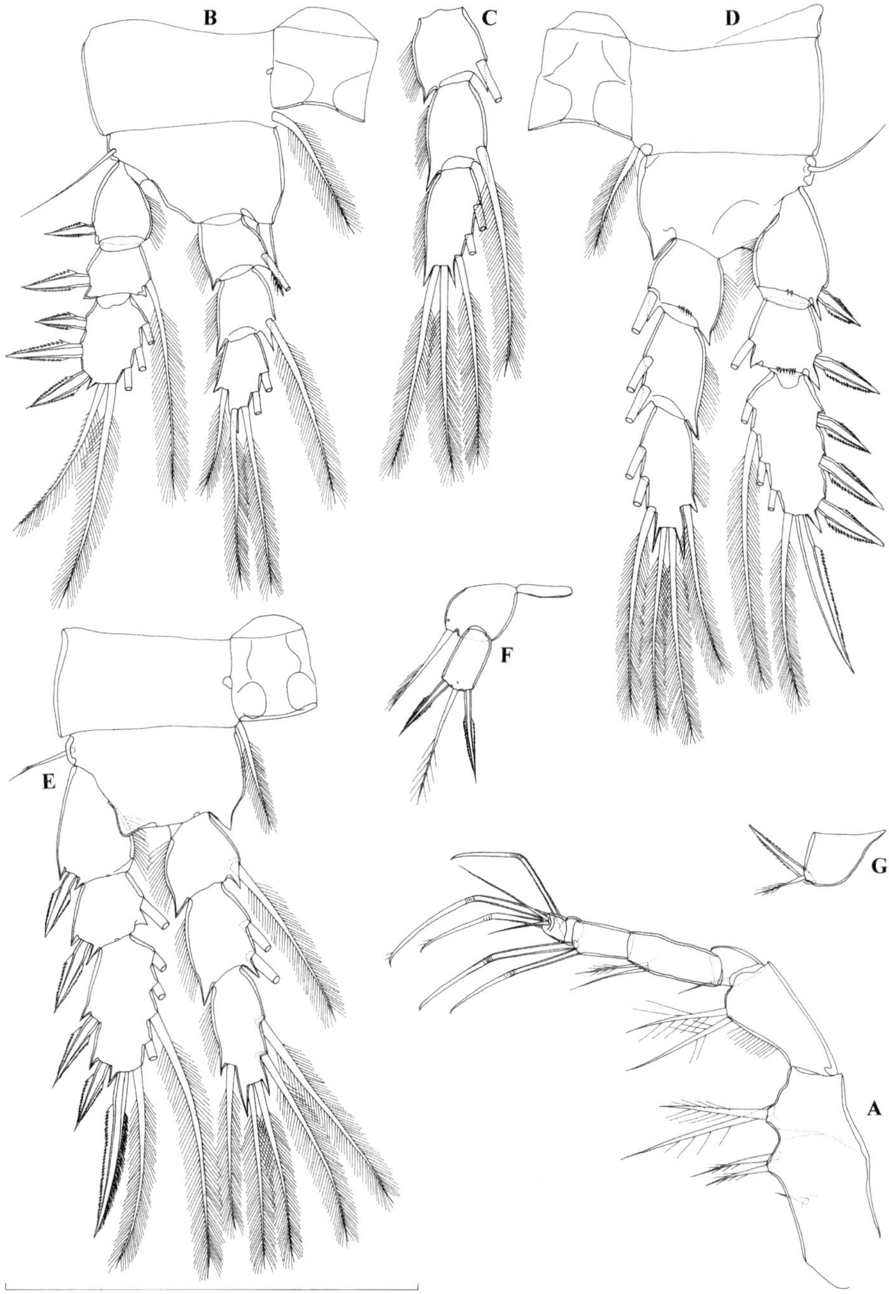

Fig. 45. *Heterocyclopina plesai* n. sp., holotype female. A, maxilliped; B, first swimming leg; C, endopod of second swimming leg; D, third swimming leg; E, fourth swimming leg; F, fifth leg; G, right sixth leg. Scale = 0.1 mm.

second endopodal segment; inner distal corners of second, third, and fourth leg pronounced as short spiniform processes; a similar, but somewhat larger and blunt process also between exopod and endopod. All exopodal and endopodal setae slender and plumose. Most exopodal spines stout and with both margins sharply serrated; however, outer apical spine on third exopodal segment of second and third leg with serrated outer and smooth inner margin. All endopodal segments with hair-like spinules along outer margin, which are also present on inner margin of first exopodal segment. Third endopodal segment of fourth swimming leg twice as long as wide; both apical setae of the same length, 1.6 times as long as the segment.

Fifth leg (figs. 44A, B, 45F) with small intercoxal sclerite, and two-segmented, comprising coxobasis and a one-segmented exopod. Intercoxal sclerite quadrangular, with slightly convex distal margin, almost five times as wide as long and unornamented. Coxobasis 1.2 times as wide as long, armed with a single unipinnate basal seta on outer margin and ornamented with a single pore at base of basal seta. Exopod quadriform, only slightly longer than coxobasis, and 1.8 times as long as its greatest width, armed with two spines and a single seta apically, and ornamented with a single cuticular pore at base of apical seta; seta bipinnate at distal part, positioned centrally and as the most apical element, almost twice as long as exopod, 1.3 times as long as inner apical spine, and about 1.6 times as long as outer apical spine; both spines strong and with sharply serrated margins.

Sixth leg (fig. 45G) relatively small and nearly triangular cuticular plate, armed with an inner strong, pinnate, and dorsally directed spine, and an outer slender and bipinnate seta; spine 1.6 times as long as seta.

Male. Unknown.

Variability. — Body length of females ranging from 0.549 mm to 0.581 mm (0.563 mm average; n = 6) and no other type of variability observed.

Etymology. — The species is named in honour of the late Dr. Corneliu Pleşa, who erected the genus *Heterocyclopina*. The name is a noun in the genitive singular.

Distribution. — This species is known from the type locality only.

Discussion. — The genus *Heterocyclopina* was erected by Pleşa (1968) to accommodate a single interstitial cyclopinid from Vietnam: *H. vietnamensis* Pleşa, 1968. He recognized its close relationship with the genus *Procyclopina* Herbst, 1955, which was described by Herbst (1955) after a single Brazilian species: *P. polyartha* Herbst, 1955. However, Pleşa (1968) recognized many important morphological differences between these two species and concluded that they could not belong to the same genus. Later, Lotufo (1995) rediscovered and redescribed *Procyclopina polyartha*, also from Brazil, and in the same paper he described another closely related Brazilian species: *P. maricopeba* Lotufo, 1995. In that same paper, Lotufo (1995) described another two Brazilian species in the genus *Procyclopina*, *P. feiticeira* Lotufo, 1995 and *P. uguaipuku* Lotufo, 1995, that were not closely related to *P. polyartha* but are very similar to *Heterocyclopina vietnamensis*. It looks very likely that he was not aware of Pleşa's (1968) paper, as he neither mentioned it in the reference section, nor did he compare his two species to *H. vietnamensis*. Martínez Arbizu (2001b) synonymized *Heterocyclopina* with *Procyclopina*, "because the diagnosis of the former genus agrees well with that of the latter genus since the description of several new species of *Procyclopina* from Brazilian psammic habitats". He did not question the taxonomic status of these Brazilian species described by Lotufo (1995). Martínez Arbizu (2001b) also stated that "*Heterocyclopina* differs from *Procyclopina* basically in having only 5 elements on distal endopodal segment of leg 4, instead of 6, and in having only 1 inner seta on middle endopodal segments of leg 2 to 4, instead of 2". The first part of this statement is not true, as all species described in either *Heterocyclopina* or *Procyclopina* have five setae on the third endopodal segment of the fourth leg, while the second part is only partially true, as *P. feiticeira* was reported to have one seta on the second endopodal segment of the second leg (see Lotufo, 1995: 249, fig. 32).

The genus *Procyclopina* differs from *Heterocyclopina* by the absence of a ventrally located copulatory pore on the female genital double somite, as well as by its short caudal rami, which are armed with only one lateral seta, inserted close to the posterior margin and dorsolaterally. It contains two Brazilian species: *Procyclopina polyartha* Herbst, 1955 (type species) and *P. maricopeba* Lotufo, 1995. I here reinstate the genus *Heterocyclopina* and consider Martínez Arbizu's (2001b) synonymization unjustified, although it was followed by Boxshall & Halsey (2004: 509). The generic diagnosis is slightly revised (see above) and the genus now includes four species: *Heterocyclopina vietnamensis* Pleşa, 1968 (type species) from Vietnam; *H. feiticeira* (Lotufo, 1995) n. comb. from Brazil; *H. uguaipuku* (Lotufo, 1995) n. comb. also from Brazil; and *H. plesai* n. sp. from the eastern coast of Australia. I also do not think *Heterocyclopina* is so closely related to *Procyclopina*. The genus *Parapseudocyclopinodes* Lindberg, 1961 to me seems to represent the closest living relative of *Heterocyclopina*, and the two could basically be distinguished only by a different armature of the fifth leg (see Herbst, 1952; Lindberg, 1961).

All four *Heterocyclopina* species could easily be distinguished from each other by a different armature formula of the swimming legs, as well as by some additional characters. However, it should be noted that Pleşa's (1968) description of *H. vietnamensis* is lacking in detail and probably has a few inaccuracies, especially in the description of the mouthparts. Martínez Arbizu (2001b) claimed to have examined additional material of "*H.* cf. *vietnamica* ... from a sandy beach in China", which displayed a clearly four-segmented mandibular exopod. I also think Pleşa (1968) omitted a few armature elements from the maxillular endopod, as well as from the maxilla and maxilliped, which is not hard to understand having in mind that some of the female specimens he studied were only 0.32 mm long. *Heterocyclopina plesai* n. sp. is probably the most primitive member of the genus, as it is the only species with the antennal exopod present (represented by two setae) and with a five-segmented mandibular exopod. The latter character is, however, unreliable, as it is very hard to say whether the first exopodal segment is a real segment or just a well-chitinized joint, as it has no armature. *Heterocyclopina plesai* has the greatest number of armature elements on the maxilliped, which could also be interpreted as a plesiomorphic character state. The autapomorphic features of the new species are the absence of the inner seta on the third endopodal segment of the second leg (fig. 45C) and the presence of aesthetascs on the mandibular endopod (fig. 43B). The aesthetascs on the mandibula were so far reported only from the exopod, and only in one cyclopinid genus: *Sensogiselina* Martínez Arbizu, 2000 (see Martínez Arbizu, 2000b). Unfortunately, no males of *Heterocyclopina plesai* were collected, so the male characters could not be compared.

Genus **Cryptocyclopina** Monchenko, 1979

Diagnosis (emended). — Small Cyclopinidae with relatively slender habitus and first pedigerous somite free or fused to cephalothorax. Urosome five-segmented in female and six-segmented in male. Abdominal somites with serrated fringes. Female genital double somite with ventral cuticular suture and a single copulatory pore situated ventrally; paired ovipores located laterally. Anal somite with broad and short anal operculum. Caudal rami cylindrical, from 2.2 to 3.8 times as long as wide, parallel, armed with six setae; lateral seta inserted at midlength. Female antennula shorter than cephalothorax and composed of a varying number of segments (from 9 to 15, but with all additional subdivisions fully expressed, it could even be interpreted as 19-segmented). Antenna four-segmented, with or without exopodal seta. Mandibula composed of coxa, basis, two-segmented endopod, and four-segmented exopod; last endopodal segment with six setae and exopodal armature formula 1.1.1.2. Maxillula composed of syncoxa, basis, and one-segmented endopod and exopod; basis armed with three setae on proximal endite and one or two setae on distal endite; exopod armed with four, and endopod with six or seven setae. Maxilla composed of praecoxa, coxa (partly fused to praecoxa), basis, and three-segmented endopod. Maxilliped composed of praecoxa, coxa (could be partly fused to praecoxa), basis, and three- to five-segmented endopod. All swimming legs with three-segmented exopods and endopods, and with only one seta on second endopodal segment; third endopodal segment setal formula 5.5.5.5 or 5.6.6.5; third exopodal segment spine formula 4.4.3.3 or 2.4.4.3, while setal formula 4.4.4.4 or 5.5.5.5. Fifth leg two-segmented, armed with outer seta on first segment and with two armature elements apically: inner a shorter spine and outer a longer seta. Female sixth leg armed with inner strong and dorsally directed spine, and outer slender seta. Male sixth leg with three elements.

Type species. — *Cryptocyclopina inopinata* Monchenko, 1979.

Other species. — *Cyclopina monchenkoi* n. sp.

17. **Cryptocyclopina monchenkoi** n. sp. (figs. 46, 47)

Material examined. — Types only: Holotype female dissected on 1 slide (WAM C37238); paratypes, 2 females (1 on 1 slide in toto (TMAG G5780), other in alcohol (WAM C37239)): Australia, Queensland, Hervey Bay, 14 August 2003, leg. T. Karanovic, 25°16.728′S 152°50.400′E.

Description. — Female (holotype). Body length, excluding caudal setae, 0.443 mm. Habitus (fig. 46A, B) very slender from dorsal view, with prosome/urosome ratio of 1.6 and greatest width at posterior end of cephalothorax. Body length/width ratio about 4.1; cephalothorax about 1.9 times as wide as genital double somite. Posterior corners of cephalothorax and first two free pedigerous somites without pronounced lateral expansions; fourth pedigerous somite slightly expanded laterally (dorsal view). Urosome almost cylindrical and fifth pedigerous somite slightly wider than genital double somite. Colour of preserved specimen milky white. Nauplius eye not visible. Rostrum pointed, directed ventrally, furnished with two large sensilla dorsally and one pore laterally on each side.

Prosome (fig. 46A, B) comprising cephalothorax and three free pedigerous somites (first pedigerous somite completely fused to cephalothorax), ovoid, and about 2.6 times as long as wide (dorsal view); length ratio of prosomal somites, beginning with cephalothorax, 100 : 23 : 18 : 10. Arthrodial membrane between prosome and urosome visible from dorsal view. Cephalothorax large, 1.7 times as long as its greatest width, representing 40% of total body length; surface of cephalic shield with several small sensilla and pores. Hyaline fringe of prosomites narrow and smooth. Fifth pedigerous somite ornamented with only two large sensilla dorsally, and with a smooth fringe both dorsally and laterally. No sclerotized joint between fifth pedigerous somite and genital double somite.

Genital double somite (fig. 46C) with laterally inflated anterior part, as long as wide (dorsal view), ornamented with a single pore dorsally. No cuticular recesses, windows, or sensilla visible. Hyaline fringe of genital double somite smooth dorsally and sharply serrated ventrally. Copulatory pore small and ovoid, situated medially at 2/5 of the double somite's length, with two ovoid spermatophores attached; copulatory duct short and very narrow, rigidly sclerotized, directed mostly anteriorly. Seminal receptacles not clearly visible. Paired ovipores located laterally at middle, covered by the reduced sixth legs. Third urosomite ornamented with two pores dorsally, and with smooth hyaline fringe dorsally, sharply serrated ventrally. Fourth urosomite without any ornamentation and with frilled and deeply serrated fringe both ventrally and dorsally. Anal somite with smooth, very broad, and slightly convex anal operculum, which represents nearly 80% of the somite's width and is situated very close to the somite's anterior margin, mostly covered by serrated fringe of fourth urosomite; ornamented with two sensilla and two pores dorsally, one pore laterally, and an uninterrupted row of small spinules along its posterior margin. Anal sinus widely open dorsally, ornamented with only two short, diagonal rows of minute spinules. Length ratio of abdominal somites, beginning with genital double somite, 100 : 50 : 47 : 33.

Caudal rami (fig. 46C) long and cylindrical, 3.8 times as long as wide (ventral view) and 1.8 times as long as anal somite, parallel, with almost no space between them; they represent 9% of total body length. Distal margin ventrally convex and ornamented with several small spinules, with medial

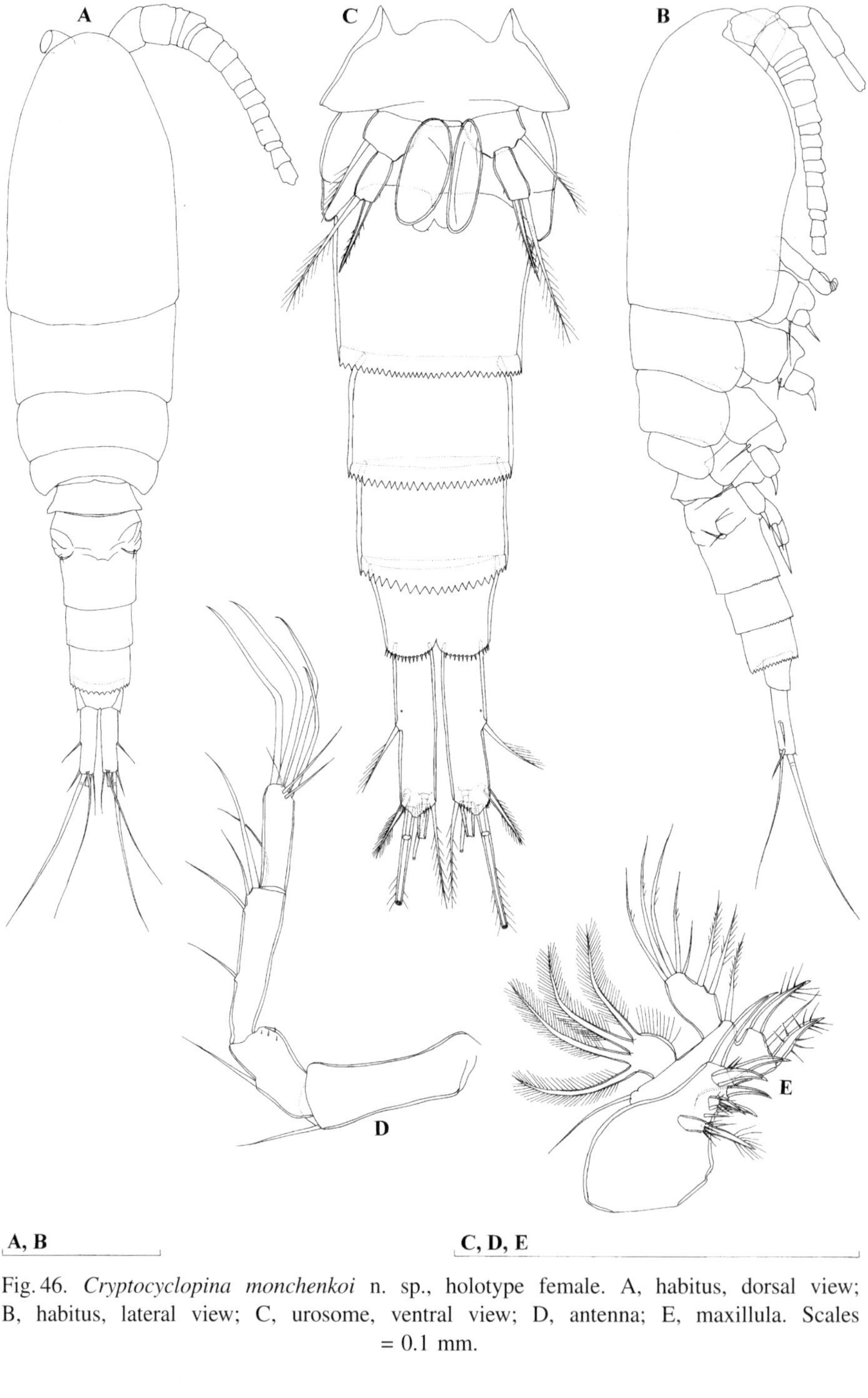

Fig. 46. *Cryptocyclopina monchenkoi* n. sp., holotype female. A, habitus, dorsal view; B, habitus, lateral view; C, urosome, ventral view; D, antenna; E, maxillula. Scales = 0.1 mm.

cuticular pore clearly visible. The only additional ornamentation consists of one dorsal, one lateral, and one ventral pore. The armature consists of one dorsal, one lateral, and four apical setae. Dorsal seta slender and very long, 2.3 times as long as ramus and only slightly shorter than outer principal apical seta, inserted on small protuberance almost reaching posterior margin, uniarticulate at base and bipinnate distally. Lateral seta slender and unipinnate, arising at middle and about half as long as ramus. Outermost apical seta stout and bipinnate, spiniform, 0.4 times as long as ramus and 0.6 times as long as innermost apical seta, which is also bipinnate but much more slender. Principal apical setae with breaking planes and bipinnate; inner seta broken on both rami.

Antennula (fig. 47A) long and stout, reaching to 3/4 of cephalothorax, 19-segmented (although sixth and seventh, eighth and ninth, tenth and eleventh, and fifteenth and sixteenth segments partly fused), unornamented, with one short aesthetasc on fourteenth and eighteenth segments subapically and one on nineteenth segment apically, and setal formula: 3.9.2.2.2.1.4.1.1.1.1.0.1.1.0.1.2.2.7. No setae with breaking planes and only one seta on seventeenth segment and three setae on nineteenth segment articulating on basal part; most setae smooth and slender (only four setae pinnate); subapical seta on sixteenth segment the longest and strongest; one seta on seventh segment spiniform but not very short. Length ratio of antennular segments, from proximal end and along caudal margins, 1 : 1.6 : 0.4 : 0.3 : 0.3 : 0.4 : 0.8 : 0.4 : 0.5 : 0.5 : 0.6 : 0.6 : 0.7 : 0.8 : 0.5 : 0.7 : 0.5 : 0.8 : 1.1.

Antenna (fig. 46D) slender, four-segmented, comprising coxobasis and three-segmented endopod, with geniculation between first and second endopodal segment; length ratio of antennal segments, 1 : 0.6 : 0.9 : 0.6. Coxobasis about 2.5 times as long as wide, unornamented, and armed with a single smooth and short seta at inner distal corner; seta representing exopod absent. First endopodal segment twice as long as wide, ornamented with several spinules on outer (caudal) margin, and armed with one smooth and short seta at middle of inner margin. Second endopodal segment about 3.9 times as long as wide, unornamented, and armed with five smooth setae (three lateral, two subapical; one subapical seta geniculate and stronger than any other seta on that segment). Third endopodal segment 3.3 times as long as wide, armed with seven apical setae, four of which are very strong and geniculate.

Labrum small and trapezoidal, with smooth and straight distal margin and ornamented with a few rows of long spinules on each side at distal lateral corner.

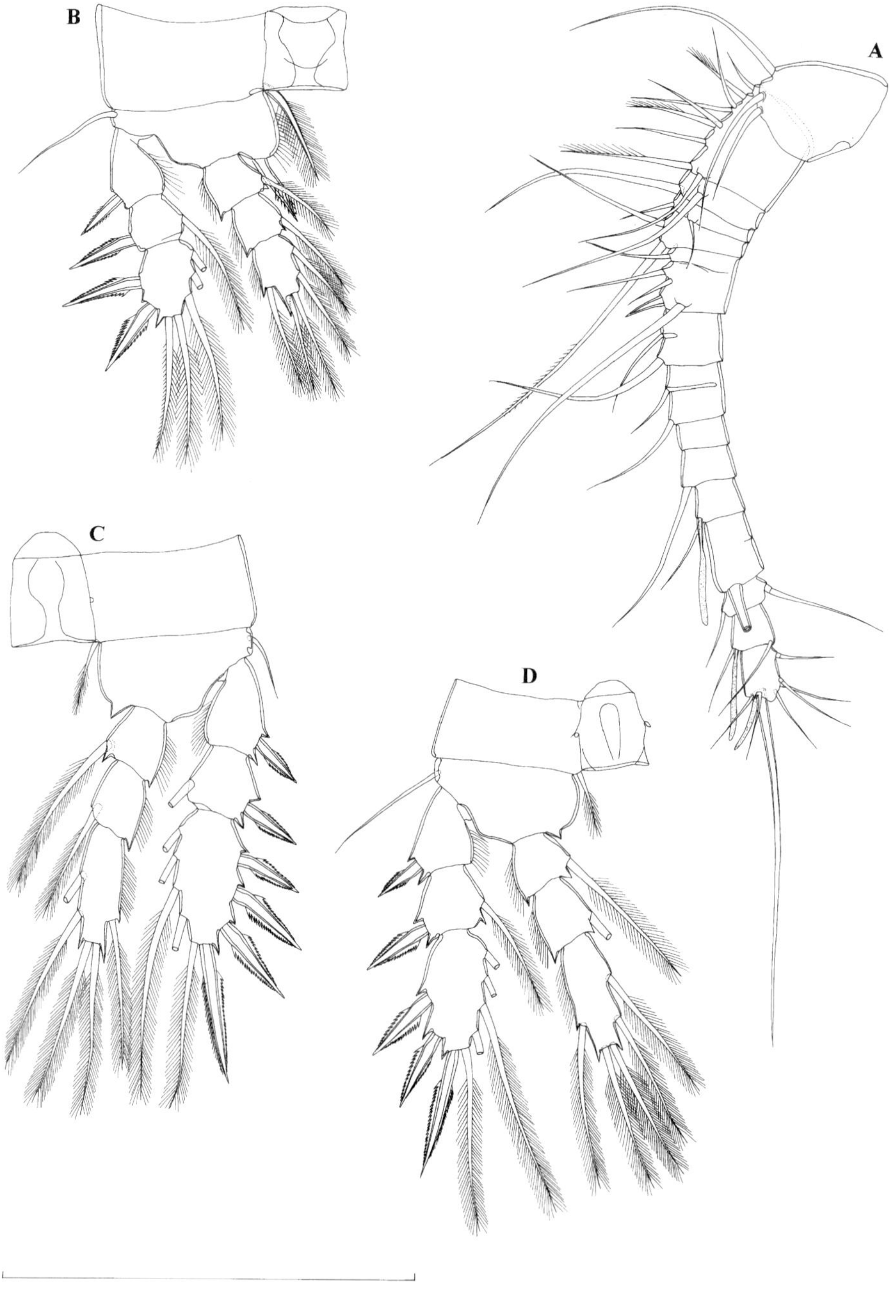

Fig. 47. *Cryptocyclopina monchenkoi* n. sp., holotype female. A, antennula; B, first swimming leg; C, third swimming leg; D, fourth swimming leg. Scale = 0.1 mm.

Mandibula comprising coxa with well developed gnathobase and distal palp; palp consisting of basis, two-segmented endopod, and four-segmented exopod. Cutting edge of coxal gnathobase divided into three regions: inner group of two complex and strong teeth; middle group of four simple teeth and several spinules (only outermost spine pinnate); and outermost group of two setae (outermost seta smooth and twice as long as inner unipinnate one). Basis large, three times as long as wide, unornamented, and armed with one plumose seta laterally. Endopod small and unornamented, with first segment attached apically to basis and armed with two plumose setae subapically; second segment about as long as the first and armed apically with six pinnate setae. Exopod slender, unornamented, and with setal formula 1.1.1.2.

Maxillula (fig. 46E) composed of syncoxa with well developed arthrite, basis, and one-segmented endopod and exopod. Arthrite of syncoxa with nine elements; four apical elements strong and spiniform, proximalmost element longest and plumose. Coxa unornamented and armed with one epipodal pinnate seta; coxal endites weakly discernible, unarmed. Basis quadriform, much longer than arthrite of syncoxa, twice as long as wide, unornamented, and armed with three setae on proximal, and one seta on distal endite; two setae on proximal endite very strong and spiniform. Endopod as long as basis' width and about 1.5 times as long as exopod, unornamented, and armed with two lateral setae on inner margin and five setae apically; all endopodal setae slender and all but one pinnate. Exopod only slightly longer than wide, with convex inner margin, which is also ornamented with a row of hair-like spinules, armed apically with four slender, plumose, and long setae.

Maxilla robust and large, six-segmented, comprising praecoxa, coxa, basis, and three-segmented endopod (although praecoxa and coxa partly fused). Praecoxa inflated, ovoid, and about as wide as long, unornamented, with four setae on proximal, and only one seta on distal endite. Coxa half as large as praecoxa, also unornamented, and armed with two pinnate setae on elongated proximal endite; distal endite highly mobile and bearing three slender and pinnate setae. Basis much shorter than coxa, expanded into spinulose, geniculate, and unipinnate claw, and armed with one minute and one large setae. First endopodal segment about as long as basis, armed with two robust and geniculate setae. Second endopodal segment about half as long as first endopodal segment, armed with two strong setae. Third endopodal segment minute, armed with one strong and two slender setae.

Maxilliped large and slender, prehensile, seven-segmented, composed of syncoxa (although praecoxa and coxa partly divided), basis, and five-segmented endopod. Syncoxa 1.8 times as long as greatest width, unornamented, with proximal praecoxal endite weakly discernible and armed with a single short seta, and distal praecoxal endite armed with three pinnate setae; the only coxal endite well-developed, armed with two strong and pinnate setae. Basis 0.7 times as long as syncoxa, twice as long as wide, much narrower at basal part, ornamented with several hair-like spinules along inner margin proximally, and armed with two pinnate setae distally on inner margin. Endopod almost twice as long as basis, unornamented, with setal formula 1.2.2.1.4; two apical setae on fifth endopodal segment, as well as all setae on fourth and third segment very strong and geniculate, and somewhat longer than in the previous species.

All swimming legs (fig. 47B, C, D) with three-segmented exopod and endopod of about the same length; their armature formula as follows (legend: inner/outer spine or seta; inner/terminal/outer):

	Exopod			Endopod		
Segments	1	2	3	1	2	3
First leg	0/1	1/1	3/2/2	1/0	1/0	2/2/1
Second leg	0/1	1/1	4/2/3	1/0	1/0	2/2/1
Third leg	0/1	1/1	4/2/3	1/0	1/0	2/2/1
Fourth leg	0/1	1/1	4/2/2	1/0	1/0	2/2/1

Third exopodal segment spine formula 2.4.4.3 and no spines on endopods. All intercoxal sclerites relatively small, quadriform, unornamented, with almost straight distal margin. Coxae also unornamented, armed with a plumose seta at inner distal corner. Basis of first leg ornamented with hair-like spinules along inner margin, basis of other swimming legs unornamented; each basis armed with a slender, smooth seta on outer margin; basis of first leg with additional robust spine on distal inner corner, which reaches 2/3 of second endopodal segment; inner distal corners of second, third, and fourth leg pronounced as short spiniform processes. All exopodal and endopodal setae slender and plumose. All exopodal spines stout and with both margins sharply serrated. All endopodal segments with hair-like spinules along outer margin, which are also present on inner margin of first exopodal segment. Third endopodal segment of fourth swimming leg 2.4 times as long as wide; both apical setae of the same length, 1.3 times as long as segment.

Fifth leg (fig. 46C) with small intercoxal sclerite and two-segmented, comprising coxobasis and a one-segmented exopod. Intercoxal sclerite quadrangular, with slightly convex distal margin, almost five times as wide as long, and unornamented. Coxobasis 1.3 times as wide as long, unornamented, and armed with a single unipinnate basal seta on outer margin. Exopod quadriform, only slightly longer than coxobasis, and twice as long as wide, ornamented with a single minute spinule on inner margin, and armed apically with inner spine and outer seta. Exopodal seta plumose, 2.8 times as long as exopod, and 1.8 times as long as exopodal spine, which is relatively slender and also bipinnate.

Sixth leg (fig. 46B) relatively small and not completely distinct, cuticular plate, armed with inner strong, pinnate, and dorsally directed spine, and outer slender and bipinnate seta; spine 1.5 times as long as the seta.

Male. Unknown.

Variability. — Body length of females ranging from 0.419 mm to 0.450 mm (0.437 mm average; n = 3) and no other type of variability observed.

Etymology. — The species is named in honour of Dr. V. I. Monchenko from the Schmalhausen Institute of Zoology (Ukraine), who described the genus *Cryptocyclopina*. The name is a noun in the genitive singular.

Distribution. — *Cryptocyclopina monchenkoi* n. sp. is known so far from the type locality only.

Discussion. — Monchenko (1979a) erected the genus *Cryptocyclopina* to accommodate a single species from the interstitial of the Caspian Sea: *Cryptocyclopina inopinata* Monchenko, 1979. Later he found this species in the Black Sea as well (Monchenko, 2006) and declared it to be a Ponto-Caspian endemic element. Although Monchenko (2006) noted some minor differences between these two populations and even redescribed the species, the line drawings he presented are exactly the same as those in Monchenko (1979a), only differently arranged, and fewer in number. This is probably the only case where a redescription contains less information than the original description, and this is rather unfortunate. For example, in both papers the antenna is illustrated with four setae on the second endopodal segment, while every other member of the family Cyclopinidae G. O. Sars, 1913 has five setae on this segment. Also, it is not clear how many armature elements there are on the first endopodal segment of the mandibula, as the proximalmost element illustrated in both papers is situated unusually proximally (an unknown condition in the family) and looks rather like a cuticular spine. Some other characters are completely unknown for *C. inopinata*, like the armature of the antennula and the position of the fifth leg on the fifth pedigerous somite (presence or absence of intercoxal sclerite, etc.), but enough information was presented by Monchenko (1979a) to justify the erection of the new genus. So far, *Cryptocyclopina* is the only known cyclopinid genus with only two armature elements on the exopod of the fifth leg, a condition so very characteristic of the nominotypical subfamily of the family Cyclopidae Dana, 1846. This does not imply a closer relationship between *Cryptocyclopina* and Cyclopinae Dana, 1846, but is a clear case of convergence.

Cryptocyclopina monchenkoi n. sp. differs from *C. inopinata* in many characters. Some of the differences, like the armature of the second endopodal segment of the antenna, or the armature of the first endopodal segment of the mandibula, can be explained as a result of an incomplete description of the latter species, but many other characters cannot. For example, even if Monchenko (1979a) could not discern one or two segments on the antennula of *C. inopinata* (and I am not implying that he did), there would still be a big difference from the 15-segmented antennula of *C. monchenkoi*, which could even be interpreted as 19-segmented if all partial sutures were fully expressed. Similarly, the maxillipedal endopod of *C. inopinata* is still significantly shorter than that of *C. monchenkoi*, even if Monchenko (1979a) possibly was not able to observe all the armature elements or the exact segmentation of the maxilliped of his tiny species. However, the conditions of the antennula and maxilliped of *C. monchenkoi* could be interpreted as plesiomorphic and thus not highly recommended for diagnostic generic characters. Furthermore, the segmentation of the antennula of *C. monchenkoi* is very hard to interpret with certainty, as it is in many other cyclopinid genera. The two species also have a different armature formula of the swimming legs, but this is a specific rather than a generic character, as no transformation of setae or spines could be observed. Each of the two species simply reduced the number of setae or spines on different legs and different segments from their common ancestor. *Cryptocyclopina monchenkoi* lacks the antennal exopod, which is represented by a single seta in *C. inopinata*, but this character could be variable even in the same population of some cyclopinid species, or even on two antennae of the same animal (see, for example, this condition in *Cyclopina amita* n. sp.). On the other hand, characters that I consider very important on the generic level in the family Cyclopinidae G. O. Sars, 1913 are the same in both *C. inopinata* and *C. monchenkoi*: the shape and armature of the caudal rami; the position of the spermatophores and copulatory pore on the female genital double somite; the segmentation of the swimming legs; and the shape of the fifth leg. Also, the mandibula, maxillula, and maxilla are very similar in both species. The segmentation of the maxillipedal endopod, and also that of the prosome (first pedigerous somite free or fused to cephalothorax), are disputable characters that may be interpreted by some to have generic value. However, a single newly described species with a four-segmented maxillipedal endopod and with a partly fused first pedigerous somite would at once fill the gap between *C. inopinata* and *C. monchenkoi*. Thus, I opt to keep both species in the same genus. The situation in the genus *Cyclopinella* G. O. Sars, 1913 (see above) nicely illustrates the need to keep the generic diagnoses in the family Cyclopinidae slightly wider than in other cyclopoid families. This is actually what we do all the time with older groups of animals in comparison with their younger relatives, in groups where systematics and phylogeny are much better known than those of copepods (reptiles and birds, fish and amphibians, etc.), and the number of plesiomorphic characters suggests that cyclopinids are indeed a very old group of copepods.

Both *Cryptocyclopina* species live in the interstitial habitat and their present distribution would suggest a Tethyan origin of the genus, but I expect the genus to be more widely distributed, and many more species to be described from marine benthic habitats in due course.

Family ABRSIIDAE n. fam.

Diagnosis. — Same as for the type genus.

Type and only genus. — *Abrsia* n. gen.

Genus **Abrsia** n. gen.

Diagnosis. — Moderately large Cyclopoida, with misophrioidiform habitus and enormous mouth appendages. First pedigerous somite free; podoplean-type tagmosis. Female urosome comprising fifth pedigerous somite, genital double somite, and three free abdominal somites. Genital double somite with a single, ventrally positioned copulatory pore, and two chitinous, beak-like structures on each side of it; copulatory duct and fused seminal receptacles forming a "T"-like structure; paired ovipores located laterally at first third of somite, covered by the reduced sixth legs. Anal somite with very broad, smooth, and convex anal operculum. Caudal rami conical, 1.8 times as long as wide, armed with six setae, and ornamented with two rows of spinules and one large lateral sensillum; all setae without breaking planes. Female antennula short and stout, 12-segmented, without aesthetascs, and with most setae very long and strong. Antenna slender, comprising coxobasis and three-segmented endopod (without exopod), with setal formula 1.1.2.6. Labrum very long and broadly rounded. Mandibula comprising coxa and well developed palp; palp consisting of allobasis, one-segmented endopod, and three-segmented exopod. Maxillula composed of syncoxa, basis, two-segmented endopod, and one-segmented exopod; exopod with four setae. Maxilla very robust and prehensile, comprising praecoxa, coxa, basis, and two-segmented endopod. Maxilliped much smaller than maxilla and less prehensile, composed of syncoxa, basis, and one-segmented endopod; endopod minute and armed with three setae. First and second swimming legs with intercoxal sclerites and composed of coxa, basis, and one-segmented exopod and endopod; first legs with one inner coxal and basal seta each, while second legs without these armature elements. Third and fourth swimming legs without intercoxal sclerites and positioned laterally; third leg composed of coxa, basis, and one-segmented exopod, while fourth leg composed of coxobasis and one-segmented exopod. All swimming legs with very long and strong setae. Fifth leg without intercoxal sclerite, composed of coxobasis and one-segmented exopod; coxobasis partly fused to somite, armed with outer basal seta; exopod armed with one long and slender seta apically and one slender spine on inner margin subapically. Female sixth leg a small cuticular plate, armed with two minute spines and an outermost short seta.

Etymology. — The generic name is dedicated to the Australian Biological Resources Study (ABRS), who financed this project for three years (Grant No. 204-05). It comprises an arbitrary combination of letters that can be treated as a Latin word and that should be conceived as a noun, gender feminine.

Type and only species. — *Abrsia misophrioides* n. sp.

18. **Abrsia misophrioides** n. sp.
(figs. 48-51)

Material examined. — Types only: Holotype ovigerous female dissected on 2 slides (WAM C37240); paratypes, 1 female + 1 copepodid (female in alcohol (WAM C37241), copepodid on 1 slide in toto (TMAG G5781)): Australia, New South Wales, Swansea, 16 August 2003, leg. T. Karanovic, 33°04.820′S 151°39.994′E.

Description. — Ovigerous female (holotype). Body length, excluding caudal setae, 0.519 mm. Habitus (fig. 48A) very robust and misophrioidiform, with enormously inflated cephalothorax and first pedigerous somite and relatively slender and cylindrical rest of body, with prosome/urosome ratio of 1.8

and greatest width at middle of cephalothorax. Mouth appendages also enormous and body impossible to position and observe from dorsal view. Body length/width ratio about 1.8 (lateral view); cephalothorax about 3.1 times as wide as genital double somite. Posterior corners of cephalothorax and free pedigerous somites without pronounced lateral expansions. Urosome cylindrical and fifth pedigerous somite slightly wider than genital double somite, while genital double somite only 1.2 times as wide as anal somite from ventral view, and even less from lateral view. Colour of preserved specimen yellowish. Nauplius eye not visible. Rostrum not developed; rostral expansion directly in contact with proximal margin of labrum.

Prosome (fig. 48) comprising cephalothorax and four free pedigerous somites, but first pedigerous somite not completely distinct from cephalothorax and inflated dorsally, almost completely covering second pedigerous somite from dorsal view. Second, third, and fourth pedigerous somites clearly distinct in dorsal and lateral view, but ventrally compacted, without clear distinctions, and with their swimming legs squeezed close to each other. Length ratio of prosomal somites from lateral view, beginning with cephalothorax, 100 : 6 : 16 : 10 : 13. Arthrodial membrane only visible from dorsal view between first and second pedigerous somites. Cephalothorax ovoid, somewhat wider than its greatest length, representing 47% of total body length; surface of cephalic shield with several small sensilla and pores. Hyaline fringe of prosomites narrow and smooth. Fifth pedigerous somite ornamented with only two large dorsolateral sensilla (fig. 49B) and with a smooth fringe dorsally and laterally. No sclerotized joint between fifth pedigerous somite and genital double somite.

Genital double somite (figs. 49A, B, 51A) short and almost cylindrical, with anterior part only 1.1 times as wide as posterior, 1.9 times as wide as long, ornamented with one lateral sensillum on each side posteriorly, with four sensilla dorsally and with two chitinous, beak-like protrusions on each side of copulatory pore ventrally. No cuticular recesses, windows, or pores visible. Hyaline fringe of genital double somite and two succeeding somites very narrow and smooth both dorsally and ventrally, with exposed arthrodial membranes between each pair of adjacent urosomites. Copulatory pore relatively small and ovoid, situated medially at 3/5 of the double somite's length; copulatory duct relatively long, bottle-shaped from both ventral and lateral view, much wider than copulatory pore, and rigidly sclerotized, directed anteriorly, and attached to relatively wide bridge between fused seminal receptacles. Seminal receptacles small, fused into curved and tubular structure,

Fig. 48. *Abrsia misophrioides* n. gen. et sp., holotype ovigerous female. A, habitus, lateral view; B, cutting edge of mandibula; C, palp of mandibula; D, maxillula. Scales = 0.1 mm.

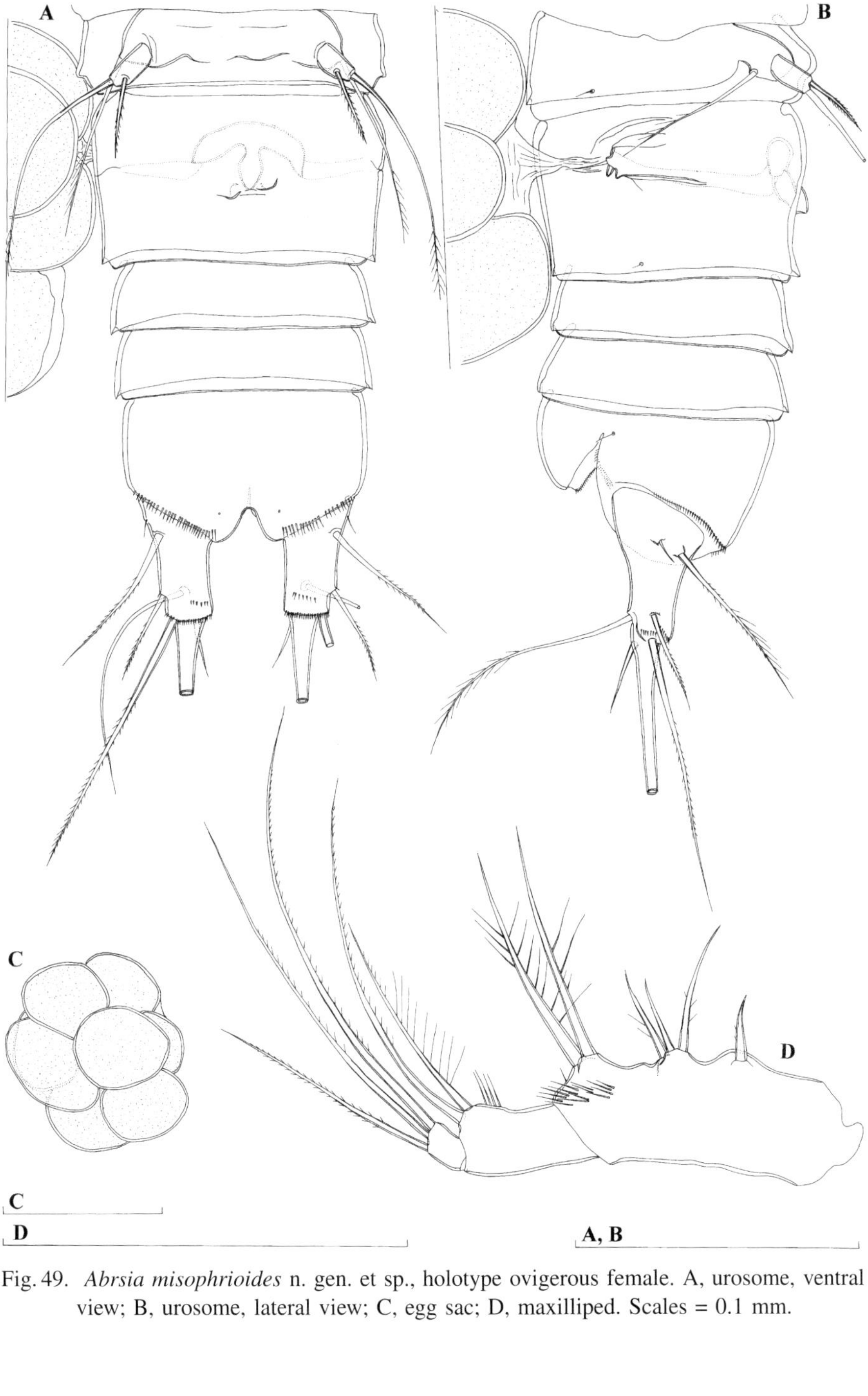

Fig. 49. *Abrsia misophrioides* n. gen. et sp., holotype ovigerous female. A, urosome, ventral view; B, urosome, lateral view; C, egg sac; D, maxilliped. Scales = 0.1 mm.

making a T-like structure with the copulatory duct from ventral view; they represent only 38% of the double somite's width. Paired ovipores located laterally at first third, covered by the reduced sixth legs, and connected to the receptacles via very long, relatively narrow, and weakly chitinized receptacle ducts; each ovipore with attached large and ovoid egg sac, that reaches anterior part of anal somite and contains eight eggs. Third and fourth urosomites without any ornamentation. Anal somite pentagonal from ventral view, with smooth, very broad, and convex anal operculum, which represents 75% of the somite's width and reaches only to the middle of the anal somite; ornamented with two large dorsal sensilla and two pores ventrally, as well as with an interrupted row of spinules along posterior margin ventrally and partly laterally. Anal sinus (fig. 51A) widely open dorsally and posteriorly, and ornamented with two long diagonal rows of minute spinules and one arched row of spinules just below anal operculum. Length ratio of abdominal somites from ventral view, beginning with genital double somite, 100 : 34 : 31 : 96.

Caudal rami (figs. 49A, B, 51A) conical, approximately 1.8 times as long as wide (ventral view) and 0.8 times as long as anal somite, parallel, with space between them more than one ramus' width; they represent 9% of total body length. Distal margin ventrally straight, with medial cuticular pore not visible, ornamented with small spinules. Additional ornamentation consisting of one lateral sensillum at anterior part, one ventral transverse row of small spinules posteriorly, and several short dorsal rows of minute spinules at posterior part. Armature consisting of one dorsal, one lateral, and four apical setae, although outermost apical seta moved somewhat laterally. Dorsal seta relatively long and slender, 1.6 times as long as ramus, inserted in a small recess at 4/5 of the ramus' length, non-articulate at base and bipinnate distally. Lateral seta somewhat stronger, arising at first fourth and somewhat ventrolaterally, also bipinnate, and about 1.2 times as long as ramus. Outermost apical seta slender and bipinnate, 0.7 times as long as ramus and 1.7 times as long as innermost apical seta, which is slender and smooth, and inserted slightly dorsally. Principal apical setae without breaking planes and bipinnate; inner seta about four times as long as outer one and almost 0.8 times as long as body.

Antennula (fig. 50A) short and stout, reaching to middle of cephalothorax, 12-segmented (although first and second segments partly fused on anterior surface, but clearly divided on posterior and caudal sides), with short subdivision on caudal margin of third and fourth segments, indicating ancestral 14-segmented stage, unornamented, without aesthetascs, and with setal formula: 1.9.3.1.0.1.0.1.2.1.2.7. No setae with breaking planes and only three

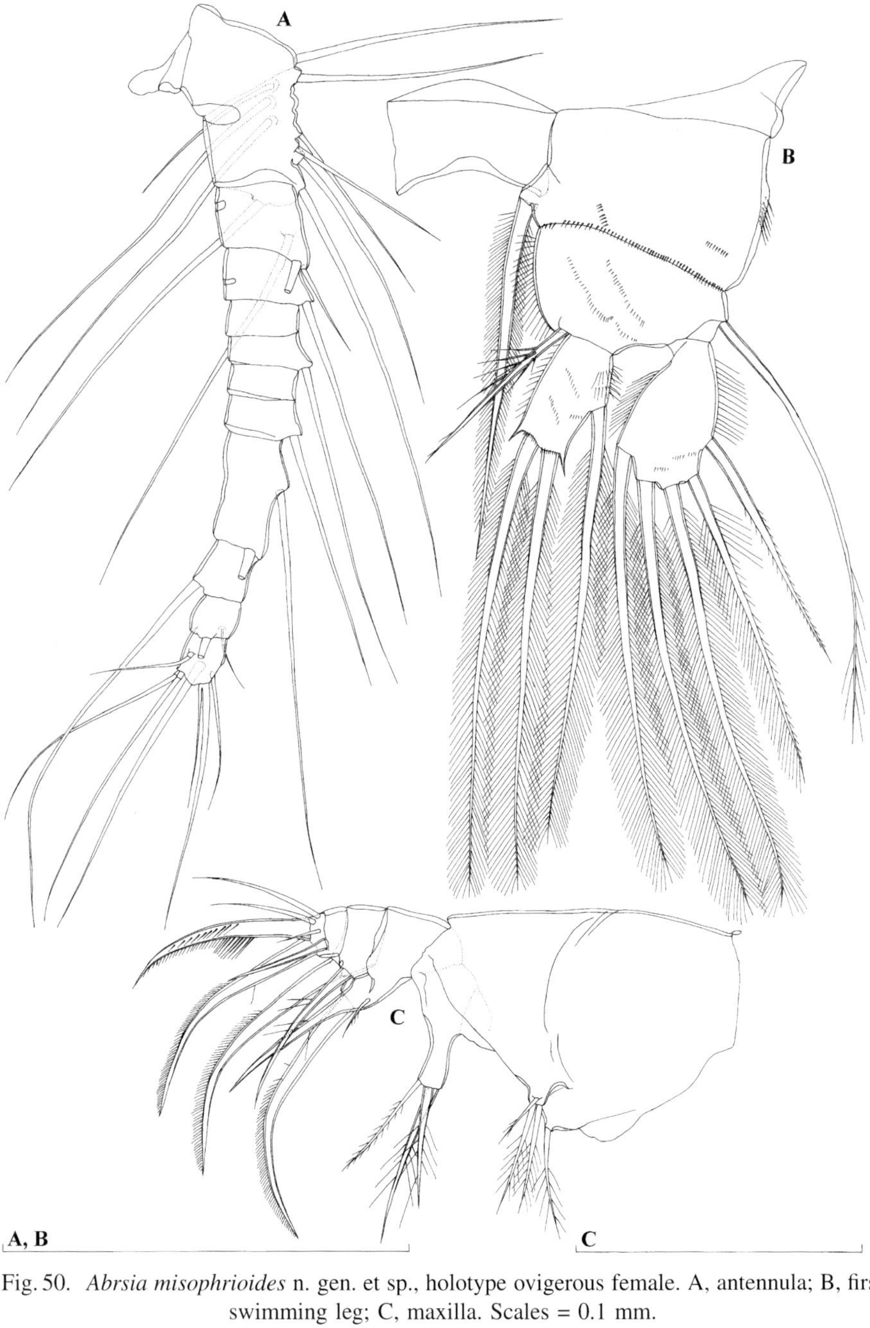

Fig. 50. *Abrsia misophrioides* n. gen. et sp., holotype ovigerous female. A, antennula; B, first swimming leg; C, maxilla. Scales = 0.1 mm.

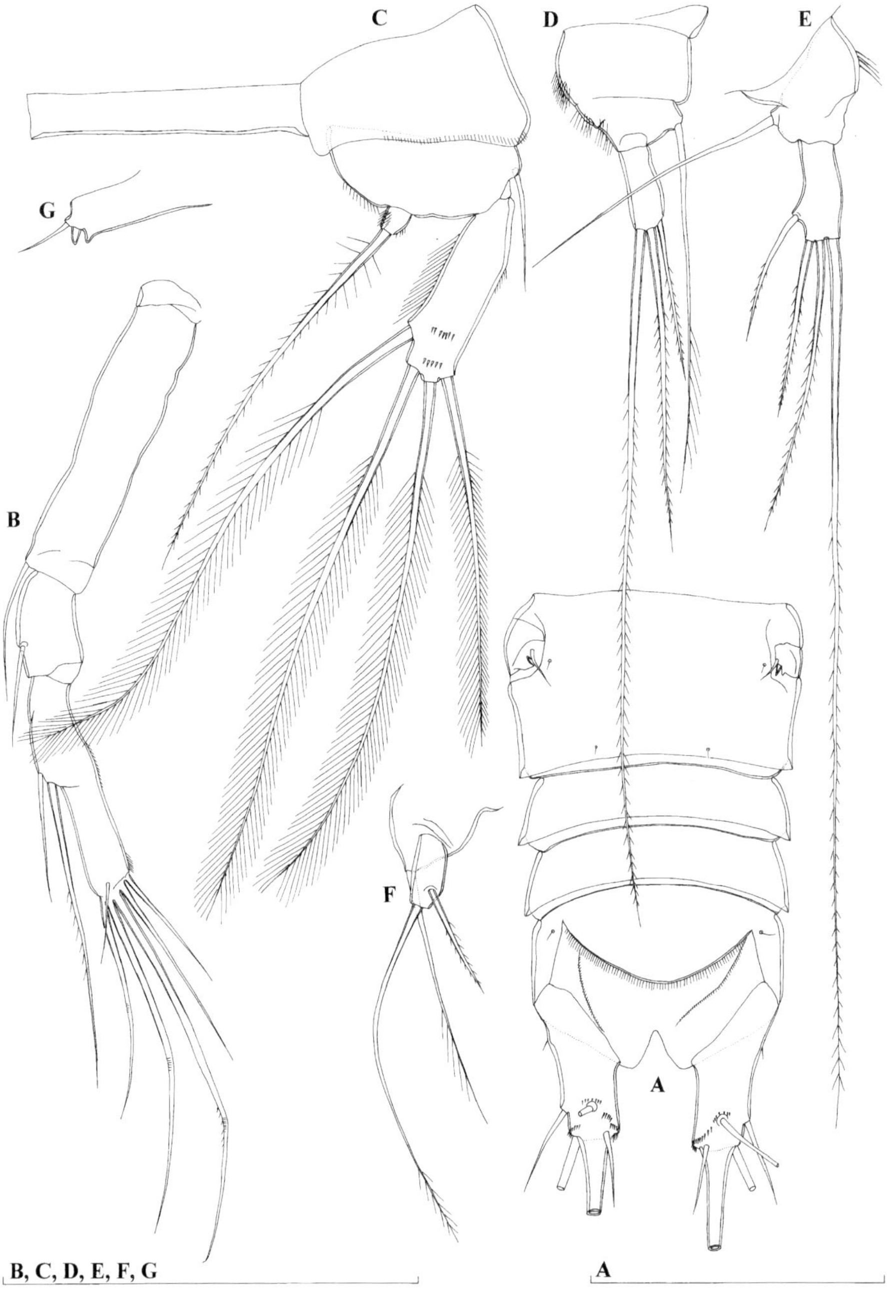

Fig. 51. *Abrsia misophrioides* n. gen. et sp., holotype ovigerous female. A, abdomen, dorsal view; B, antenna; C, second swimming leg; D, third swimming leg; E, fourth swimming leg; F, right fifth leg, ventral view; G, sixth leg. Scales = 0.1 mm.

setae on twelfth segment articulating on basal part; most setae very long and all setae smooth and slender. Length ratio of antennular segments, from proximal end and along caudal margins, 1 : 2 : 1.7 : 1.4 : 1 : 0.8 : 0.9 : 1 : 3 : 1.5 : 1.1 : 1.4.

Antenna (fig. 51B) slender and four-segmented, comprising coxobasis and three-segmented endopod (although second and third endopodal segments partly fused), with geniculation between first and second endopodal segments; length ratio of antennal segments, 1 : 0.4 : 0.4 : 0.4. Coxobasis about 3.8 times as long as wide, unornamented, and armed with a single smooth seta at inner distal corner; seta representing exopod absent. First endopodal segment 1.6 times as long as wide, also unornamented, and armed with one smooth seta at middle. Second endopodal segment about 1.9 times as long as wide, ornamented with small spinules along outer (caudal) margin, and armed with two setae subapically; one seta smooth and short, the other unipinnate, strong, and almost as long as the entire endopod. Third endopodal segment 2.5 times as long as wide, ornamented as previous segment, and armed with six apical setae, three of which are very strong and geniculate.

Labrum very long and broadly rounded, more than 0.8 times as long as antennula.

Mandibula (fig. 48B, C) comprising coxa with well developed gnathobase and distal palp; palp consisting of allobasis (first endopodal segment fused to basis), one-segmented endopod, and three-segmented exopod. Cutting edge of coxal gnathobase divided into three regions: inner group of three complex and strong teeth and five very long spinules; middle group of five simple teeth and 12 smaller spinules; and outermost group of two short and unipinnate setae. Allobasis large, 3.5 times as long as wide, unornamented, and armed with one seta laterally and two setae subapically (armature of former first endopodal segment). Endopod slender and small, quadriform, 0.7 times as long as the basis' width, and 2.3 times as long as wide, armed apically with two setae and one lateral seta on each side (four setae altogether). Exopod slender, about as long as basis's width, unornamented, and with setal formula 1.1.2. All setae on allobasis, endopod, and exopod slender and plumose.

Maxillula (fig. 48D) composed of syncoxa with well developed arthrite, basis, two-segmented endopod, and one-segmented exopod, although endopod and exopod not clearly distinct from basis. Arthrite of syncoxa with 11 unequal lateral elements and one smooth minute seta on posterior surface; four apicalmost elements strong and geniculate, proximalmost element plumose. Coxa unornamented and armed with two epipodal setae, one of which actually moved to basal part of basis; coxal endites weakly discernible, unarmed. Basis elongated, only slightly longer than arthrite of syncoxa, almost 2.4 times as long as wide, unornamented, and armed with three setae on only endite; two of these are very strong, heavily ornamented, and spiniform; third seta slender and bipinnate, and even slightly shorter than the other two. Endopod as long as the basis' width and about as long as wide; also unornamented and armed with three lateral setae on inner margin of first segment (proximalmost bipinnate, other two plumose) and two long plumose setae apically on second segment. Exopod about as long as wide and slightly shorter than endopod, unornamented, and armed apically with four slender, plumose, and very long setae.

Maxilla (fig. 50C) very robust and large, five-segmented, comprising praecoxa, coxa, basis, and two-segmented endopod (although praecoxa and coxa partly fused). Praecoxa inflated and ovoid, 1.2 times as wide as long, unornamented, and with one plumose seta on its only endite. Coxa about as long as praecoxa but less inflated, also unornamented, armed with three pinnate setae on elongated proximal endite; distal endite highly mobile and bearing three pinnate setae. Basis half as long as coxa, expanded into spinulose claw, which is almost straight and ornamented with only three long spinules, and armed with one minute seta and one strong and geniculate seta, which is unipinnate along its convex margin and 1.5 times as long as the claw. First endopodal segment shorter than basis, armed with two robust and geniculate setae. Second endopodal segment about half as long as first endopodal segment, and armed with three strong and three slender setae. One seta on first endopodal segment and one seta on second endopodal segment geniculate and very similar to strong basal seta.

Maxilliped (fig. 49D) much smaller than maxilla, more slender and less prehensile, three-segmented, composed of syncoxa, basis, and one-segmented endopod. Syncoxa 2.5 times as long as its greatest width, ornamented with a bunch of strong spinules near inner distal corner, with proximal praecoxal endite weakly discernible and unarmed, and distal praecoxal endite armed with a single smooth seta; the two coxal endites well-developed, proximal one with three, and distal with two strong setae. Basis twice as long as wide and half as long as syncoxa, cylindrical, ornamented with several large spinules along inner margin at middle, and armed with two long and unipinnate setae distally on inner margin. Endopod minute, 0.3 times as long as basis and 1.4 times as long as wide, unornamented, and armed apically with three very long, slender, and pinnate setae. Longest seta on endopod about as long as the entire maxilliped.

First swimming leg (fig. 50B) with intercoxal sclerite and composed of coxa, basis, and one-segmented exopod and endopod. Intercoxal sclerite quadriform, unornamented, and with concave distal margin. Coxa armed with strong and long seta on inner distal corner, which is about as long as the entire leg, and ornamented with small spinules along its posterior margins, as well as with several spinules on outer margin and two short rows of spinules on anterior surface. Basis armed with slender and heavily ornamented spine at distal inner corner and with outer slender seta, ornamented with hair-like spinules along inner margin and with two diagonal rows of spinules on anterior surface. Exopod somewhat longer than basis, 1.8 times as long as wide and much narrower at basal part, ornamented with hair-like spinules along both inner and outer margins and with two short rows of minute spinules on anterior surface, and armed apically with three long plumose setae, laterally with one slender spine, and subapically with one element, which looks like a transitional stage between a spine and a seta. Endopod quadriform, about as long as basis and 1.6 times as long as wide, with pronounced distal lateral corners, armed with three long plumose setae (two apically and one on outer margin), and ornamented similarly to exopod. All exopodal and endopodal setae longer than entire first leg.

Second swimming leg (fig. 51C) also with intercoxal sclerite and composed of coxa, basis, and one-segmented exopod and endopod. Intercoxal sclerite very wide and short, about six times as wide as long, unornamented, and with straight distal margin. Coxa quadriform, unarmed, and ornamented with only a posterior row of small spinules. Basis somewhat shorter and narrower than coxa, semicircular, ornamented with hair-like spinules along inner margin, and armed with an outer smooth, short, and slender seta. Exopod about as long as coxa and basis combined, longer than exopod of first leg, and more than three times as long as wide, armed with four long and plumose setae apically, and ornamented with hair-like spinules along its inner margin, several spinules on outer margin, and two transverse rows of stronger spinules on anterior surface distally. Endopod minute, 0.35 times as long as basis and slightly longer than wide, conical, armed apically with a single long and bipinnate seta, and ornamented with hair-like spinules on both inner and outer margins. Longest exopodal setae 1.5 times as long as entire second leg.

Third swimming leg (fig. 51D) very small, without intercoxal sclerite, positioned ventrolaterally on somite, and composed of coxa, basis, and one-segmented exopod. The only ornamentation consists of hair-like spinules along inner margins of coxa and basis. Coxa unarmed, while basis with long and slender outer basal seta. Exopod 2.3 times as long as wide and 1.6 times as long as basis, armed apically with three bipinnate elements (setae or possibly slender spines); innermost element more than three times as long as entire third leg, 2.1 times as long as middle element, and 4.3 times as long as outermost element.

Fourth swimming leg (fig. 51E) even smaller than third leg, also without intercoxal sclerite, positioned almost laterally on somite, composed of coxobasis and one-segmented exopod. Coxobasis ornamented with several long spinules on inner margin and armed with slender and long basal seta on outer margin. Exopod about as long as coxobasis and twice as long as wide, slightly inflated distally, unornamented, and armed with three apical elements and one element on outer margin; innermost apical element almost 4.0 times as long as entire fourth leg, 3.0 times as long as middle apical element, 4.9 times as long as outermost apical element, and 6.6 times as long as lateral element.

Fifth leg (figs. 49A, B, 51F) inserted ventrolaterally, unornamented, without intercoxal sclerite, and two-segmented, comprising coxobasis and one-segmented exopod, but coxobasis basally fused to somite. Coxobasis armed with long and unipinnate outer basal seta. Exopod armed with one long and slender seta apically and one slender spine on inner margin subapically; apical seta plumose at distal end, 4.6 times as long as exopod, and 2.9 times as long as subapical spine, which is bipinnate.

Sixth leg (fig. 51G) a not clearly distinct and small cuticular plate, unornamented, and armed with two minute and smooth spines (innermost spine fused basally, middle spine distinct) and an outermost smooth and somewhat longer seta.

Male. Unknown.

Variability. — Body length of females ranging from 0.506 mm to 0.519 mm (0.513 mm average; n = 2); no other type of variability was observed.

Etymology. — The specific name refers to the species' misophrioidiform habitus, which is most unusual in cyclopoid copepods. It is composed of the stem of the copepod order Misophrioida Brady, 1878 and the Latin suffix for likeness "-oides", and is consequently an adjective with one termination.

Distribution. — At present, *Abrsia misophrioides* n. gen., n. sp. is known from the type locality only.

Discussion. — In their overview of copepod morphology, Huys & Boxshall (1991) recognized ten copepod orders. Ho et al. (2003) established another new order, Thaumatopsylloida, for the previously troublesome family Thaumatopsyllidae G. O. Sars, 1913. This was not accepted by Boxshall & Halsey (2004), who regarded this family to belong to the Cyclopoida Rafinesque, 1815, following Huys & Boxshall (1991). The latter authors studied this family in detail (as Thespesiopsyllidae Wilson, 1924) and transferred it to the Cyclopoida from the order Monstrilloida Dana, 1849. In my opinion, the argument is on Ho's side and the Thaumatopsylloida G. O. Sars, 1913 is a valid order. In any case, it is at least as valid as the orders Platycopioida G. O. Sars, 1911, Mormonilloida Giesbrecht, 1893, or Gelyelloida Rouch & Lescher-Moutoué, 1977 are, for example. Boxshall & Halsey (2004) also considered the Cyclopoida to be paraphyletic, and merged all families of the copepod order Poecilostomatoida Thorell, 1859 into the former order. This was supported by Huys et al. (2006), notwithstanding the fact that in their rDNA-based analysis the poecilostomatoid taxa were resolved as a sister group to the Cyclopoida. Ho et al. (2006) rejected this view and re-examined the phylogenetic relationship of 61 poecilostomatoid families, omitting only the Bradophilidae Marchenkov, 2002. Again, I think the argument is on Ho's side and the Poecilostomatoida is a valid monophyletic order, and I treated it as such here (see above). In the end, it is all the same if we accept poecilostomatoids as a valid order or if we use the term "poecilostome complex of families" as Huys et al. (2006: 407) did. I here list all 11 copepod orders, with their correct authors, as this was not done properly in Huys & Boxshall (1991), Damkaer (1996), nor in Boxshall & Halsey (2004), all constituting most important recent reference works: Calanoida Dana, 1846; Cyclopoida Rafinesque, 1815; Gelyelloida Rouch & Lescher-Moutoué, 1977; Harpacticoida Dana, 1846; Misophrioida Brady, 1878; Monstrilloida Dana, 1849; Mormonilloida Giesbrecht, 1893; Platycopioida G. O. Sars, 1911; Poecilostomatoida Thorell, 1859; Siphonostomatoida Burmeister, 1835; and Thaumatopsylloida G. O. Sars, 1913. Note that I previously did not share L. B. Holthuis & J. C. von Vaupel Klein's view (in Karanovic, 2004a) that Rafinesque (1815; fide Holthuis, 1954) could be accepted as the author of the family Cyclopidae and consequently of the order Cyclopoida, as this reference was obscure and hard to verify for me, and the family name was apparently not latinized by Rafinesque or connected with his name after latinization for almost 140 years.

Abrsia n. gen. undoubtedly belongs to the order Cyclopoida, as it has a podoplean-type tagmosis, a single copulatory pore on the genital double somite, an antenna without an exopod, as well as mouthparts that are very similar to those of more primitive cyclopoid families. The order Cyclopoida today has around 1360 valid species, classified into 203 genera and the following 17 families: Archinotodelphyidae Lang, 1949; Ascidicolidae Thorell, 1859; Botryllophilidae G. O. Sars, 1921; Buproridae Thorell, 1859; Chordeumiidae Boxshall, 1988; Cucumaricolidae Bouligand & Delamare Deboutteville, 1959; Cyclopidae Rafinesque, 1815; Cyclopinidae G. O. Sars, 1913; Enterognathidae Illg & Dudley, 1980; Enteropsidae Aurivillius, 1885; Fratiidae Ho, Conrad & López-González, 1998; Lernaeidae Cobbold, 1879; Mantridae Leigh-Sharpe, 1934; Notodelphyidae Dana, 1853; Oithonidae Dana, 1853; Ozmanidae Ho & Thatcher, 1989; and Speleoithonidae Da Rocha & Iliffe, 1991. The family Archinotodelphyidae today contains five species in two genera (Boxshall & Halsey, 2004), all living in association with ascidians, and could be distinguished from the Abrsiidae n. fam. by many characters, including: cyclopiform body, slender caudal rami, well developed rostrum, claw-like setae on the antenna, four-segmented maxilliped, swimming legs with intercoxal sclerites and three-segmented rami, and four to five setae on the fifth leg. The Ascidicolidae (three species in two genera), Botryllophilidae (54 species in seven genera), Buproridae (three species in the type genus), Enterognathidae (five species in three genera), and Enteropsidae (33 species in five genera) used to be subfamilies of the Ascidicolidae s.l.; they are mostly associates or parasites of ascidians (except the Enterognathidae, which are parasites of crinoids, holothurians, or hemichordates), and differ from the Abrsiidae by their transformed and fleshy body, numerous reductions in the antennula, antenna, and mouthparts, as well as by their expanded and dorsolaterally located female fifth leg. The Chordeumiidae count 12 species in six genera, all endoparasites of brittle stars, and originally used to be placed in the Siphonostomatoida; they differ from the Abrsiidae also by many characters, including: swollen and transformed body, non-discrete caudal rami, unsegmented or absent antennula, papilliform or absent antenna, absent mandibula, maxillula, and maxilliped, reduced or absent maxilla, unsegmented or absent swimming legs, and lobate and dorsolaterally located female fifth legs. The Cucumaricolidae have only two species in the type genus, both parasites of sea cucumbers, and were originally placed in the Poecilostomatoida; they are also highly transformed and have numerous reductions in most appendages (including absent mandibula, maxillula, maxilla, and fourth and fifth legs), which can easily distinguish them from my new family. The Cyclopidae today count around 780 species in 60 genera, mostly free-living in freshwater (with some in coastal and brackish waters), and could be distinguished from the Abrsiidae by the following characters: cyclopiform body, cylindrical caudal rami, minute or absent mandibular palp, reduced maxillular palp, swimming legs with intercoxal sclerites and with two- or three-segmented rami (rarely one-segmented). The Cyclopinidae are free-living forms, mostly inhabiting marine benthos, and count around 122 species in 41 genera (including those described here); they differ from the Abrsiidae by the following characters: cyclopiform body, cylindrical caudal rami, at least two aesthetascs on the antennula; five setae on the second endopodal segment of the antenna, one-segmented endopod of the maxillula, four- to seven-segmented maxilliped, swimming legs with intercoxal sclerites and with three- or two-segmented rami, and usually a more complex fifth leg. The family Fratiidae is still monospecific, living in ascidians, and differs from the Abrsiidae by the following: cylindrical caudal rami, eight-segmented antennula armed with four aesthetascs, four setae on the second endopodal segment of the antenna, two-segmented mandibular palp, bilobed maxillula, three-segmented maxilla, four-segmented maxilliped, swimming legs with intercoxal sclerites and three- or two-segmented rami, and four elements on the second segment of the fifth leg. The Lernaeidae count 118 species in 17 genera, all of which are parasites of freshwater fishes; they differ from the Abrsiidae by their fleshy and highly transformed body, as well as by many reductions and transformations of their appendages. The Mantridae have three species in as many genera and they are all associates of bivalve molluscs; they could be distinguished from the Abrsiidae by the following characters: cyclopiform body, 15-segmented antennula, antenna with a separate coxa, with five setae on the second endopodal segment, and a massive claw on the third, five or six setae on the maxillipedal endopod, swimming legs with intercoxal sclerites and with three-segmented rami, and four setae on the second segment of the fifth leg. The Notodelphyidae count 146 species in 46 genera, mostly associates or parasites of ascidians (rarely occurring on octocorals) and also more or less transformed, but those that exhibit a more primitive morphology differ from the Abrsiidae by the following: cyclopiform body with the fourth and fifth pedigerous somites modified as an incubatory region, swimming legs with intercoxal sclerites and three-segmented rami, an up to 15-segmented antennula with at least three aesthetascs, a modified claw on top of the antenna and five setae on its second endopodal segment. The Oithonidae are elements of the marine plankton and count 51 species in three genera; they differ from the Abrsiidae by the following: cyclopiform body, paired copulatory pores inside genital apertures, cylindrical caudal rami, very long and slender antennula, five setae on the second endopodal segment of the antenna, one to three robust elements on the mandibular allobasis and the mandibular endopod very small, a one-segmented maxillular endopod, a four-segmented maxilliped, swimming legs with intercoxal sclerites and three- or two-segmented rami, and a fifth leg with the first segment completely incorporated into the somite and a very slender second segment. The Ozmanidae are monospecific, the single species being a parasite of a freshwater snail, and differ from the Abrsiidae by many characters, including many reductions of the mouth appendages. The Speleoithonidae contain three species in the type genus, all inhabiting anchialine caves; they could be distinguished from the Abrsiidae by the following characters: cyclopiform body, long and 18-segmented antennula, four or five setae on the second endopodal segment of the antenna, a two-segmented mandibular exopod, the maxillular endopod absent, a six-segmented maxilla, a four-segmented maxilliped, swimming legs with intercoxal sclerites and three-segmented rami, and the female fifth leg with intercoxal sclerite and with only one apical seta on the slender second segment. I should mention here that I am not very convinced the Speleoithonidae deserve a separate family status, as all the characters that distinguish them from the Oithonidae are actually plesiomorphies (including a single copulatory pore).

The Abrsiidae could easily be distinguished from the other cyclopoid families by a combination of plesiomorphic and apomorphic characters. The most important are the following: misophrioidiform habitus with enormously inflated cephalothorax and first pedigerous somite, and a relatively slender and cylindrical rest of the body; conical caudal rami; disproportionately large mouth appendages, especially the maxilla; absence of aesthetascs on the antennula; two-segmented endopod of the maxillula; two chitinous, beak-like structures on each side of the copulatory pore; first and second swimming legs with both rami one-segmented; third and fourth swimming legs without intercoxal sclerites, positioned close to each other, and almost laterally; third swimming leg with one-segmented exopod and with completely reduced endopod; and fourth swimming leg with fused coxa and basis, without an endopod, and with a one-segmented exopod. The Misophrioida have a distensible gut and exoskeletal modifications to allow prosomal swelling, which comprise special adaptations to the adoption of gorging as a feeding strategy (Huys & Boxshall, 1991). Whether this is also the case in the Abrsiidae remains to be discovered, but it would seem very plausible at this point.

Although the habitus in fig. 48 looks as if it has been pressed with the cover-slip, that is not the case; in fact, the mouth appendages are so enormous that it is impossible to position the body and observe a specimen from a dorsal view. A two-segmented maxillular endopod is very rare in the Cyclopoida and it was so far reported only for *Nearchinotodelphys indicus* Ummerkutty, 1960 from the family Mantridae, and possibly also for *Cyclopinoides elegans* (T. Scott, 1894) from the family Cyclopinidae (see Ummerkutty, 1960; Huys & Boxshall, 1991). But, this certainly is a plesiomorphic character state, as it is known for some Harpacticoida and Misophrioida, while some Calanoida even have a three-segmented maxillular endopod. An antennula without any aesthetascs has not been confirmed so far in the order Cyclopoida, but it is known in some Mormonilloida, where it certainly originated independently. Equally unique are the two beak-like structures on each side of the copulatory pore (fig. 49A, B). The only thing similar was reported by Huys (1988) for *Gelyella droguei* Rouch & Lescher-Moutoué, 1977 from the order Gelyelloida, for which he speculated that these structures might represent the sixth legs. As the sixth legs in the Abrsiidae are clearly visible and separate from the beak-like structures, it is probable that these structures originated independently in the Gelyelloida and Cyclopoida, and may not even be homologous structures. Reductions of the swimming legs' armature and segmentation are quite common in the Cyclopoida, especially in parasitic forms, but they are never achieved in the same way as in the Abrsiidae. A superficially similar second leg could be found in the above-mentioned *Gelyella droguei*, but in this species the intercoxal sclerite is absent and the coxae are fused (see Rouch & Lescher-Moutoué, 1977; Huys, 1988). Species from the order Thaumatopsylloida have a very reduced fourth leg (sometimes even uniramous), without an intercoxal sclerite, but they have other swimming legs with three-segmented rami and the intercoxal sclerites present, among other differences (see Fosshagen, 1970; McKinnon, 1994; Ho et al., 2003).

Because of the primitive mouth appendages, the Abrsiidae display most similarities with the primitive cyclopoid families Archinotodelphyidae, Cyclopinidae, Mantridae, and the most primitive genera of the Notodelphyidae, but these are plesiomorphic character states and thus provide very little information about their phylogenetic relationships. Although it may seem at a first glance that the Abrsiidae have originated from these primitive cyclopoids, neither of the above-mentioned families (or any other family) could be identified as its ancestor. Also, neither of the more advanced and specialized families could be seen as originating from the Abrsiidae. That is why this unusual Australian cyclopoid occupies a pretty isolated position, and is consequently elevated here to the levels of a new genus and a new family.

Family CYCLOPIDAE Rafinesque, 1815
Subfamily HALICYCLOPINAE Kiefer, 1927
Genus **Neocyclops** Gurney, 1927
[= *Protoneocyclops* Petkovski, 1986]

19. **Neocyclops australiensis** n. sp.
(figs. 52-56)

Material examined. — Types: Holotype female dissected on 2 slides (WAM C37242); paratype female dissected on 1 slide (TMAG G5782): Australia, Western Australia, Cape Leeuwin (near Augusta), 1 December 2002, leg. T. Karanovic, 34°22′S 115°10′E.

Other material: Australia, Western Australia, Withcliffe (near Margaret River), beach near Conto Spring, 19 February 2002, leg. T. Karanovic and S. Eberhard, 34°02′S 115°00′E: 1 male + 4 females + 1 copepodid (male dissected on 2 slides (WAM C37244), 2 females dissected on 1 slide each (WAM C37245 and WAM C37246), others in alcohol (WAM C37247)).

Australia, Western Australia, Busselton, jetty, 8 December 2002, leg. T. Karanovic, 33°39′S 115°20′E: 1 copepodid in alcohol (WAM C37307).

Australia, Western Australia, Perth, Munster, Jervoise Bay, 8 December 2002, leg. T. Karanovic, 32°08′S 115°46′E: 1 male + 5 females + 2 copepodids (1 male + 1 female on 1 slide in toto (TMAG G5783), others in alcohol (WAM C37248)).

Australia, Western Australia, Shark Bay, Shell Beach, 2 August 2003, leg. T. Karanovic, 26°13.032′S 113°46.632′E: 1 male dissected on 1 slide (WAM C37306) + 1 female dissected on 2 slides (WAM C37249).

Australia, Western Australia, Exmouth, 3 August 2003, leg. T. Karanovic, 21°56.726′S 114°08.383′E: 1 female in alcohol (WAM C37250).

Australia, Northern Territory, Dundee Beach, 7 August 2003, leg. T. Karanovic, 12°46.555′S 130°21.377′E: 1 female dissected on 1 slide (WAM C37251).

Australia, Queensland, Yeppoon, 13 August 2003, leg. T. Karanovic, 23°08.582′S 150°45.659′E: 15 females (1 ovigerous) (1 female dissected on 1 slide (WAM C37252), 2 females on 1 slide in toto (TMAG G5784), others in alcohol (WAM C37253)).

Australia, Queensland, Gladstone, Barney Point Beach, 14 August 2003, leg. T. Karanovic, 23°50.859′S 151°16.378′E: 1 female dissected on 1 slide (WAM C37254).

Australia, Queensland, Hervey Bay, 14 August 2003, leg. T. Karanovic, 25°16.728′S 152°50.400′E: 1 female in alcohol (WAM C37255).

Australia, New South Wales, Wollongong, Boat Harbour, 17 August 2003, leg. T. Karanovic, 34°25.326′S 150°54.296′E: 11 males + 5 females + 3 copepodids (1 female dissected on 1 slide (WAM C37256), others in alcohol (WAM C37257)).

Australia, South Australia, Stansbury, 26 December 2002, leg. T. Karanovic, 34°54.524′S 137°47.873′E: 1 copepodid in alcohol (WAM C37308).

Australia, South Australia, Smoky Bay, 24 December 2002, leg. T. Karanovic, 32°22.735′S 133°56.006′E: 1 male + 1 female in alcohol (WAM C37258).

Description. — Female (holotype). Body length, excluding caudal setae, 0.731 mm. Habitus (fig. 52A) typical cyclopoid shape, very robust, with prosome/urosome ratio of 1.2 and greatest width at posterior end of cephalothorax. Body length/width ratio about 2.6; cephalothorax about 2.4 times as wide as genital double somite. Posterior corners of cephalothorax somewhat expanded laterally; other free pedigerous somites without pronounced lateral expansions. Fifth pedigerous somite narrower than genital double somite. Colour of preserved specimen milky white. Nauplius eye not visible. Rostrum (fig. 54G) well developed, broadly rounded, and furnished with two large sensilla.

Cephalothorax (fig. 52A) almost triangular, slightly longer than its greatest width, representing 33% of total body length. Surface of cephalic shield with several large sensilla; no other ornamentation visible. Hyaline fringe of prosomites narrow and smooth. Large arthrodial membrane between cephalothorax and second pedigerous somite (first free somite) very thin and without any resemblance to first pedigerous somite or any attachment area of muscles. First pedigerous somite completely fused to cephalosome. Fifth pedigerous somite ornamented with only two large dorsal sensilla and with smooth fringe dorsally and ventrally (fig. 52A). Sclerotized joint between fifth pedigerous somite and genital double somite well developed, almost as a pseudosomite (fig. 52B).

Genital double somite (figs. 52B, 53A) with lateral expansions bearing spiniform projections at first third, about 0.8 times as long as wide, ornamented with characteristic pattern of nine pores ventrally and 15 pores dorsally. Hyaline fringe of genital double somite and two succeeding somites narrow, not frilled, and completely smooth. No cuticular recesses or sensilla visible. Copulatory pore small, ovoid, situated just before 1/3 of the somite's length, with two spermatophores attached; copulatory duct narrow and rigidly sclerotized, with first half directed dorsally and second half directed anteriorly, almost L-shaped from lateral view (fig. 52C). Seminal receptacle (fig. 52B) complex, curved, and tubular, but entirely situated anteriorly of copulatory pore; central part, connected to copulatory duct, almost cylindrical, positioned close to ventral cuticular surface and 2.6 times as wide as long; constricted part on each side about 0.3 times as wide as central part, and turning backwards (dorsally); lateral arms expanded, almost as wide as central part, and connected to dorsolateral genital areas through another constriction (fig. 52C). Ovipores (fig. 53A) situated somewhat dorsolaterally, covered by the reduced sixth legs, bearing two minute spines and one pinnate seta.

Fig. 52. *Neocyclops australiensis* n. sp., holotype female. A, habitus, dorsal view; B, abdomen, ventral view; C, seminal receptacle, lateral view; D, maxillula; E, maxilla; F, maxilliped. Scales = 0.1 mm.

Third urosomite ornamented with four ventral and six dorsal pores in posterior half, while fourth urosomite with four ventral and only two dorsal pores; their hyaline fringes smooth and narrow. Anal somite (figs. 52B, 53A) with smooth, broad, and slightly convex anal operculum, which represents 72% of the somite's width and reaches only to 3/5 of the somite's length; ornamented with two large sensilla and two pores dorsally, one pore laterally on each side, two pores ventrally, and a transverse row of spinules on posterior margin. Anal sinus widely open dorsally, ornamented with four diagonal (two on each side) and one arched transversal (just below anal operculum) rows of minute spinules (fig. 53A).

Caudal rami (figs. 52B, 53A) cylindrical, 2.5 times as long as wide (ventral view), parallel, with space between them about one half of a ramus' width; they represent 8% of total body length. Distal margin ventrally smooth and convex, with medial cuticular pore hardly visible. Ornamentation consisting of several small spinules at base of outermost apical seta and four ventral, one lateral, and three dorsal pores. Dorsal seta very long, almost 1.4 times as long as ramus, inserted on a small protuberance reaching border of ramus, uniarticulate at base and unipinnate distally. Lateral seta smallest, arising at 3/4, positioned dorsolaterally, uniplumose, and about 0.4 times as long as a ramus' width. Outermost apical seta stout and spiniform, bipinnate, about half the ramus length and 1.4 times as long as innermost apical seta, which is also bipinnate, but more slender than outermost one. Principal apical setae with breaking planes and bipinnate; inner seta about 1.4 times as long as outer and 0.4 times as long as body.

Antennula (fig. 53B) slender, reaching to 2/3 of cephalothorax, 12-segmented, unornamented, with one slender aesthetasc on ninth, one on eleventh, and one on twelfth segment, and setal formula: 8.4.2.6.4.2.2.3.2.2.2.7. No setae with breaking planes and only one seta on tenth, one on eleventh, and three on twelfth segment articulating on basal part. One seta on sixth segment spiniform and very short; all other setae slender and most setae smooth, only 2 setae (less than 5%) pinnate at their distal parts. Length ratio of antennular segments, from proximal end and along caudal margins, 1 : 0.4 : 0.2 : 0.4 : 0.3 : 0.2 : 0.5 : 1 : 0.6 : 0.4 : 0.5 : 0.7.

Antenna (fig. 53C) four-segmented, comprising coxobasis and three-segmented endopod; all endopodal segments of about the same length, coxobasis somewhat longer. Coxobasis about 1.8 times as long as wide, unornamented, and armed with two smooth setae at distal inner corner; seta representing exopod absent. First endopodal segment armed with one smooth seta at middle, ornamented with longitudinal, short row of long spinules along external

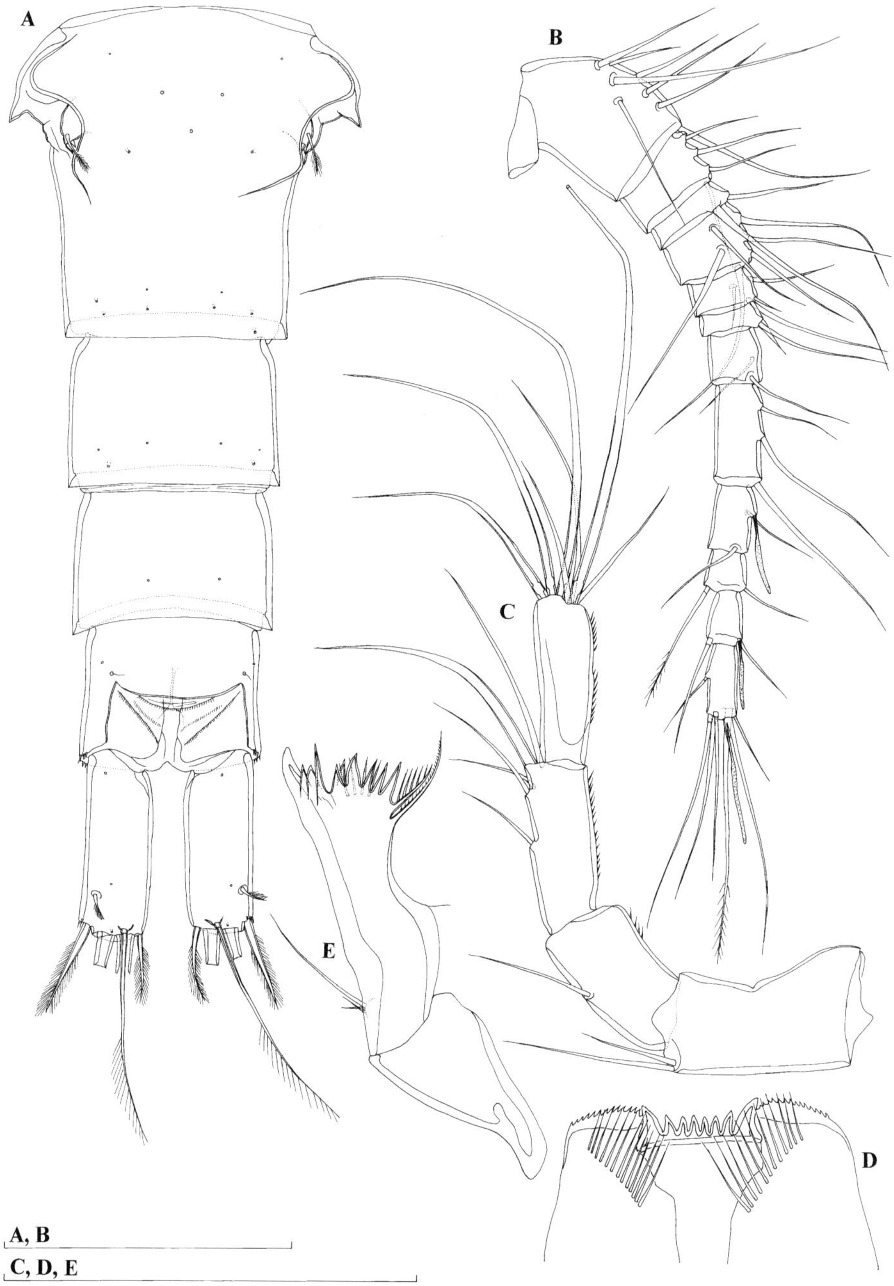

Fig. 53. *Neocyclops australiensis* n. sp., holotype female. A, abdomen, dorsal view; B, antennula; C, antenna; D, labrum; E, mandibula. Scales = 0.1 mm.

margin. Second endopodal segment about 2.4 times as long as wide, ornamented with longitudinal row of spinules along external margin, and armed with five smooth setae (three lateral, two subapical; one subapical seta somewhat longer and much stronger than any other seta on that segment). Third endopodal segment 2.8 times as long as wide, armed with seven smooth apical setae (four of which strong, geniculate, and with breaking planes) and ornamented as the previous segment.

Labrum (fig. 53D) trapezoidal, but with very broad cutting edge and distolateral corners almost on the same plane, ornamented with two diagonal rows of 12 (13) long and slender spinules on ventral surface, and with a finely serrated distolateral margin. Cutting edge slightly concave, with eight strong and sharp teeth (outermost ones strongest). No other ornamentation visible.

Mandibula (fig. 53E) with palp reduced to small protuberance, which bears two minute and one much longer, smooth setae (but longest seta not reaching base of gnathobasal teeth). Cutting edge of coxal gnathobase clearly divided into three regions: inner group of four complex smooth teeth, innermost tooth strongest, outermost sharpest; middle group of six simple smooth teeth and four spinules; outer group of one strongly unipinnate tooth and outermost unipinnate seta. Gnathobasal seta about 0.8 times as long as longest seta on reduced palp.

Maxillula (fig. 52D) composed of well developed praecoxa and two-segmented palp. Arthrite of praecoxa with four apical, very strong spines; three of these basally fused to arthrite, geniculate, and strongly unipinnate; also an unfused spine, not geniculate and bipinnate. Praecoxa armed with seven armature elements along inner margin, the longest one plumose. Palp with completely divided, but minute endopod, which bears two apical and one lateral pinnate setae, and is armed laterally with one exopodal smooth seta and apically with two slender and smooth setae, and one robust unipinnate spine.

Maxilla (fig. 52E) four-segmented but praecoxa fused to coxa on posterior surface. Proximal endite of praecoxa armed with two setae (one plumose, other pinnate distally); distal endite very small, unarmed. Proximal endite of coxa with one short pinnate seta; distal endite highly mobile and bearing one proximal, completely fused, and very stout seta, and one distal, smaller seta. Seta fused to endite ornamented with five elongated spinules. Basis expanded into spinulose claw and armed with two setae; strong seta slightly longer than claw. Endopod armed with three spinulose and unipinnate, and two slender and smooth setae. Longest endopodal spinulose seta about 1.3 times as long as strong seta on basis; shortest endopodal spinulose seta about 1.2 times as long as longer endopodal smooth seta. All three endopodal spinulose setae, as well as claw and strong seta on basis, with most proximal spinules longer than succeeding spinules in row.

Maxilliped (fig. 52F) slender, four-segmented, composed of syncoxa, basis, and two-segmented endopod. Syncoxa 1.8 times as long as its greatest width, unornamented, and armed with three pinnate setae; middle seta strongest, 1.8 times as long as most distal one, and almost twice as long as most proximal seta. Basis 1.7 times as long as wide, ornamented with one arched row of spinules at middle, one longitudinal row of spinules along outer margin, and several spinules at base of setae, and armed with two subequal setae, each with two elongated spinules. First endopodal segment unornamented and armed with two pinnate setae; distal one the strongest on the maxilliped and about 1.5 times as long as proximal one. Second endopodal segment very small, unornamented, but armed with three smooth setae; innermost seta the longest on the maxilliped, about 1.4 times as long as middle one, and more than 3.0 times as long as outermost seta.

All swimming legs (fig. 54A, B, C, D) with three-segmented exopod and endopod; their armature formula as follows (legend: inner/outer spine or seta; inner/terminal/outer):

	Exopod			Endopod		
Segments	1	2	3	1	2	3
First leg	1/1	1/1	3/2/3	1/0	2/0	3/2/1
Second leg	1/1	1/1	4/2/3	1/0	2/0	3/2/1
Third leg	1/1	1/1	4/2/3	1/0	2/0	3/2/1
Fourth leg	1/1	1/1	4/2/2	1/0	2/0	2/2/1

Third exopodal segment spine formula 3.4.4.3 and third endopodal segment spine formula 2.3.3.3. All intercoxal sclerites without ornamentation and with rounded protrusions at both sides of distal margin; that of first leg most pronounced (fig. 54A). Coxae also unornamented (except for short row of minute spinules along posterior margin of third leg), and armed with a plumose seta at inner distal corner; this seta shortest on fourth leg (fig. 54D). Spine inserted at inner, protruded corner of basis of first leg reaching halfway the second endopodal segment, stout and bipinnate (fig. 54A). Basis of fourth swimming leg unornamented, those of other legs with a row of hair-like spinules along inner margin and with a few minute spinules along distal

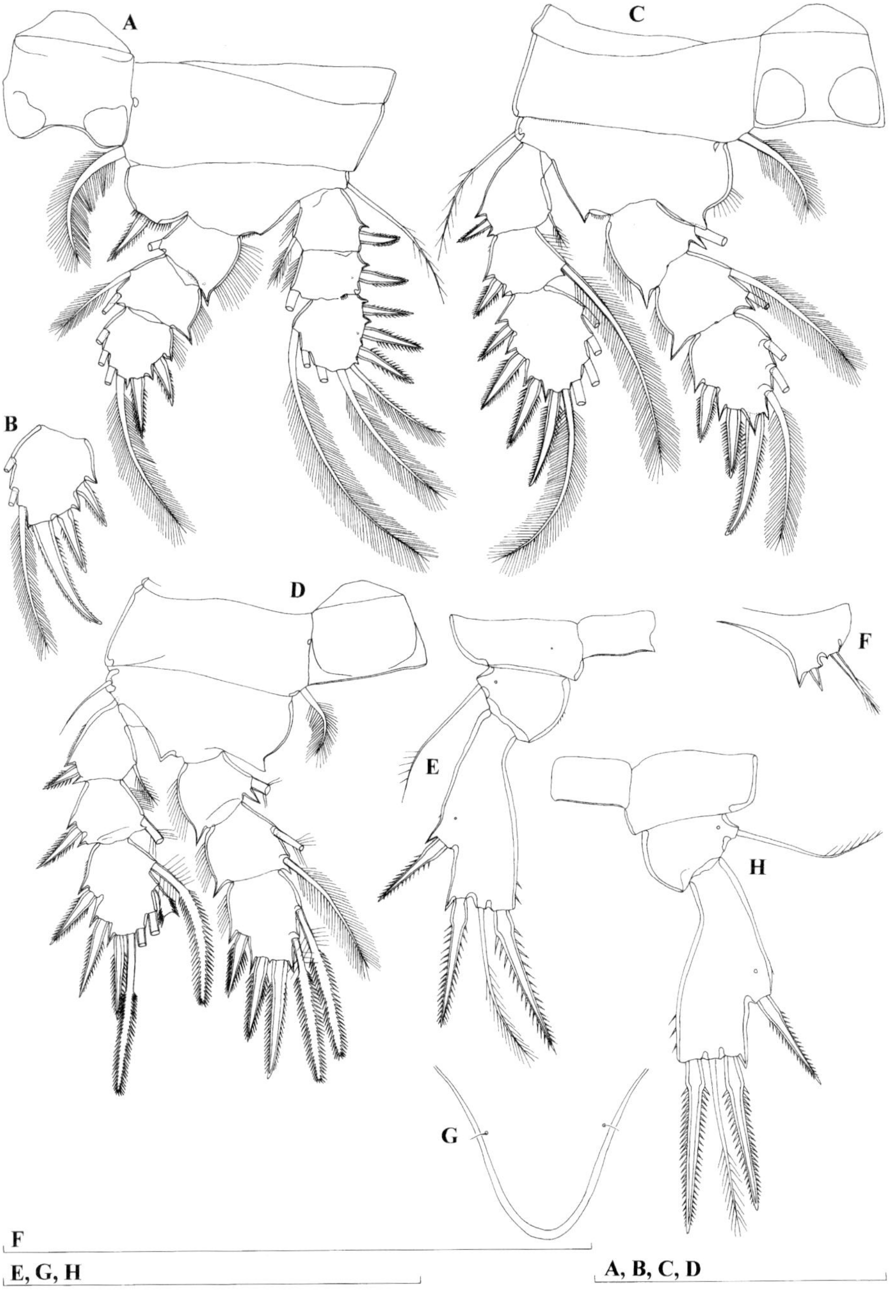

Fig. 54. *Neocyclops australiensis* n. sp., A-G, holotype female; H, female from Withcliffe (0.792 mm). A, first swimming leg; B, third endopodal segment of second swimming leg; C, third swimming leg; D, fourth swimming leg; E, fifth leg; F, sixth leg; G, rostrum; H, fifth leg. Scales = 0.1 mm.

margin close to endopod; inner distal corners of second, third, and fourth leg pronounced as blunt, spiniform processes; each basis armed with a slender seta on outer margin, which is smooth in fourth leg and pinnate in the others. Most exopodal and endopodal setae slender and plumose; outer apical seta on third exopodal segment of first leg pinnate along outer margin, plumose along inner (fig. 54A); all setae on third endopodal and exopodal segments of fourth leg modified into very strong, spiniform armature elements, with a few hair-like spinules at base and sharply serrated lateral margins (fig. 54D). All exopodal and endopodal spines stout and bipinnate. All endopodal segments with hair-like spinules along outer margins, which are also present on inner margin of second exopodal segment of all legs and on first exopodal segment of first leg. Cuticular pores visible only on first leg exopod and endopod. Third endopodal segment of fourth swimming leg about 1.2 times as long as wide; inner apical spine about 1.4 times as long as outer one (fig. 54D).

Fifth leg (fig. 54E) with intercoxal sclerite and three-segmented, comprising coxa, basis, and a one-segmented exopod. Intercoxal sclerite quadrangular, with almost straight distal margin, about twice as wide as long, and unornamented. Coxa unarmed, ornamented with a single cuticular pore closer to the inner margin, and almost 2.5 times as wide as long. Basis nearly triangular, armed with a single unipinnate seta on short outer margin; ornamented with cuticular pore at base of seta and with several minute spinules along convex and much longer inner margin; basis about 1.4 times as wide as its greatest length. Exopod 2.1 times as long as its greatest width, much narrower at basal part, armed with three spines and a single seta, and ornamented with one pore at base of proximal spine, two spinules at distal end of inner margin, and one minute spinule on outer margin; seta plumose, positioned centrally and as the most apical element, only slightly longer than inner apical spine, about 1.3 times as long as outer apical spine, and 1.9 times as long as lateral spine; all spines with swollen basal part and very strongly bipinnate.

Fig. 55. *Neocyclops australiensis* n. sp., A-D, female from Shell Beach (0.786 mm); E-K, female from Dundee Beach (0.633 mm); L-P, female from Barney Point (0.705 mm). A, habitus, dorsal view; B, right caudal ramus, ventral view; C, third endopodal segment of fourth swimming leg; D, fifth leg; E, habitus, dorsal view; F, left caudal ramus, dorsal view; G, eighth and ninth segments of antennula; H, basis of antenna; I, cutting edge of labrum; J, fifth leg; K, sixth leg; L, right caudal ramus, dorsal view; M, eighth and ninth segments of antennula; N, basis of antenna; O, third endopodal segment of fourth swimming leg; P, fifth leg. Scales = 0.1 mm.

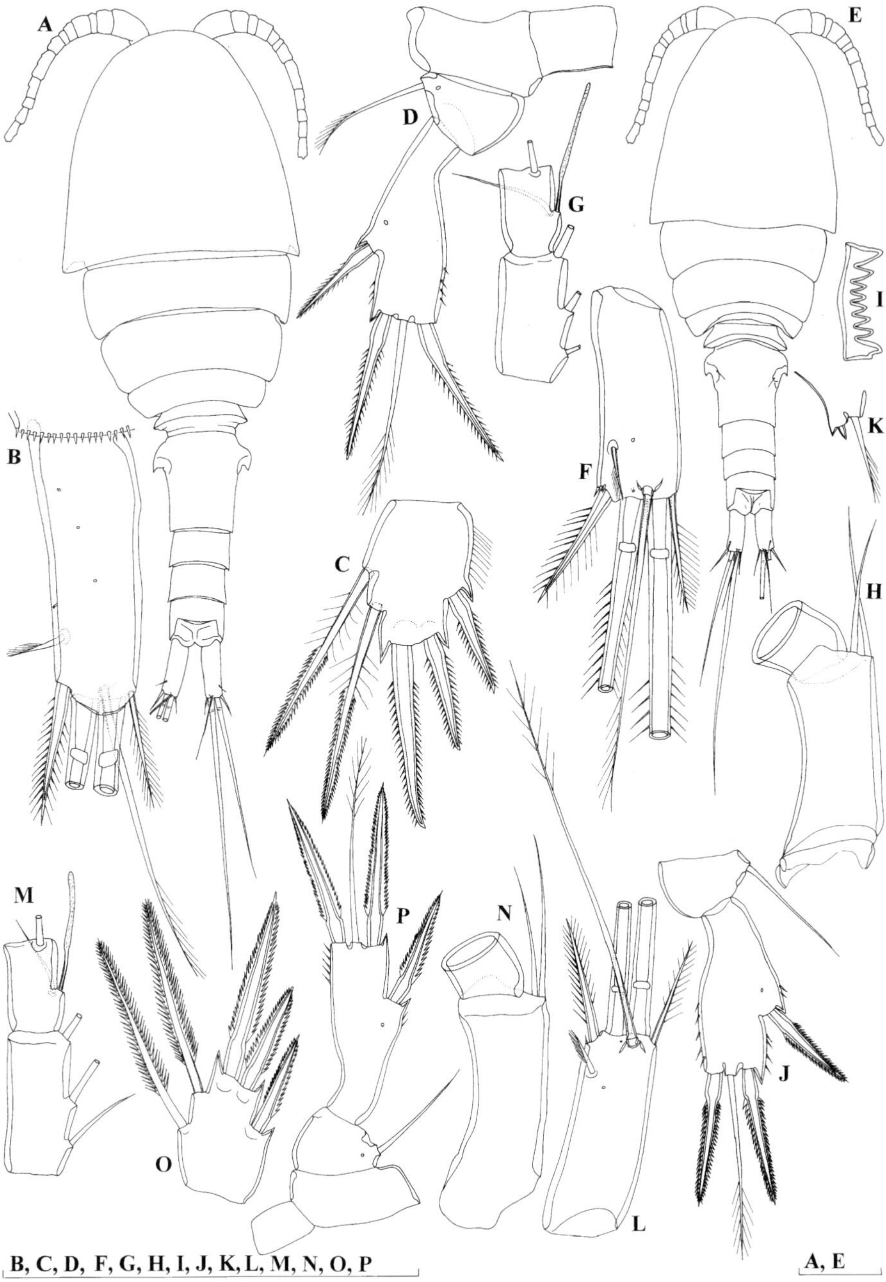
A
B
C
D
E
F
G
H
I
J
K
L
M
N
O
P
B, C, D, F, G, H, I, J, K, L, M, N, O, P
A, E

Sixth leg (fig. 54F) not clearly distinct, triangular plate, armed with two almost equal, smooth spines (inner one fused to plate) and one pinnate, much longer seta.

Male (allotype). Body length, excluding caudal setae, 0.473 mm. Habitus (fig. 56A) less robust than that of female, with prosome/urosome ratio of 1.2, and greatest width near posterior end of cephalothorax. Body length/width ratio about 2.9; cephalothorax 2.1 times as wide as genital somite.

Cephalothorax (fig. 56A) slightly longer than its greatest width, with posterior corners not pronounced laterally. Ornamentation of prosomites, colour, and nauplius eye similar to those of female. Hyaline fringes of all somites smooth.

Genital somite 1.6 times as wide as long (dorsal view), ornamented with four pores dorsally, two pores ventrally, and four pores and one sensillum laterally (fig. 56B); third and fourth urosomites ornamented with two pores ventrally, two dorsally, and two laterally on each side; fifth urosomite with two dorsal pores and one lateral pore on each side. Anal somite similar to that of female.

Caudal rami (fig. 56A, B) about 2.4 times as long as wide, parallel, with space between them less than half of a ramus' width. Armature and ornamentation similar to those of female, except anterior dorsal pore absent.

Antennula (fig. 56B) of length almost equal to that of cephalothorax, 14-segmented, geniculate, with geniculation between twelfth and thirteenth segments. Aesthetasc formula: 3.0.0.0.1.0.0.1.0.0.1.0.0.2. All aesthetascs slender and not very long. Setal formula: 8.4.4.0.1.4.1.2.4.2.1.1.1.10. All setae on tenth, eleventh, and twelfth segments, and one seta on ninth segment very short and spiniform. Only seta on eleventh and one seta on tenth segments pinnate; all other setae smooth.

Antenna (fig. 56D) with exopodal seta on coxobasis present, otherwise similar to that of female.

Labrum, mandibula, maxillula, maxilla, maxilliped, and swimming legs similar to those in female.

Fifth leg (fig. 56E, F) with incompletely divided exopod and with additional seta on inner margin of first exopodal segment; intercoxal sclerite narrower, and apical seta on second exopodal segment shorter than in female; pore on coxa absent, pore on exopod present on one leg (fig. 56F), absent on opposite appendage (fig. 56E). First exopodal segment twice as long as second one.

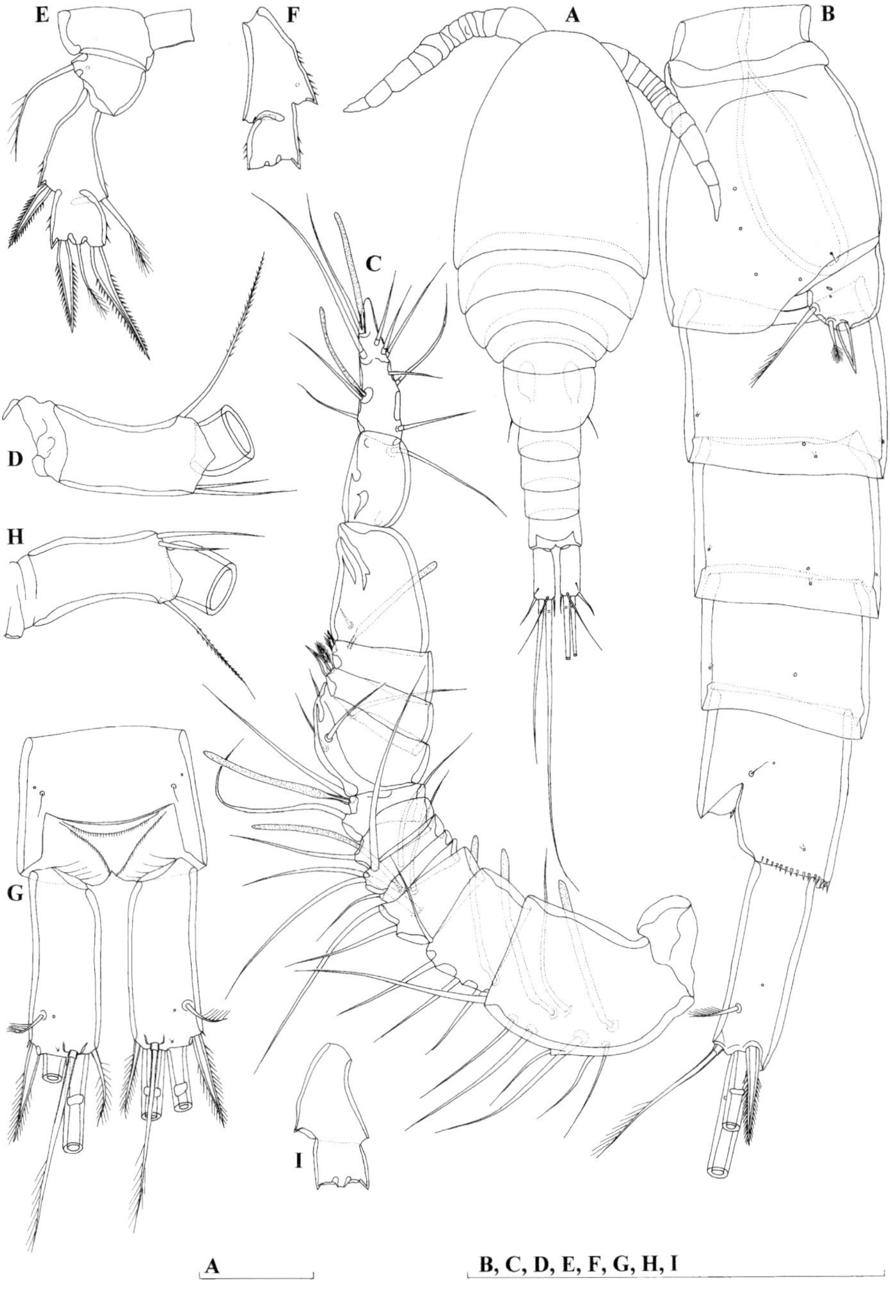

Fig. 56. *Neocyclops australiensis* n. sp., A-F, allotype male; G-I, male from Shell Beach (0.515 mm). A, habitus, dorsal view; B, abdomen, lateral view; C, antennula; D, basis of antenna; E, fifth leg; F, exopod of opposite fifth leg without armature; G, anal somite and caudal rami, dorsal view; H, basis of antenna; I, exopod of fifth leg without armature. Scales = 0.1 mm.

Sixth leg (fig. 56B) a distinct, large plate, armed with one smooth spine and two pinnate setae; ornamented with two cuticular pores. Outermost seta 2.5 times as long as middle seta and twice as long as innermost spine.

Variability. — Body length of females ranging from 0.538 mm to 0.896 mm (0.705 mm average; n = 13), while in males it ranges from 0.455 mm to 0.515 mm (0.481 mm average; n = 3).

The cuticular pore pattern is variable and even asymmetrical on the same specimen. For example, the holotype female has three pores on the left posterior part of the genital double somite (ventral view) and only two on the right side (fig. 52B). The allotype male has one cuticular pore on the exopod of the right fifth leg, while the left one is without pore (fig. 56E, F). However, the species is very conservative, considering its truly wide distribution in Australia. Some small differences were noticed in the female habitus (fig. 55A, E), proportions and ornamentation of the caudal rami (fig. 55B, F, L), proportions of the antennular segments (fig. 55G, M), proportions of the third endopodal segment of the fourth swimming leg (fig. 55C, O), and proportions and ornamentation of the fifth leg (figs. 54H, 55D, P). The length of the exopodal seta on the antenna is variable in males (fig. 56D, H), as is the articulation between the first and second exopodal segments of the fifth leg (fig. 56E, F, I), although the segments are never completely fused or divided. Generally, however, most characters are very stable, including the number of teeth in the cutting edge of the labrum (fig. 55I), the armature of the antennal coxobasis in females (fig. 55H, N), the female sixth leg (fig. 55K), and the shape and ornamentation of the caudal rami in males (fig. 56G).

Etymology. — The specific name refers to the species' wide Australian distribution. The name is an adjective for place, made with the Latin suffix "-ensis".

Distribution. — *Neocyclops australiensis* n. sp. has a very wide Australian distribution and is recorded in all states examined, except in Victoria. It occurs on Shell Beach (Shark Bay, Western Australia) sympatrically with two other new species: *N. sharkbayensis* n. sp. and *N. trajani* n. sp. Also, *N. australiensis* was found sympatrically with *N. tropicus* n. sp. at four different locations: Exmouth (Western Australia), Dundee Beach (Northern Territory), and Yeppoon and Hervey Bay (both in Queensland).

Discussion. — Gurney (1927b) erected the genus *Neocyclops* to accommodate the new species from the Suez Canal he described earlier the same year (Gurney, 1927a) as *Halicyclops salinarum*. He (Gurney, 1927b) also corrected and amended the species description, as the original one was based "on a single damaged specimen". Sewell (1949) described the genus *Eurycyclops* to accommodate two new species he described in the same publication from the Maldives: *E. magna* and *E. parva*. He realized that his new genus was closely related to *Neocyclops*, but he found enough justification for it in the fact that his two new species had a 12-segmented antennula (while the antennula of *N. salinarum* is eight-segmented) and a single seta as a mandibular exopod (which seemed to be absent in *N. salinarum*). Herbst (1952), on the other hand, was completely unaware of Gurney's (1927b) and Sewell's (1949) publications, when he erected the genus *Pareuryte*, for his new German species, *P. remanei*, collected by Prof. Remane, although a short generic diagnosis for *Neocyclops* and a short species description of *N. salinarum* were given by Kiefer (1929b), a reference listed in Herbst's (1952) paper. He compared this new genus only with the genera *Euryte* Philippi, 1843 (today in a completely separate subfamily) and *Cyclopinella* G. O. Sars, 1913 (which was even then in a separate family), but his new species was morphologically very similar to those described by Sewell (1949). Three years later, Herbst (1955) described another new *Pareuryte* species from Brazil (*P. vicina*), while at the same time describing another closely related species in the genus *Neocyclops* (as *N. medius*). He was now obviously aware of a close relationship between these two genera, but not yet of Sewell's (1949) paper and the genus *Eurycyclops*. He distinguished *Pareuryte* from *Neocyclops* on the basis of the presence or absence of the mandibular palp and the armature of the male fifth leg, although he did not find any males of his new *Pareuryte vicina*. Aguesse & Dussart (1957) found and redescribed *N. salinarum* from the French coast of the Mediterranean, but they did not make any comments on the generic confusion around this species. Kiefer (1960b) described a new species from the Balearic Islands as *Pareuryte mediterranea*, comparing it only to the two species described by Herbst (1952, 1955) and without making any comments about the genera *Neocyclops* or *Eurycyclops*. It was Pleşa (1961) who first drew attention to a close relationship between *Neocyclops*, *Eurycyclops*, and *Pareuryte*, when he described another new species from Ghana as *Eurycyclops affinis*. He noted about the genus *Eurycyclops* that "it is not unlikely that this genus may be synonymous with the genus *Neocyclops* described by Gurney (1927) but certainly not with *Pareuryte* Herbst". However, he listed all the species described until then in any of the three genera (as "the described species which are very similar to one another”), pending a revision of this complex. Two years later, Pleşa (1963) more confidently united all the species previously described in *Neocyclops*, *Eurycyclops*, or *Pareuryte* in one genus, recognizing only the genus *Eurycyclops* as valid, and provided a key to the six species and subspecies known until then (he considered *E. remanei*, *E. vicinus*, and *E. mediterraneus* as subspecies, not distinct species as they were originally described). Herbst (1964) acknowledged a close relationship of *Neocyclops*, *Eurycyclops*, and *Pareuryte* (although without acknowledging the previous work of Pleşa, 1961, 1963), proposing them as subgenera of the genus *Neocyclops*. Vervoort (1964) accepted that “the genus *Pareuryte* Herbst is evidently identical with Sewell’s genus *Eurycyclops*”, but considered *Neocyclops* and *Eurycyclops* as two valid and separate genera. Wells (1967) admitted that he “followed Herbst (1964, p.168) in considering *Neocyclops* Gurney, 1927, *Eurycyclops* Sewell, 1949 and *Pareuryte* Herbst, 1952 as synonymous, but with each forming a subgenus”, while Monchenko (1974) considered the segmentation of the antennula to be a valid character in separating *Neocyclops* from *Eurycyclops*. Pleşa (1973, 1981) finally recognized that the genera *Neocyclops*, *Eurycyclops*, and *Pareuryte* are synonyms, and he (Pleşa, 1981: 21) clearly stated that: “C’est ainsi que, pour des raisons de priorité, le nom correct qui doit être retenu est celui donné par Gurney: *Neocyclops*”. He also provided a key to nine species and subspecies. This was accepted by Pesce (1985), Dussart & Defaye (1985), Fiers (1986), Herbst (1986), Petkovski (1986), and all subsequent workers.

Petkovski (1986) revised the genus *Neocyclops* and divided it into two subgenera (*Neocyclops* s. str. and *Protoneocyclops* Petkovski, 1986) based on the segmentation of the male fifth leg. He established a new species, *N.* (*P.*) *herbsti*, for the specimens from the Red Sea, mentioned by Herbst (1964) as *N. salinarum* (Gurney), as well as a new species, *N.* (*P.*) *wellsi*, for the specimens from Mozambique, mentioned by Wells (1967) as *N. affinis* (Pleşa). Petkovski (1986) accepted another 10 species and subspecies as valid, providing a key to all 12 taxa described until then in the genus *Neocyclops*: *N.* (*N.*) *affinis* (Pleşa, 1961); *N.* (*P.*) *herbsti* Petkovski, 1986; *N. improvisus* Pleşa, 1973; *N. magnus* (Sewell, 1949); *N.* (*P.*) *mediterraneus* (Kiefer, 1960); *N.* (*N.*) *medius* Herbst, 1955; *N. parvus* (Sewell, 1949); *N.* (*N.*) *remanei remanei* (Herbst, 1952); *N.* (*N.*) *remanei vicinus* (Herbst, 1955); *N.* (*N.*) *salinarum* (Gurney, 1927); *N.* (*P.*) *stocki* Pesce, 1985; and *N.* (*P.*) *wellsi* Petkovski, 1986. As can be seen from this list, three species could not be assigned to any subgenus as their males were unknown. The systematic concept of the genus *Neocyclops* presented by Petkovski (1986) was not challenged so far, but his subgenera have to be abandoned in the light of new discoveries presented in this monograph. Some further criticism is given in the discussion section for *Neocyclops petkovskii* De Laurentiis et al., 1997 (see below). Since Petkovski (1986), the following additions to the genus *Neocyclops* have been made: Pesce & Galassi (1987) recorded *N. mediterraneus* for the first time from Italy; Pesce & Galassi (1993) recorded *N. medius* and *N. stocki* from several new localities in the West Indies and also described one new species from Curaçao as *N.* (*P.*) *geltrudeae*; Lotufo & Da Rocha (1993b) reported *N. medius* and *N. vicinus* from several Brazilian beaches; Da Rocha & Iliffe (1993) reported *N. stocki* from an anchialine cave in Bermuda; Da Rocha (1995a) described a new species from Belize as *N.* (*P.*) *ferrarii*; Jaume & Boxshall (1996a) redescribed a male of *N. mediterraneus* from the Balearic Islands; and De Laurentiis et al. (1997) described the first representative from Australia as a new species, *N.* (*N.*) *petkovskii*. Together with five new Australian species described in this monograph and three others established as new as a result of taxonomic analyses of published data, the genus *Neocyclops* now numbers 24 members and a key to aid in their identification is presented in the discussion section for *N. petkovskii* (see below).

Neocyclops australiensis n. sp., unlike any other species of the genus for which males are known, has the exopod of the male fifth leg incompletely divided, thus making it impossible to assign this species to any of the two subgenera established by Petkovski (1986). This, together with the situation in *N. petkovskii* (see below), made me abandon the subgeneric subdivision of the genus *Neocyclops*. As noted by Da Rocha (1995a), the great difficulty in studying the taxonomy of this genus is the scarcity of diagnostic characters distinguishing species (i.e., a very conservative morphology). A significant variability of some species, as well as incomplete descriptions, further complicate the situation. However, I did not have any problems in distinguishing the six Australian species presented here (although their distributions often overlapped), because they were studied in detail. But distinguishing them from other described species was not so easy, as many characters could not be compared. *Neocyclops australiensis* differs from the other five Australian species by many characters, but most notably by the proportions and armature of its caudal rami. It also has very sharp lateral projections on the genital double somite and a very robust habitus. In many situations in which it occurred sympatrically with other congeners, it was possible to tell they were distinct species under the dissecting microscope, even before one could be sure of their generic placement. This is the only species that could be found all around Australia (recorded in all states investigated, except Victoria, and from tropical regions to the shores of the cold Southern Ocean) and it was found sympatrically with three other species: *N. sharkbayensis* n. sp., *N. trajani* n. sp., and *N. tropicus* n. sp. However, its variability is not more significant than that of more localized species. For example, all the females of *N. australiensis* examined lack the exopod of the antenna, while this is a variable feature in *N. tropicus*. The sexual dimorphism in the armature of the antenna, besides in *N. australiensis* and some specimens of *N. tropicus*, was confirmed so far only for *N. vicinus* (Herbst, 1955) by Lotufo & Da Rocha (1993b). However, this condition may occur more widely, as males of some species are still unknown and those of others were rarely studied in detail. *Neocyclops australiensis* seems to be most closely related to *N. remanei*, *N. vicinus*, and *N. wellsi*, but can be distinguished from these by the proportions and armature of the caudal rami, as well as by some additional small differences in the antenna or the fifth leg. *Neocyclops affinis* (Pleşa, 1961), described from Ghana by Pleşa (1961), has caudal rami very similar to those of *N. australiensis*, but its dorsal seta is longer than half the length of the principal outer apical seta, while in *N. australiensis* the principal apical seta is proportionately much longer. The two species could be additionally distinguished by the armature formula of the swimming legs, the shape of the male fifth leg, as well as by the size of the lateral projections on the female genital double somite. The homonym of *N. affinis* (Pleşa, 1961), described by Dussart (1974) from Ethiopia, I consider to be a separate species (see the discussion section for *N. petkovskii*).

20. **Neocyclops sharkbayensis** n. sp.
(figs. 57-59)

Material examined. — Types: Holotype female dissected on 2 slides (WAM C37259); paratypes, 5 females (2 dissected on 1 slide each (WAM C37260 and WAM C37261), 1 on 1 slide in toto (TMAG G5785), others in alcohol (WAM C37262)): Australia, Western Australia, Shark Bay, Nanga Bay Resort, 2 August 2003, leg. T. Karanovic, 26°15.277′S 113°48.333′E.

Other material: Australia, Western Australia, Shark Bay, Shell Beach, 2 August 2003, leg. T. Karanovic, 26°13.032′S 113°46.632′E: 3 females + 2 copepodids in alcohol (WAM C37263).

Description. — Female (holotype). Body length, excluding caudal setae, 0.762 mm. Habitus (fig. 57A) typical cyclopoid shape, very robust, with prosome/urosome ratio of 1.2, and greatest width at posterior end of cephalothorax. Body length/width ratio about 2.9; cephalothorax about 2.5 times as wide as genital double somite. Posterior corners of cephalothorax somewhat expanded laterally; other free pedigerous somites without pronounced lateral expansions. Fifth pedigerous somite as wide as genital double somite. Colour of preserved specimen milky white, but much more transparent than in previous species. Nauplius eye not visible. Rostrum well developed, broadly rounded, and furnished with two large sensilla.

Cephalothorax (fig. 57A) almost triangular, slightly shorter than its greatest width, representing 33% of total body length. Surface of cephalic shield with several large sensilla; no other ornamentation visible. Hyaline fringe of prosomites narrow and smooth. Large arthrodial membrane between cephalothorax and second pedigerous somite (first free somite) very thin; first pedigerous somite completely fused to cephalosome. Fifth pedigerous somite ornamented with only two large dorsal sensilla and with a smooth fringe dorsally and ventrally. Sclerotized joint between fifth pedigerous somite and genital double somite well developed, almost as a pseudosomite (fig. 57B).

Genital double somite (fig. 57B) with blunt lateral expansions at first third, slightly longer than wide, without pattern of pores ventrally, ornamented instead with two sensilla and a posterior row of minute spinules. Hyaline fringe of genital double somite and two succeeding somites narrow, not frilled, and completely smooth. No cuticular recesses visible. Copulatory pore small, ovoid, situated just after 1/3 of the somite's length; copulatory duct narrow and rigidly sclerotized, with first half directed dorsally and second half directed anteriorly, L-shaped from lateral view. Seminal receptacle (fig. 57B) very similar to that of previous species, with somewhat smaller lateral arms. Ovipores situated somewhat dorsolaterally, covered by the reduced sixth legs, bearing two minute spines and one pinnate seta. Third urosomite ornamented with four ventral, one lateral, and two dorsal pores in posterior half, as well as with a posterior row of minute spinules ventrally. Fourth urosomite with two ventral, one lateral, and two dorsal pores, and also with a posterior row of minute spinules. Anal somite (fig. 57B) with smooth, very broad, and convex anal operculum, which represents 77% of the somite's width and reaches to only 3/5 of the somite's length; ornamented with two large sensilla and two pores dorsally, one pore laterally on each side, four pores ventrally, and a transverse row of spinules on the posterior margin. Anal sinus widely open dorsally, ornamented with a few diagonal rows of minute spinules.

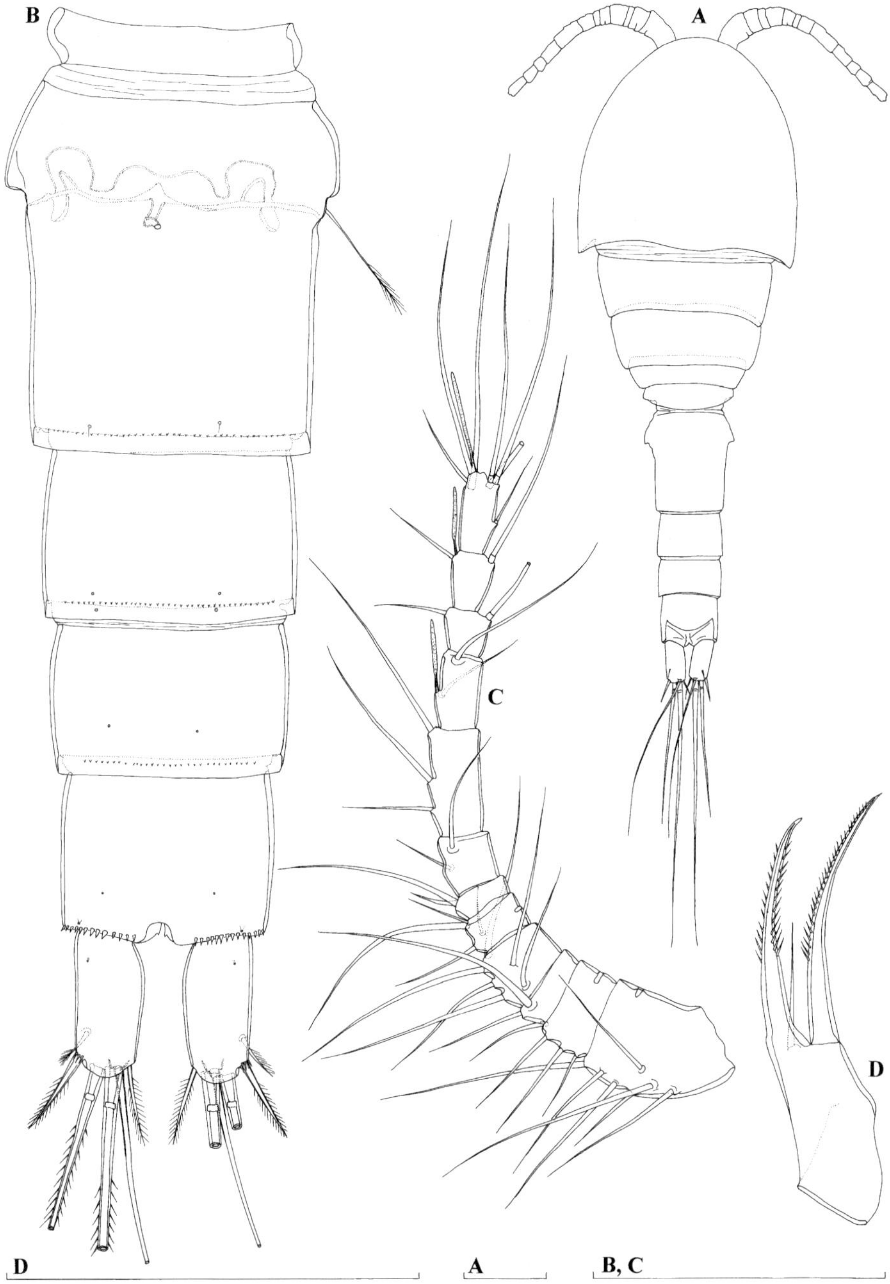

Fig. 57. *Neocyclops sharkbayensis* n. sp., holotype female. A, habitus, dorsal view; B, abdomen, ventral view; C, antennula; D, basis of maxilla. Scales = 0.1 mm.

Caudal rami (fig. 57B) cylindrical, twice as long as wide (ventral view), parallel, with space between them about one half of a ramus' width; they represent 6% of total body length. Distal margin ventrally smooth and convex, with medial cuticular pore hardly visible. The ornamentation consists of several small spinules at the base of the outermost apical seta and one ventral, one lateral, and one dorsal pore. Dorsal seta extremely long, almost 4.0 times as long as ramus and slightly longer than outer principal apical seta, inserted on small protuberance reaching border of ramus, uniarticulate at base and unipinnate distally. Lateral seta smallest, arising at 3/4, positioned dorsolaterally, uniplumose, and about 0.7 times as long as a ramus' width. Outermost apical seta stout and spiniform, bipinnate, about half the ramus' length, and only slightly longer than innermost apical seta, which is also bipinnate, but more slender than outermost one. Principal apical setae with breaking planes and bipinnate; inner seta about twice as long as outer, and 0.4 times as long as body.

Antennula (fig. 57C) slender, reaching to 2/3 of cephalothorax, 12-segmented, unornamented, with one slender aesthetasc on ninth, one on eleventh, and one on twelfth segment, and setal formula: 7.4.2.6.4.2.2.3.2.2.2.7. No setae with breaking planes and only one seta on tenth, one on eleventh, and three on twelfth segment articulating on basal part. One seta on sixth segment spiniform and short, all other setae slender, and all setae smooth. Atavistic remnants of articulation present on first, second, and fifth segments caudally. Length ratio of antennular segments, from proximal end and along caudal margins, 1 : 0.5 : 0.2 : 0.5 : 0.4 : 0.2 : 0.5 : 1.2 : 0.8 : 0.5 : 0.6 : 0.9.

Antenna (fig. 58A) four-segmented, ornamented with small spinules along caudal margin, comprising coxobasis and a three-segmented endopod; all endopodal segments of about the same length, coxobasis somewhat longer. Coxobasis about 2.3 times as long as wide, armed with two subequal smooth setae at distal inner corner; seta representing exopod absent. First endopodal segment armed with one smooth seta at middle. Second endopodal segment about 2.7 times as long as wide, armed with five smooth setae (three lateral, two subapical; one subapical seta somewhat longer and much stronger than any other seta on that segment). Third endopodal segment 3.6 times as long as wide, armed with seven smooth apical setae, four of which strong, geniculate, and with breaking planes.

Labrum (fig. 58B) trapezoidal, with very broad cutting edge, ornamented with two diagonal rows of 13 (14) long and slender spinules on ventral surface, with two groups of smaller spinules on dorsal surface, and with

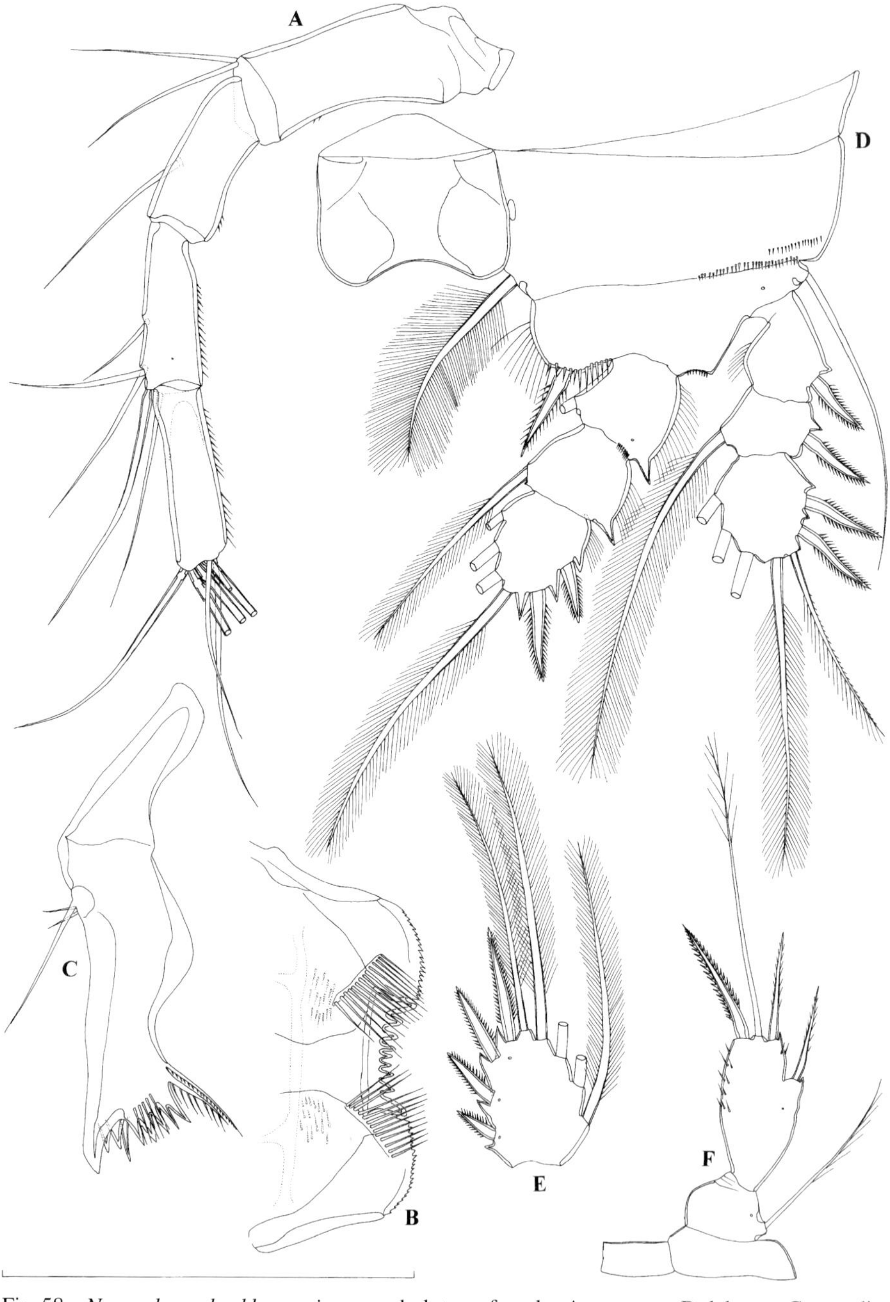

Fig. 58. *Neocyclops sharkbayensis* n. sp., holotype female. A, antenna; B, labrum; C, mandibula; D, first swimming leg; E, third exopodal segment of third swimming leg; F, fifth leg. Scale = 0.1 mm.

finely serrated distolateral margin. Cutting edge slightly concave, with nine strong and sharp teeth (outermost ones strongest).

Mandibula (fig. 58C) with palp reduced to small protuberance, which bears two minute and one much longer smooth setae (but longest seta not reaching base of gnathobasal teeth). Cutting edge of coxal gnathobase clearly divided into three regions: inner group of four complex smooth teeth, innermost tooth strongest, outermost sharpest; middle group of five simple smooth teeth and five spinules; outer group of one strongly unipinnate tooth and outermost unipinnate seta. Gnathobasal seta about 0.7 times as long as longest seta on reduced palp.

Maxillula, maxilla (fig. 57D), and maxilliped the same as in the previous species.

All swimming legs (figs. 58D, E, 59A, B) with three-segmented exopod and all, except the fourth, with a three-segmented endopod; endopod of fourth leg with ancestral second and third segments completely fused on both anterior and posterior surface, although inner and outer margins with pointed expansions clearly marking ancestral segmentation; their armature formula as follows (legend: inner/outer spine or seta; inner/terminal/outer):

	Exopod			Endopod		
Segments	1	2	3	1	2	3
First leg	0/1	1/1	3/2/2	1/0	1/0	3/2/1
Second leg	0/1	1/1	4/2/3	1/0	1/0	3/2/1
Third leg	0/1	1/1	4/2/3	1/0	1/0	3/2/1
Fourth leg	0/1	1/1	4/2/2	1/0	3/2/1	-

Third exopodal segment spine formula 2.4.4.3 and third endopodal segment spine formula 2.3.3.3. All intercoxal sclerites without ornamentation and with rounded protrusions at both sides on distal margin; that of first leg most pronounced. Coxae of second to fourth legs also unornamented, that of first leg with two transverse rows of small spinules near distal margin; all coxae armed with a plumose seta at inner distal corner. Spine inserted at inner, protruded corner of basis of first leg reaching 3/4 of second endopodal segment, stout and bipinnate. Basis of each leg with a row of hair-like spinules along inner margin and with a few minute spinules along distal margin close to endopod; inner distal corners of second, third, and fourth leg pronounced as blunt spiniform processes; each basis armed with a slender and very long seta on outer margin; basis of first leg additionally ornamented with a row of strong spinules at base of inner spine. Most exopodal and endopodal setae

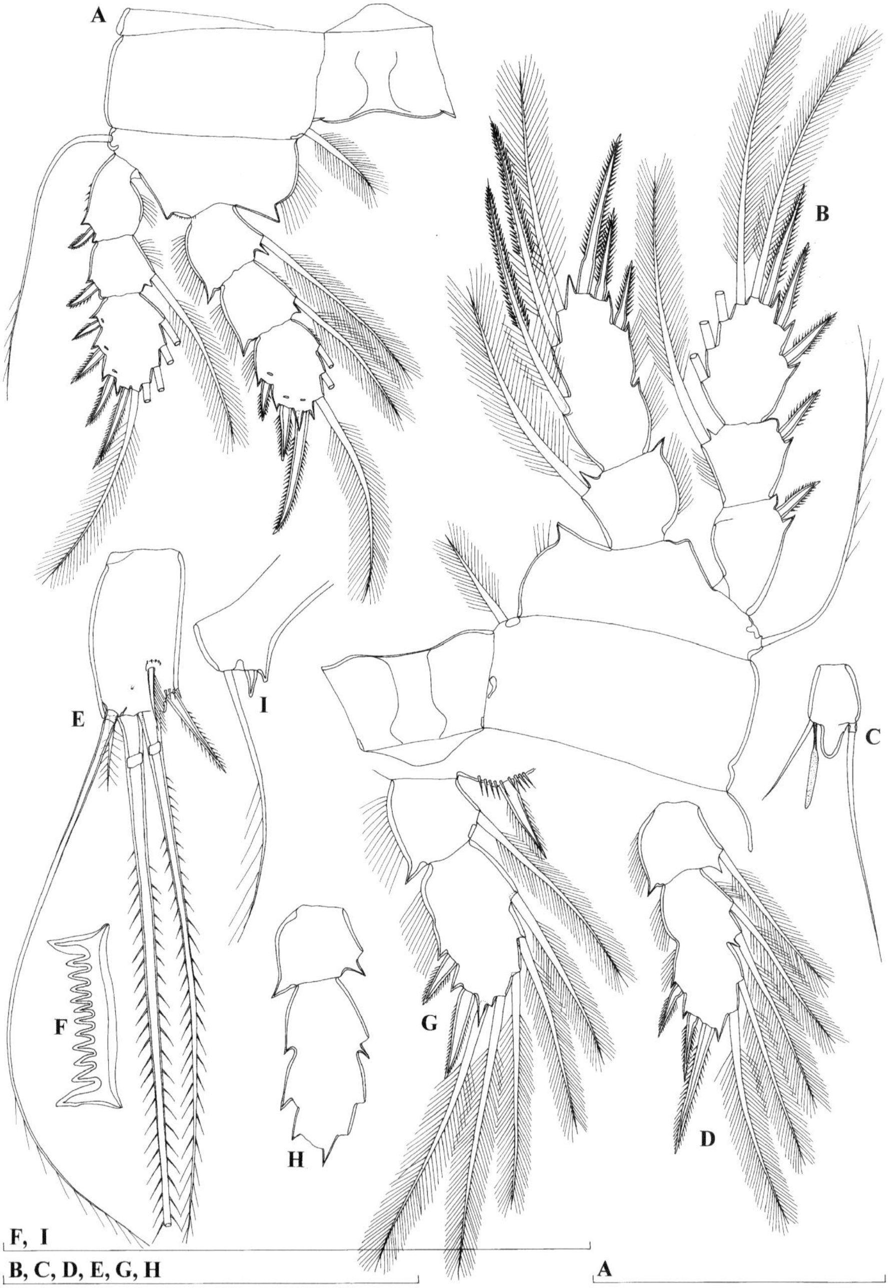

Fig. 59. *Neocyclops sharkbayensis* n. sp., A-B, holotype female; C-D, paratype female (0.635 mm); E-I, paratype female (0.642 mm). A, first swimming leg; B, fourth swimming leg; C, terminal segment of antennula; D, endopod of fourth swimming leg; E, right caudal ramus, dorsal view; F, cutting edge of labrum; G, endopod of first swimming leg; H, endopod of fourth swimming leg without armature; I, sixth leg. Scales = 0.1 mm.

slender and plumose; outer apical seta on third exopodal segment of first leg pinnate along outer margin, plumose along inner; two setae on second endopodal segment of fourth leg (one on ancestral second segment and proximal seta on ancestral third segment) modified into very strong, spiniform armature elements, with a few hair-like spinules at their base and sharply serrated lateral margins. All exopodal and endopodal spines stout and bipinnate. All endopodal segments with hair-like spinules along outer margins, which are also present on inner margin of some exopodal segments. Cuticular pores visible on third exopodal and endopodal segments of second and third leg (three pores each), as well as on basis and first endopodal segment of first leg. Second endopodal segment of fourth swimming leg about twice as long as wide; inner apical spine about 1.8 times as long as outer one.

Fifth leg (fig. 58F) with intercoxal sclerite and three-segmented, comprising coxa, basis, and a one-segmented exopod. Intercoxal sclerite quadrangular, with almost straight distal margin, about twice as wide as long, and unornamented. Coxa unarmed, unornamented, and almost 2.7 times as wide as long. Basis much narrower than coxa, armed with a single unipinnate seta on outer margin; ornamented with cuticular pore at base of seta and about 0.8 times as long as its greatest width. Exopod 1.8 times as long as its greatest width, much narrower at basal part, armed with three spines and a single seta, and ornamented with one pore at base of proximal spine and two longitudinal rows of spinules along inner margin, and two spinules on outer margin distally; seta plumose, positioned centrally, and as the most apical element, 2.3 times as long as inner apical spine, about 2.8 times as long as outer apical spine, and 3.1 times as long as lateral spine; innermost apical spine with swollen basal part and much stronger than the other two spines.

Sixth leg not clearly distinct, triangular plate, armed with two almost equal, smooth spines (inner one fused to plate) and one pinnate, much longer seta.

Male. Unknown.

Variability. — Body length of females ranging from 0.635 mm to 0.762 mm (0.694 mm average; n = 6). One paratype female has an abnormal antennula (fig. 59C), with the opposite appendage normally built, and also the endopod of the fourth swimming leg with two setae on the ancestral second segment and no setae modified (fig. 59D). Another paratype female has a slightly shorter caudal ramus (fig. 59E), more teeth in the cutting edge of the labrum (fig. 59F), a two-segmented endopod of one first leg (fig. 59G) (other leg normally built), and a sixth leg with the innermost spine very short (fig. 59I); this female has the same endopod of the fourth swimming leg as the holotype female. Unfortunately, no males of this species were collected.

Etymology. — The species is named after Shark Bay in Western Australia, as its distribution seems to be restricted to this World Heritage Area. The specific name is an adjective for place, made with the Latin suffix "-ensis".

Distribution. — The species was recorded on only two beaches in Shark Bay, Western Australia. In both localities it lives sympatrically with *N. trajani* n. sp., while on the Shell Beach they also live together with the previously mentioned *N. australiensis* n. sp.

Discussion. — *Neocyclops sharkbayensis* n. sp. could be distinguished from all previously described species of the genus *Neocyclops* by its reduced armature formula of the swimming legs. Namely, it lacks an inner seta on the first exopodal segment of all legs, it has only one seta on the second endopodal segment of all legs, and it bears only two spines on the third exopodal segment of the first leg. Another unusual character of this species is its extremely long dorsal caudal seta, which is 4.0 times as long as the caudal ramus. Only two other species have this seta similarly long: *N. pilbarensis* n. sp. and *N. geltrudeae* Pesce & Galassi, 1993. The first species is described as new below, and seems to be restricted to the Pilbara region of Western Australia (some 500 km north of Shark Bay), while the second species was described from Curaçao (West Indies) by Pesce & Galassi (1993). Both species differ markedly from *N. sharkbayensis*, so the unusually long caudal seta seems to be nothing more than a convergence. *Neocyclops sharkbayensis* is the only Australian representative of its genus without a known male, so the male characters can not be compared.

21. **Neocyclops trajani** n. sp.
(figs. 60-62)

Material examined. — Types: Holotype female dissected on 2 slides (WAM C37264); allotype male dissected on 2 slides (WAM C37265); paratypes, 1 male + 9 females + 3 copepodids (2 females dissected on 1 slide each (WAM C37266 and WAM C37267), 1 male + 1 female on 1 slide in toto (TMAG G5786), others in alcohol (WAM C37268)): Australia, Western Australia, Shark Bay, Nanga Bay Resort, 2 August 2003, leg. T. Karanovic, 26°15.277′S 113°48.333′E.

Other material: Australia, Western Australia, Shark Bay, Shell Beach, 2 August 2003, leg. T. Karanovic, 26°13.032′S 113°46.632′E: 5 females (1 ovigerous) (1 on 1 slide in toto (TMAG G5787), others in alcohol (WAM C37269)).

Australia, Western Australia, Carnavon, town beach, 2 August 2003, leg. T. Karanovic, 24°52.959′S 113°39.302′E: 1 juvenile male in alcohol (WAM C37270).

Description. — Female (holotype). Body length, excluding caudal setae, 0.565 mm. Habitus (fig. 60A) not very robust, with prosome/urosome ratio of 1.3, and greatest width at posterior end of cephalothorax. Body length/width ratio about 2.7; cephalothorax about 2.3 times as wide as genital double somite. Posterior corners of cephalothorax not expanded laterally; other free pedigerous somites also without pronounced lateral expansions. Fifth pedigerous somite as wide as genital double somite. Colour of preserved specimen milky white to transparent. Nauplius eye not visible. Rostrum broadly rounded and furnished with two large sensilla.

Cephalothorax (fig. 60A) slightly shorter than its greatest width, representing 36% of total body length. Surface of cephalic shield with several large sensilla; no other ornamentation visible. Hyaline fringe of prosomites narrow and smooth. First pedigerous somite completely fused to cephalosome. Arthrodial membrane between prosome and urosome well developed. Fifth pedigerous somite ornamented with only two large dorsal sensilla, and with a smooth fringe dorsally and ventrally. Sclerotized joint between fifth pedigerous somite and genital double somite well developed, almost as a pseudosomite.

Genital double somite (figs. 60B, 61A) with lateral expansions (but without spiniform projections) at first third, about 0.9 times as long as wide (ventral view), and ornamented with a characteristic pattern of 14 pores ventrally, four pores laterally, and nine pores and two sensilla dorsally. Hyaline fringe of genital double somite and two succeeding somites narrow, not frilled, and completely smooth. No cuticular recesses or windows visible. Copulatory pore very small, ovoid, situated just after 1/3 of the somite's length; copulatory duct simple, narrow, and rigidly sclerotized, directed dorsally and anteriorly. Seminal receptacle (fig. 60B) complex, curved, and tubular, but entirely situated anteriorly of copulatory pore; the central part connected to the copulatory duct is cylindrical but arched, positioned close to ventral cuticular surface; constricted part on each side only slightly narrower than central part and turning backwards (dorsally); lateral arms expanded, wider than central part, and connected to dorsolateral genital areas without another constriction. Ovipores (fig. 61A) situated somewhat dorsolaterally, covered by the reduced sixth legs, bearing two minute spines and one pinnate seta. Third urosomite ornamented with eight ventral, two lateral, and four dorsal pores in posterior half, while fourth urosomite with four ventral and only two

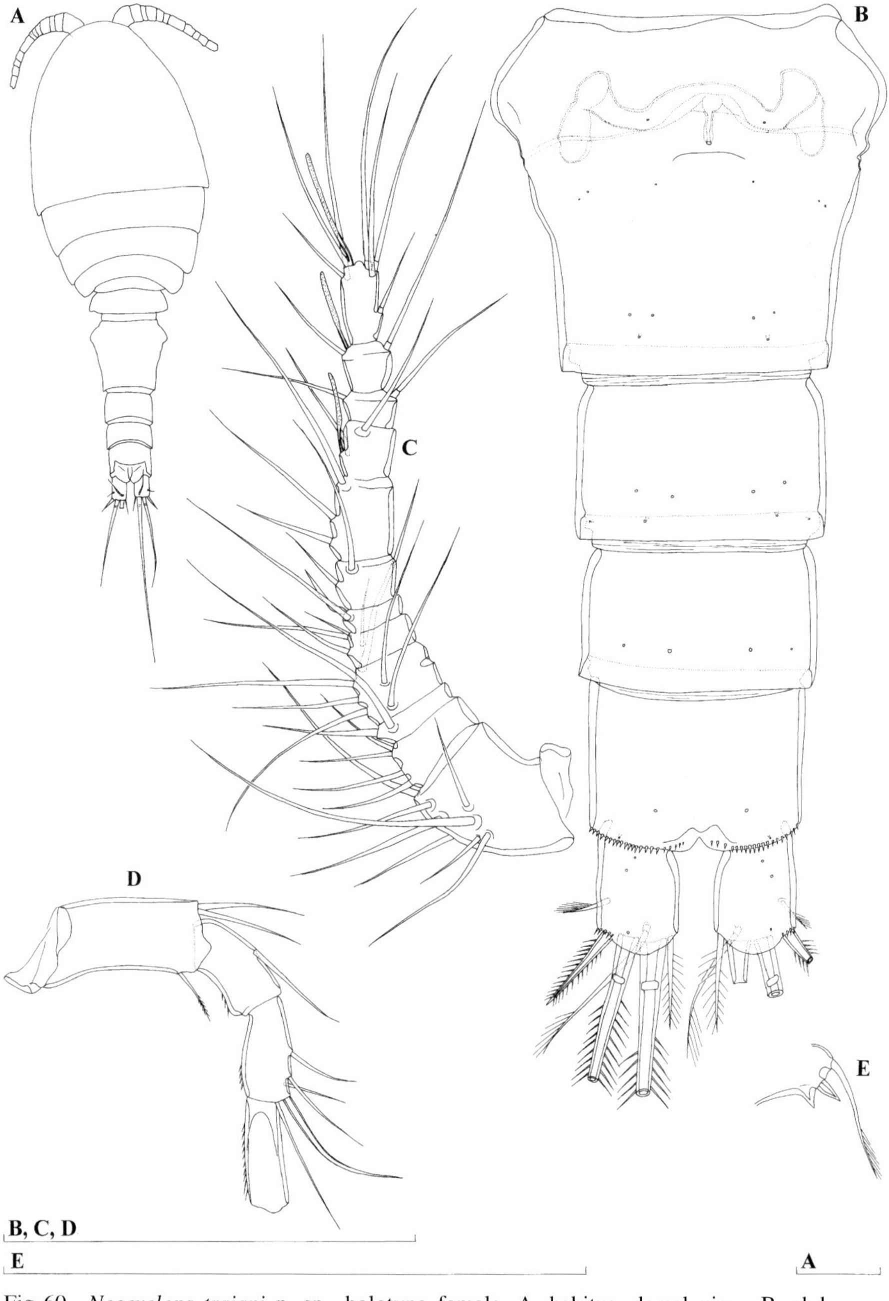

Fig. 60. *Neocyclops trajani* n. sp., holotype female. A, habitus, dorsal view; B, abdomen, ventral view; C, antennula; D, antenna; E, sixth leg. Scales = 0.1 mm.

dorsal pores; the hyaline fringes of both somites smooth and narrow. Anal somite (figs. 60B, 61A) with smooth, broad, and convex anal operculum, which represents 73% of the somite's width and reaches to only 3/5 of the somite's length; ornamented with two large sensilla and two pores dorsally, one pore laterally on each side, four pores ventrally, and a transverse row of spinules on posterior margin. Anal sinus widely open dorsally, ornamented with two diagonal (one on each side) and one arched transversal (just below anal operculum) rows of minute spinules (fig. 61A).

Caudal rami (figs. 60B, 61A) cylindrical but very short, 1.3 times as long as wide (ventral view), parallel, with space between them about one half of a ramus' width; they represent less than 5% of total body length. Distal margin ventrally smooth and convex, with medial cuticular pore hardly visible. Ornamentation consisting of several small spinules at base of outermost apical seta and three ventral, two lateral, and three dorsal pores. Dorsal seta almost 1.5 times as long as ramus, inserted in small recess close to posterior margin of ramus, uniarticulate at base and unipinnate distally. Lateral seta the smallest, arising at 2/3, positioned dorsolaterally, uniplumose, and about 0.5 times as long as a ramus' width. Outermost apical seta stout and spiniform, bipinnate, slightly shorter than ramus, and about as long as innermost apical seta, which is also bipinnate but more slender than outermost one. Principal apical setae with breaking planes and strongly bipinnate; inner seta about 1.8 times as long as outer one and 0.3 times as long as body.

Antennula (fig. 60C) slender, hardly reaching to half of cephalothorax, 12-segmented, unornamented, with one slender aesthetasc on ninth, one on eleventh, and one on twelfth segment, and setal formula: 8.4.2.6.4.2.2.3.2.2.2.7. Atavistic remnant of articulation present caudally on fourth segment. No setae with breaking planes and only one seta on tenth, one on eleventh, and three on twelfth segment articulating on basal part. One seta on sixth segment spiniform but not short; all other setae slender and smooth. Length ratio of antennular segments, from proximal end and along caudal margins, 1 : 0.5 : 0.2 : 0.7 : 0.4 : 0.3 : 0.6 : 1.1 : 0.7 : 0.5 : 0.6 : 1.1.

Antenna (fig. 60D) four-segmented, comprising coxobasis and a three-segmented endopod; all endopodal segments of about the same length, coxobasis longer. Coxobasis about 2.3 times as long as wide, unornamented, and armed with two smooth, subequal setae at distal inner corner, and a bipinnate short seta representing the exopod; exopodal seta 0.3 times as long as coxobasis and half as long as the other two coxobasal setae. First endopodal segment armed with one smooth seta at middle, ornamented with

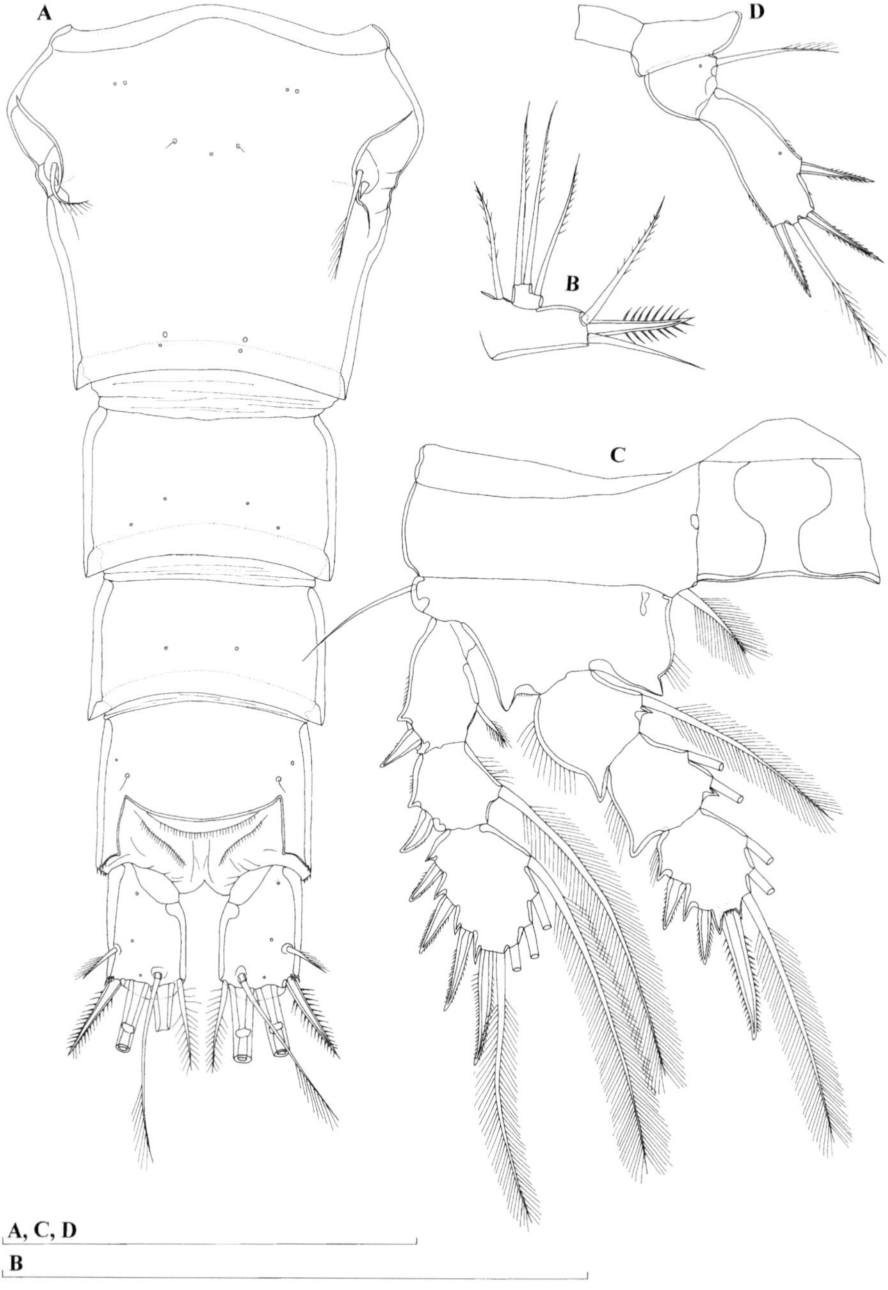

Fig. 61. *Neocyclops trajani* n. sp., holotype female. A, abdomen, dorsal view; B, palp of maxillula; C, third swimming leg; D, fifth leg. Scales = 0.1 mm.

a longitudinal short row of spinules along external margin. Second endopodal segment about twice as long as wide, ornamented with longitudinal row of spinules along external margin, armed with five smooth setae (three lateral, two subapical; one subapical seta much stronger than any other seta on that segment). Third endopodal segment 2.7 times as long as wide, armed with seven smooth apical setae (four of which strong, geniculate, and with breaking planes), and ornamented as previous segment.

Labrum trapezoidal, with very broad cutting edge, ornamented with two diagonal rows of 13 long and slender spinules on ventral surface, with two groups of smaller spinules on dorsal surface, and with a finely serrated distolateral margin. Cutting edge slightly concave, with nine strong and sharp teeth (outermost ones strongest).

Mandibula with palp reduced to a small protuberance, which bears two minute and one much longer smooth setae (but longest seta not reaching base of gnathobasal teeth). Cutting edge of coxal gnathobase clearly divided into three regions: inner group of four complex smooth teeth, innermost tooth strongest, outermost sharpest; middle group of six simple smooth teeth and four spinules; outer group of one strongly unipinnate tooth and outermost unipinnate seta. Gnathobasal seta somewhat shorter than longest seta on reduced palp.

Maxillula (fig. 61B) composed of well developed praecoxa and two-segmented palp. Arthrite of praecoxa with four apical, very strong spines; three of these basally fused to arthrite, geniculate, and strongly unipinnate; also an unfused spine, not geniculate and bipinnate. Praecoxa armed with seven armature elements along inner margin, longest one plumose. Palp with completely divided endopod, which bears two apical and one lateral unipinnate setae, and is armed laterally with one exopodal bipinnate seta, and apically with two slender setae (one bipinnate, other smooth) and one robust bipinnate spine.

Maxilla four-segmented but praecoxa fused to coxa on posterior surface. Proximal endite of praecoxa armed with two setae (one plumose, other pinnate distally); distal endite very small, unarmed. Proximal endite of coxa with one short pinnate seta; distal endite highly mobile and bearing one proximal, completely fused, and very stout seta, and one distal, smaller seta. Seta fused to endite ornamented with several elongated spinules. Basis expanded into spinulose claw and armed with two setae; strong seta slightly longer than the claw. Endopod armed with three spinulose and unipinnate, and two slender and smooth setae. Proportions of armature elements very similar to those of two previous species.

Maxilliped also very similar to that of two previous species.

All swimming legs (figs. 61C, 62A) with three-segmented exopod and endopod, but articulation between second and third endopodal segments of fourth leg incomplete; their armature formula as follows (legend: inner/outer spine or seta; inner/terminal/outer):

	Exopod			Endopod		
Segments	1	2	3	1	2	3
First leg	1/1	1/1	3/2/3	1/0	2/0	3/2/1
Second leg	1/1	1/1	4/2/3	1/0	2/0	3/2/1
Third leg	1/1	1/1	4/2/3	1/0	2/0	3/2/1
Fourth leg	1/1	1/1	4/2/2	1/0	2/0	2/2/1

Third exopodal segment spine formula 3.4.4.3 and third endopodal segment spine formula 2.3.3.3. All intercoxal sclerites without ornamentation and with rounded protrusions at both sides of distal margin; that of first leg most pronounced. Coxae also unornamented, armed with plumose seta at inner distal corner. Spine inserted at inner, protruded corner of basis of first leg reaching halfway second endopodal segment, stout and bipinnate. Basis of all legs with a row of hair-like spinules along inner margin and with a few minute spinules along distal margin close to endopod; inner distal corners of second, third, and fourth leg pronounced as blunt spiniform processes; each basis armed with a slender smooth seta on outer margin. All exopodal and endopodal setae slender and plumose; outer apical seta on third exopodal segment of first leg pinnate along outer margin, plumose along inner; no modified setae on fourth leg. All exopodal and endopodal spines stout and bipinnate. All endopodal segments with hair-like spinules along outer margins, which are also present on inner margin of second exopodal segment of all legs. Cuticular pores visible only on fourth leg, exopod and endopod. Third endopodal segment of fourth swimming leg about as long as wide; inner apical spine about 1.8 times as long as outer one.

Fifth leg (fig. 61D) with intercoxal sclerite, three-segmented, comprising coxa, basis, and a one-segmented exopod. Intercoxal sclerite quadrangular, with almost straight distal margin, almost twice as wide as long, and unornamented. Coxa unarmed and unornamented, 2.4 times as wide as long. Basis nearly triangular, armed with a single unipinnate seta on short outer margin; ornamented with cuticular pore at base of seta; inner margin much longer, rounded, and unornamented; basis about as long as its greatest width. Exopod 2.4 times as long as its greatest width, with somewhat narrower basal part, armed with three spines and a single seta, and ornamented with one pore at base of proximal spine and several spinules along outer and inner margins distally; seta plumose, positioned centrally and as the most apical element, twice as long as inner apical spine, about 1.8 times as long as outer apical spine, and 2.1 times as long as lateral spine; all spines stout and bipinnate but without swollen basal part.

Sixth leg (fig. 60E) not clearly distinct, triangular plate, armed with two almost equal, smooth spines (inner one fused to plate) and one unipinnate, much longer seta.

Male (allotype). Body length, excluding caudal setae, 0.426 mm. Habitus (fig. 62B) slender, with prosome/urosome ratio of 1.3 and greatest width near posterior end of cephalothorax. Body length/width ratio about 3; cephalothorax 2.1 times as wide as genital somite.

Cephalothorax (fig. 62B) slightly longer than its greatest width, with posterior corners not pronounced laterally. Ornamentation of prosomites, colour, and nauplius eye similar to those of female. Hyaline fringes of all somites smooth.

Genital somite 1.7 times as wide as long (dorsal view), ornamented with four pores dorsally, two pores ventrally, and five pores laterally (fig. 62C); third and fourth urosomites ornamented with two pores ventrally, two dorsally, and four laterally on each side; fifth urosomite with two dorsal pores and two lateral pores on each side. Anal somite similar to that of female.

Caudal rami (fig. 62C) about 1.4 times as long as wide, parallel, with space between them less than half of a ramus' width. Armature and ornamentation similar as in female.

Antennula (fig. 62D) slightly shorter than cephalothorax, 14-segmented, geniculate, with geniculation between twelfth and thirteenth segments. Aesthetasc formula: 3.0.0.0.1.0.0.1.0.0.0.0.0.2. All aesthetascs slender and not very long; that on eighth segment longest. Setal formula: 8.4.4.0.1.4.1.4.4.2.2.1.1.10. All setae on tenth, eleventh, and twelfth segments very short and spiniform. Only two setae on eleventh and one seta on tenth segment pinnate; all other setae smooth.

Antenna (fig. 62E) with exopodal seta on coxobasis much longer than in female: about 2.2 times as long as other two coxobasal setae and 1.2 times as long as coxobasis; otherwise similar to that of female.

Labrum, mandibula, maxillula, maxilla, maxilliped, and swimming legs similar to those of female.

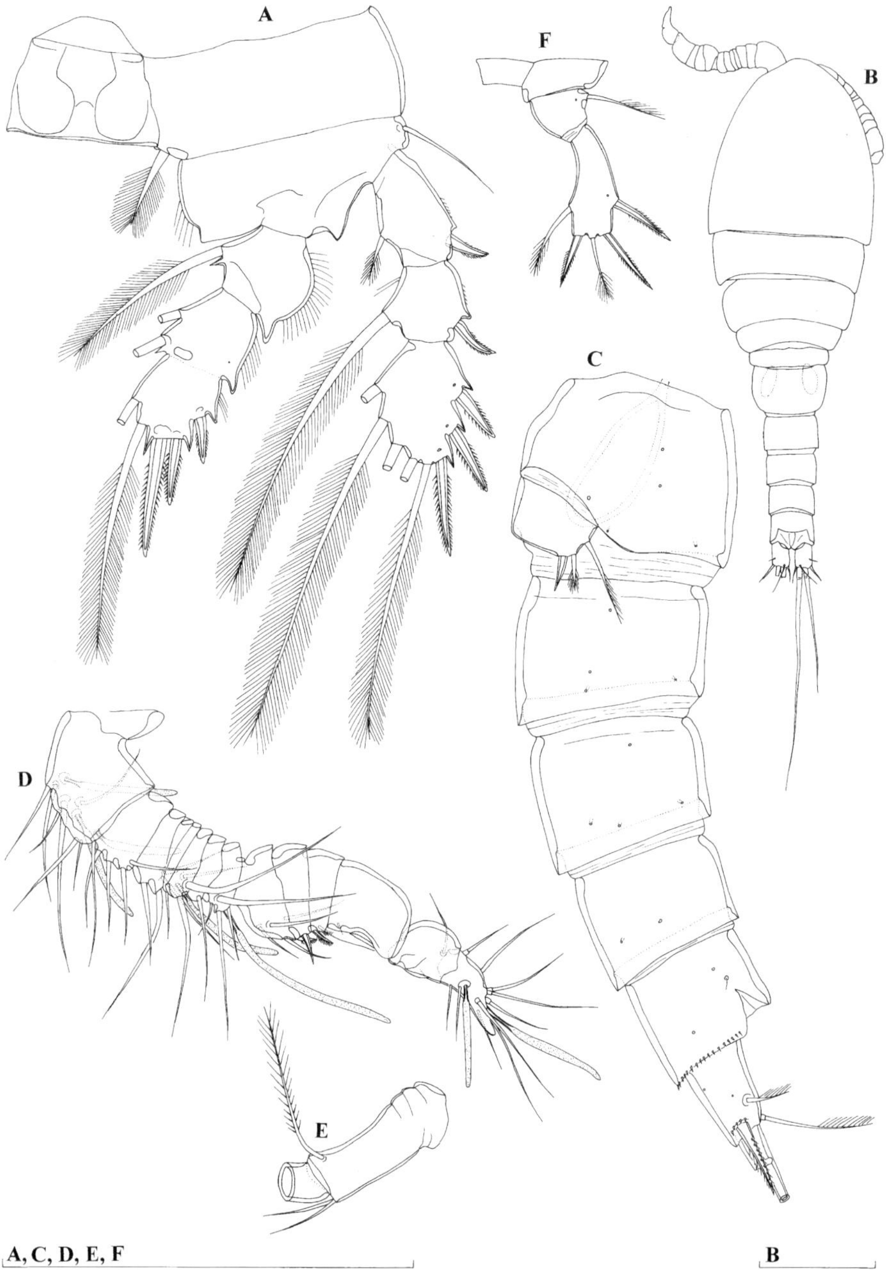

Fig. 62. *Neocyclops trajani* n. sp., A, holotype female; B-F, allotype male. A, fourth swimming leg; B, habitus, dorsal view; C, abdomen, lateral view; D, antennula; E, basis of antenna; F, fifth leg. Scales = 0.1 mm.

Fifth leg (fig. 62F) with a one-segmented exopod (without any traces of subdivision), but with additional seta on its inner margin; intercoxal sclerite narrower and apical exopodal seta much shorter than in female, about as long as outermost apical spine; both inner and outer exopodal margins smooth. Basis similar to that of female, as is the ratio of the exopodal spines. Exopod about 2.3 times as long as its greatest width.

Sixth leg (fig. 62C) a distinct, large plate, armed with one smooth spine and two pinnate setae, unornamented. Outermost seta 2.5 times as long as middle seta and innermost spine.

Variability. — Body length of females ranging from 0.462 mm to 0.573 mm (0.523 mm average; n = 11), while in males it ranges from 0.426 mm to 0.435 mm (0.431 mm average; n = 2). The atavistic remnant of articulation caudally on the fourth antennular segment is present in the holotype female on both sides, but most other females are without this feature. No other type of variability was observed.

Etymology. — This species is named in honour of Dr. Trajan K. Petkovski, curator emeritus of the Museum of Natural History, Skopje, Macedonia, as a recognition of his outstanding contribution to the taxonomy and systematics of the genus *Neocyclops*. The specific name is a noun in the genitive singular.

Distribution. — *Neocyclops trajani* n. sp., just like the previously described *N. sharkbayensis* n. sp., could be found in Shark Bay, Western Australia, and is also found some 80 km north of Shark Bay, in Carnarvon. In Shark Bay it lives sympatrically with *N. sharkbayensis* on both beaches they were recorded from, while on one of those beaches these two species also live together with *N. australiensis* n. sp.

Discussion. — *Neocyclops trajani* n. sp. could be distinguished from its five Australian congeners, as well as from all other described species of *Neocyclops*, by its very short caudal rami, which are only 1.3 times as long as wide. It is most closely similar to *N. tropicus* n. sp. (see below), which also has short caudal rami (1.5 times as long as wide), and a three-segmented male fifth leg. However, the caudal rami of the latter species are always constricted at the base, ornamented dorsally with several large spinules, and bear a much longer dorsal seta. The two species also differ in the cuticular pore pattern and armature of the fourth and fifth legs, but these characters can be very variable in *N. tropicus* and are not always reliable. All other species of *Neocyclops* have the caudal rami index between 1.7 and 3.5, but most have this index > 2.

22. **Neocyclops tropicus** n. sp. (figs. 63-66)

Material examined. — Types: Holotype female dissected on 1 slide (WAM C37271); paratypes, 11 females + 4 copepodids (1 female dissected on 1 slide (WAM C37272), 2 females on 1 slide in toto (TMAG G5788), others in alcohol (WAM C37273)): Australia, Western Australia, Dampier, 4 August 2003, leg. T. Karanovic, 20°40.051′E 116°42.244′E.

Other material: Australia, Western Australia, Exmouth, 3 August 2003, leg. T. Karanovic, 21°56.726′S 114°08.383′E: 1 male + 2 females + 1 copepodid (1 female dissected on one slide (WAM C37310); others in alcohol (WAM C37274)).

Australia, Western Australia, Exmouth Gulf, Pebble Beach, 3 August 2003, leg. T. Karanovic, 22°02.479′S 114°06.889′E: 1 male + 1 female + 1 copepodid (female dissected on 1 slide (WAM C37275), others in alcohol (WAM C37276)).

Australia, Western Australia, Karratha, jetty, 4 August 2003, leg. T. Karanovic, 20°43.298′S 116°53.458′E: 4 females (1 ovigerous) + 1 copepodid in alcohol (WAM C37277).

Australia, Western Australia, Point Samson, 4 August 2003, leg. T. Karanovic, 20°37.516′S 117°11.795′E: 1 female in alcohol (WAM C37278).

Australia, Western Australia, Port Hedland, town beach, 4 August 2003, leg. T. Karanovic, 20°18.553′S 118°35.359′E: 11 females (2 ovigerous) (2 females on 1 slide in toto (TMAG G5789), others in alcohol (WAM C37279)).

Australia, Northern Territory, Dundee Beach, 7 August 2003, leg. T. Karanovic, 12°46.555′S 130°21.377′E: 1 female dissected on 1 slide (WAM C37280) + 1 copepodid in alcohol (WAM C37281).

Australia, Northern Territory, Darwin, Mindil Beach, 7 August 2003, leg. T. Karanovic, 12°26.754′S 130°49.856′E: 1 female in alcohol (WAM C37282).

Australia, Queensland, Alva (near Ayr), 12 August 2003, leg. T. Karanovic, 19°27.232′S 147°28.924′E: 1 female dissected on 1 slide (WAM C37283).

Australia, Queensland, Midge Point (near Proserpine), 13 August 2003, leg. T. Karanovic, 20°38.983′S 148°43.338′E: 1 copepodid in alcohol (WAM C37284).

Australia, Queensland, Yeppoon, 13 August 2003, leg. T. Karanovic, 23°08.582′S 150°45.659′E: 7 females (1 ovigerous) (1 female dissected on 1 slide (WAM C37285), others in alcohol (WAM C37286)).

Australia, Queensland, Hervey Bay, 14 August 2003, leg. T. Karanovic, 25°16.728′S 152°50.400′E: 1 male + 5 females (4 ovigerous) + 1 copepodid (1 male (WAM C37287) and 1 female (WAM C37288) dissected on 1 slide each, others in alcohol (WAM C37289)).

Description. — Female (holotype). Body length, excluding caudal setae, 0.546 mm. Habitus (fig. 63A) slender, with prosome/urosome ratio of 1.2 and greatest width at posterior end of cephalothorax. Body length/width ratio about 3.1; cephalothorax about 2.1 times as wide as genital double somite. Posterior corners of cephalothorax somewhat expanded laterally; other free pedigerous somites without pronounced lateral expansions. Fifth pedigerous somite narrower than genital double somite. Colour of preserved specimen milky white. Nauplius eye not visible. Rostrum broadly rounded and furnished with two large sensilla.

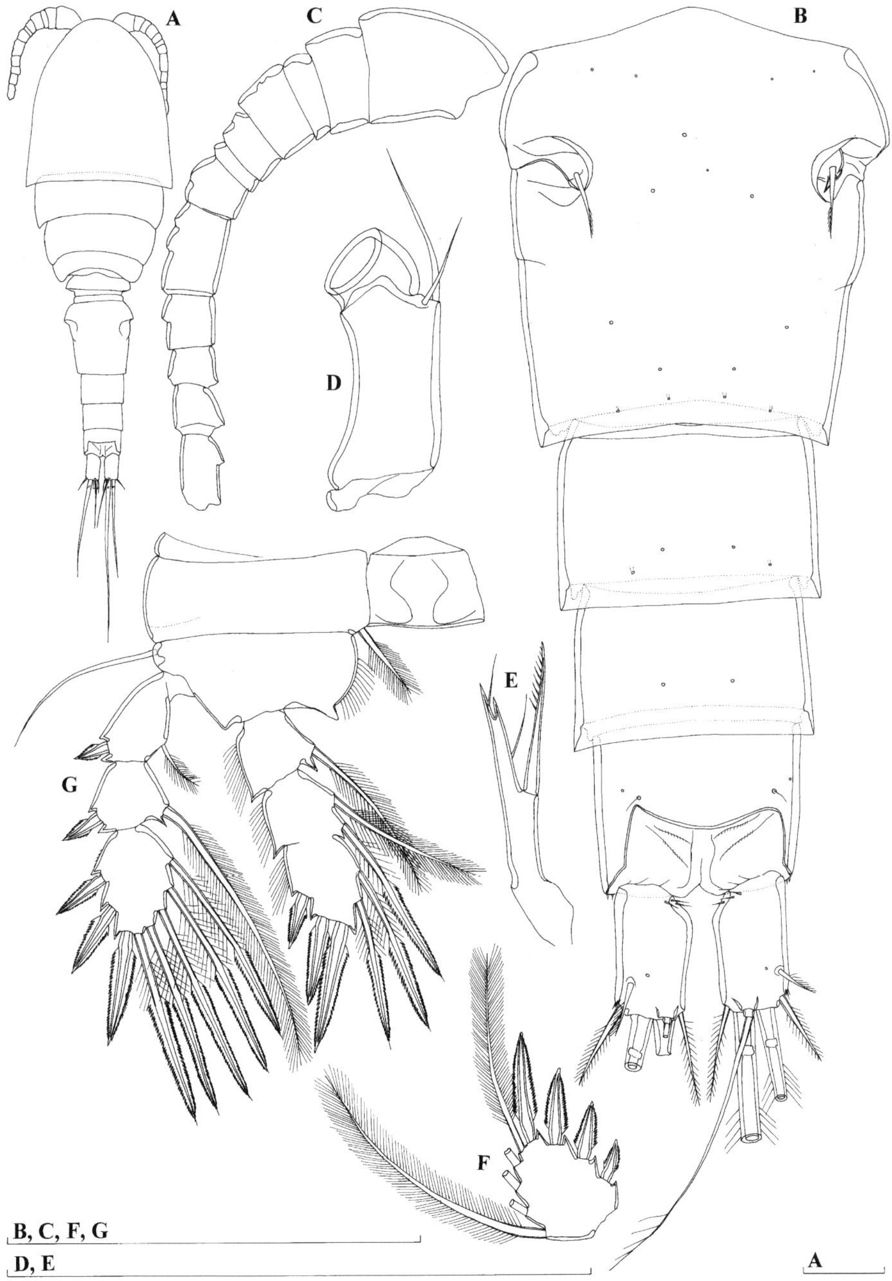

Fig. 63. *Neocyclops tropicus* n. sp., holotype female. A, habitus, dorsal view; B, abdomen, dorsal view; C, antennula without armature; D, basis of antenna; E, distal coxal endite of maxilla; F, third exopodal segment of third swimming leg; G, fourth swimming leg. Scales = 0.1 mm.

Cephalothorax (fig. 63A) 1.1 times as long as its greatest width, representing 35% of total body length. Surface of cephalic shield with several large sensilla, as are surfaces of other prosomites; no other ornamentation visible. Hyaline fringe of prosomites narrow and smooth. First pedigerous somite completely fused to cephalosome. Sclerotized joint between fifth pedigerous somite and genital double somite well developed, almost as a pseudosomite.

Genital double somite (figs. 63B, 64A) with small lateral expansions (but without spiniform projections) at first third, about as long as wide (ventral view), ornamented with characteristic pattern of five pores ventrally and 16 pores dorsally. Hyaline fringe of genital double somite and two succeeding somites narrow, not frilled, and completely smooth. No cuticular recesses, windows, or sensilla visible. Copulatory pore small (but larger than in previous species), ovoid, situated anterior to 1/3 of somite's length; copulatory duct straight, not so narrow, and rigidly sclerotized, directed anteriorly. Seminal receptacle (fig. 64A) complex, curved and tubular, not wholly situated anteriorly of copulatory pore; the central part connected to the copulatory duct is arched, positioned close to ventral cuticular surface, with its lateral parts positioned posteriorly of copulatory pore; constricted part on each side about 0.3 times as wide as central part and turning backwards (dorsally); lateral arms expanded, somewhat wider than central part, and connected to dorsolateral genital areas through another constriction. Ovipores (fig. 63B) situated somewhat dorsolaterally, covered by the reduced sixth legs, bearing two minute spines and one pinnate seta. Third urosomite ornamented with four ventral and four dorsal pores in posterior half, while fourth urosomite with only two ventral and two dorsal pores. Anal somite (figs. 63B, 64A) with smooth, broad, and convex anal operculum, which represents 70% of the somite's width and reaches to only 3/5 of the somite's length; ornamented with two large sensilla and two pores dorsally, one pore laterally on each side, four pores ventrally, and a transverse row of spinules on the posterior margin. Anal sinus widely open dorsally, ornamented with four diagonal (two on each side) rows of minute spinules (fig. 63B).

Caudal rami (figs. 63B, 64A) cylindrical, but with constricted base, 1.5 times as long as wide (ventral view), divergent, with space between them about one ramus' width; they represent 6% of total body length. Distal margin ventrally convex, ornamented with a row of minute spinules, with medial cuticular pore hardly visible. The additional ornamentation consists of several spinules at base of outermost apical seta, three larger spinules at proximal part of ramus dorsally, and two ventral, one lateral, and one dorsal pore. Dor-

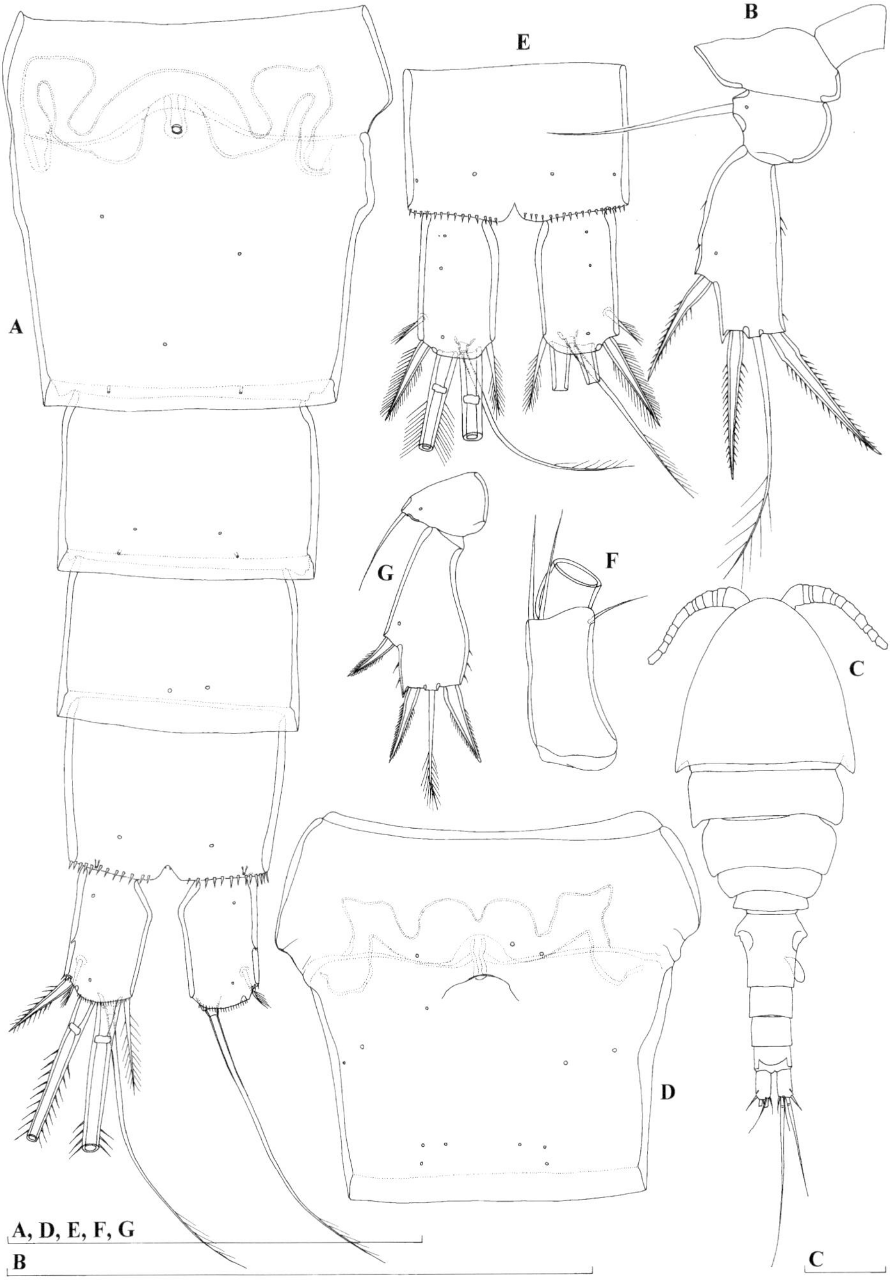

Fig. 64. *Neocyclops tropicus* n. sp., A-B, holotype female; C-G, female from Alva (0.592 mm). A, abdomen, ventral view; B, fifth leg; C, habitus, dorsal view; D, genital double somite, ventral view; E, anal somite and caudal rami, ventral view; F, basis of antenna; G, fifth leg. Scales = 0.1 mm.

sal seta very long, almost 2.4 times as long as ramus and half as long as outer principal apical seta, inserted on a small protuberance that reaches posterior border of ramus, uniarticulate at base, and unipinnate distally. Lateral seta smallest, arising at 3/4, positioned dorsolaterally, uniplumose, and about 0.7 times as long as a ramus' width. Outermost apical seta stout and spiniform, bipinnate, about 0.6 times as long as ramus, and 0.8 times as long as innermost apical seta, which is also bipinnate, but more slender than outermost one. Principal apical setae with breaking planes and strongly bipinnate; inner seta about 1.7 times as long as outer one and 0.35 times as long as body.

Antennula (fig. 63C) not very slender, reaching to 2/3 of cephalothorax, 12-segmented, unornamented, with one slender aesthetasc on ninth, one on eleventh, and one on twelfth segment, and setal formula: 8.4.2.6.4.2.2.3.2.2.2.7. No setae with breaking planes and only one seta on tenth, one on eleventh, and three on twelfth segment articulating on basal part. One seta on sixth segment spiniform and very short; all other setae slender and all setae smooth. Length ratio of antennular segments, from proximal end and along caudal margins, 1 : 0.5 : 0.2 : 0.5 : 0.3 : 0.2 : 0.4 : 0.9 : 0.6 : 0.5 : 0.6 : 0.8.

Antenna (fig. 53C) four-segmented, comprising coxobasis and a three-segmented endopod; all endopodal segments of about the same length, coxobasis somewhat longer. Coxobasis twice as long as wide, unornamented, and armed with two smooth setae at distal inner corner, inner seta 1.7 times as long as outer one; seta representing exopod absent. First endopodal segment armed with one smooth seta at middle, and ornamented with a longitudinal short row of long spinules along external margin. Second endopodal segment about 2.5 times as long as wide, ornamented with longitudinal row of spinules along external margin, and armed with five smooth setae (three lateral, two subapical; one subapical seta somewhat longer and much stronger than any other seta on that segment). Third endopodal segment 2.8 times as long as wide, armed with seven smooth apical setae (four of which strong, geniculate, and with breaking planes) and ornamented as previous segment.

Labrum trapezoidal, but with very broad cutting edge and distolateral corners almost on the same plane, ornamented with two diagonal rows of 12 long and slender spinules on ventral surface and with a finely serrated distolateral margin. Cutting edge slightly concave, with nine strong and sharp teeth (outermost ones strongest). No other ornamentation visible.

Mandibula with palp reduced to a small protuberance, which bears two minute and one much longer smooth setae (but longest seta not reaching base of gnathobasal teeth). Cutting edge of coxal gnathobase clearly divided into three regions: inner group of four complex smooth teeth, innermost tooth strongest, outermost sharpest; middle group of six simple smooth teeth and four spinules; outer group of one strongly unipinnate tooth and outermost unipinnate seta. Gnathobasal seta somewhat shorter than longest seta on reduced palp.

Maxillula composed of well developed praecoxa and two-segmented palp. Arthrite of praecoxa with four apical, very strong spines; three of these basally fused to arthrite, geniculate, and strongly unipinnate; the unfused spine not geniculate, but bipinnate. Praecoxa armed with seven armature elements along inner margin, the longest one plumose. Palp with completely divided, small endopod, which bears two apical and one lateral pinnate setae, and is armed laterally with one smooth exopodal seta and apically with two slender and smooth setae and one robust unipinnate spine.

Maxilla (fig. 63E) four-segmented but praecoxa fused to coxa on posterior surface. Proximal endite of praecoxa armed with two setae (one plumose, the other pinnate distally); distal endite very small, unarmed. Proximal endite of coxa with one short pinnate seta; distal endite highly mobile and bearing one proximal, completely fused, and very stout seta, and one distal, smaller seta. Seta fused to endite ornamented with two elongated spinules and bifurcate. Basis expanded into a spinulose claw and armed with two setae; strong seta slightly longer than claw. Endopod armed with three spinulose and unipinnate, and two slender and smooth setae. Proportions of armature elements similar to those in three previous species.

Maxilliped also very similar to those of three previous species.

All swimming legs (fig. 63F, G) with three-segmented exopod and endopod, although second and third endopodal segments of fourth leg not completely divided; their armature formula as follows (legend: inner/outer spine or seta; inner/terminal/outer):

	Exopod			Endopod		
Segments	1	2	3	1	2	3
First leg	1/1	1/1	3/2/3	1/0	2/0	3/2/1
Second leg	1/1	1/1	4/2/3	1/0	2/0	3/2/1
Third leg	1/1	1/1	4/2/3	1/0	2/0	3/2/1
Fourth leg	1/1	1/1	4/2/2	1/0	2/0	2/2/1

Third exopodal segment spine formula 3.4.4.3 and third endopodal segment spine formula 2.3.3.3. All intercoxal sclerites without ornamentation and with almost straight distal margin. Coxae also unornamented, armed with a plumose seta at inner distal corner. Spine inserted at inner, protruded corner of basis of first leg reaching halfway second endopodal segment, stout and bipinnate. Basis of each leg with a row of hair-like spinules along inner margin; inner distal corners of second, third, and fourth leg pronounced as blunt spiniform processes; each basis armed with a slender smooth seta on outer margin. Most exopodal and endopodal setae slender and plumose; outer apical seta on third exopodal segment of first leg pinnate along outer margin, plumose along inner; all setae on third endopodal and exopodal segments and distal seta on second endopodal segment of fourth leg modified into very strong, spiniform armature elements, with a few hair-like spinules at the base, and sharply serrated distal lateral margins (fig. 63G). All exopodal and endopodal spines stout and with sharply serrated margins. All endopodal segments with hair-like spinules along outer margins. Cuticular pores visible only on first leg exopod and endopod. Third endopodal segment of fourth swimming leg about 1.2 times as long as wide; inner apical spine about 1.5 times as long as outer one.

Fifth leg (fig. 64B) with intercoxal sclerite and three-segmented, comprising coxa, basis, and a one-segmented exopod. Intercoxal sclerite trapezoidal, with slightly concave distal margin, almost twice as wide as long, and unornamented. Coxa unarmed and unornamented, about twice as wide as long. Basis about 1.5 times as wide as its greatest length, armed with a single smooth seta on outer margin, ornamented with only a cuticular pore at base of seta. Exopod 2.1 times as long as its greatest width, much narrower at basal part, armed with three spines and a single seta, and ornamented with one pore at base of proximal spine, three spinules along inner margin, and two spinules on outer margin; seta plumose, positioned centrally and as the most apical element, 1.5 times as long as inner apical spine, about 1.7 times as long as outer apical spine, and twice as long as lateral spine; all spines very strongly bipinnate, innermost apical one with swollen basal part.

Sixth leg not clearly distinct, triangular plate, armed with two almost equal, smooth spines (inner one fused to plate) and one pinnate, much longer seta.

Male from Hervey Bay. Body length, excluding caudal setae, 0.379 mm. Habitus (fig. 66A) more robust than in female, with prosome/urosome ratio of 1.2, and greatest width at posterior end of cephalothorax. Body length/width ratio about 3.3; cephalothorax 2.2 times as wide as genital somite.

Cephalothorax (fig. 66A) 1.2 times as long as its greatest width, with posterior corners not pronounced laterally. Ornamentation of prosomites, colour,

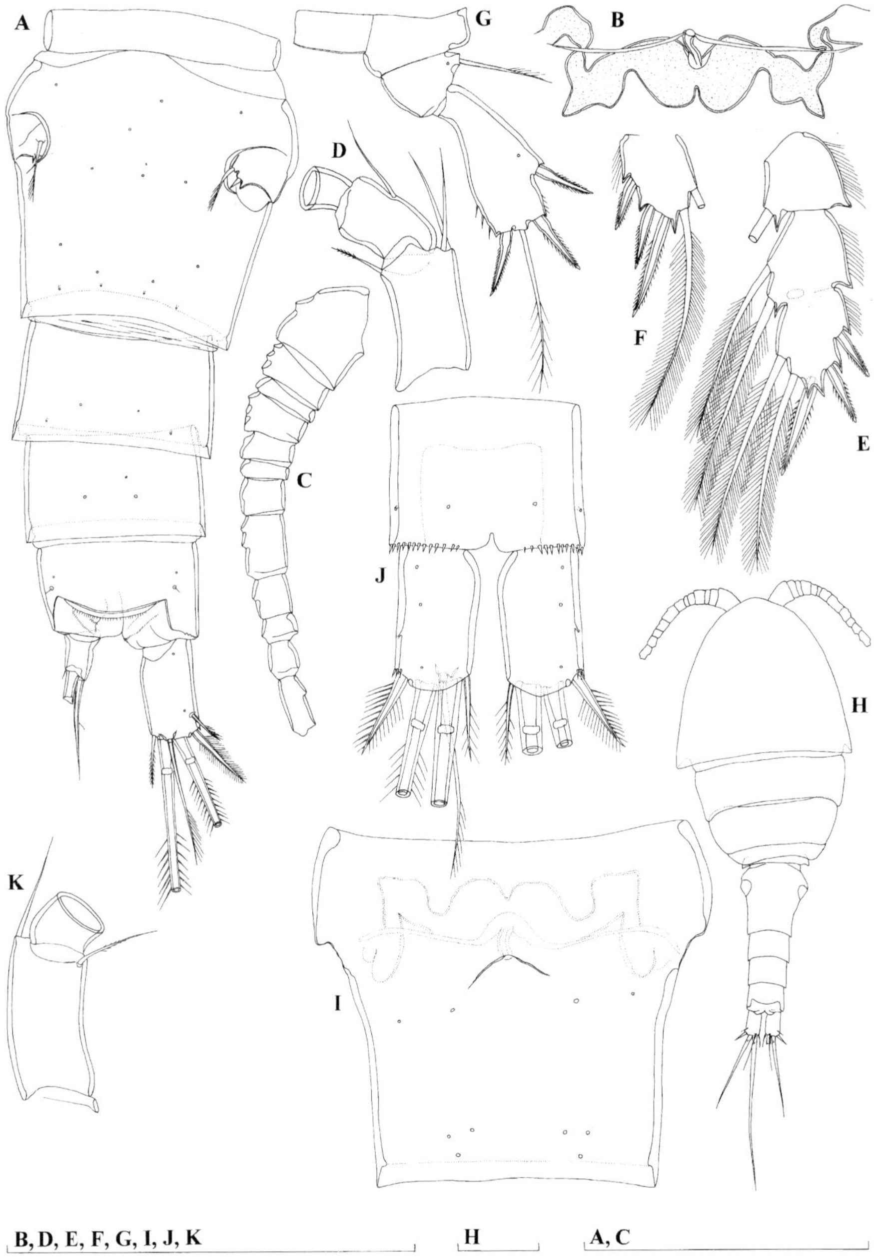

Fig. 65. *Neocyclops tropicus* n. sp., A-G, female from Exmouth (0.532 mm); H, female from Exmouth (0.527 mm); I-K, female from Dundee Beach (0.525 mm). A, abdomen, dorsal view; B, seminal receptacle, ventral view; C, antennula without armature; D, antenna; E, endopod of fourth swimming leg; F, third endopodal segment of fourth swimming leg; G, fifth leg; H, habitus, dorsal view; I, genital double somite, ventral view; J, anal somite and caudal rami, ventral view; K, basis of antenna. Scales = 0.1 mm.

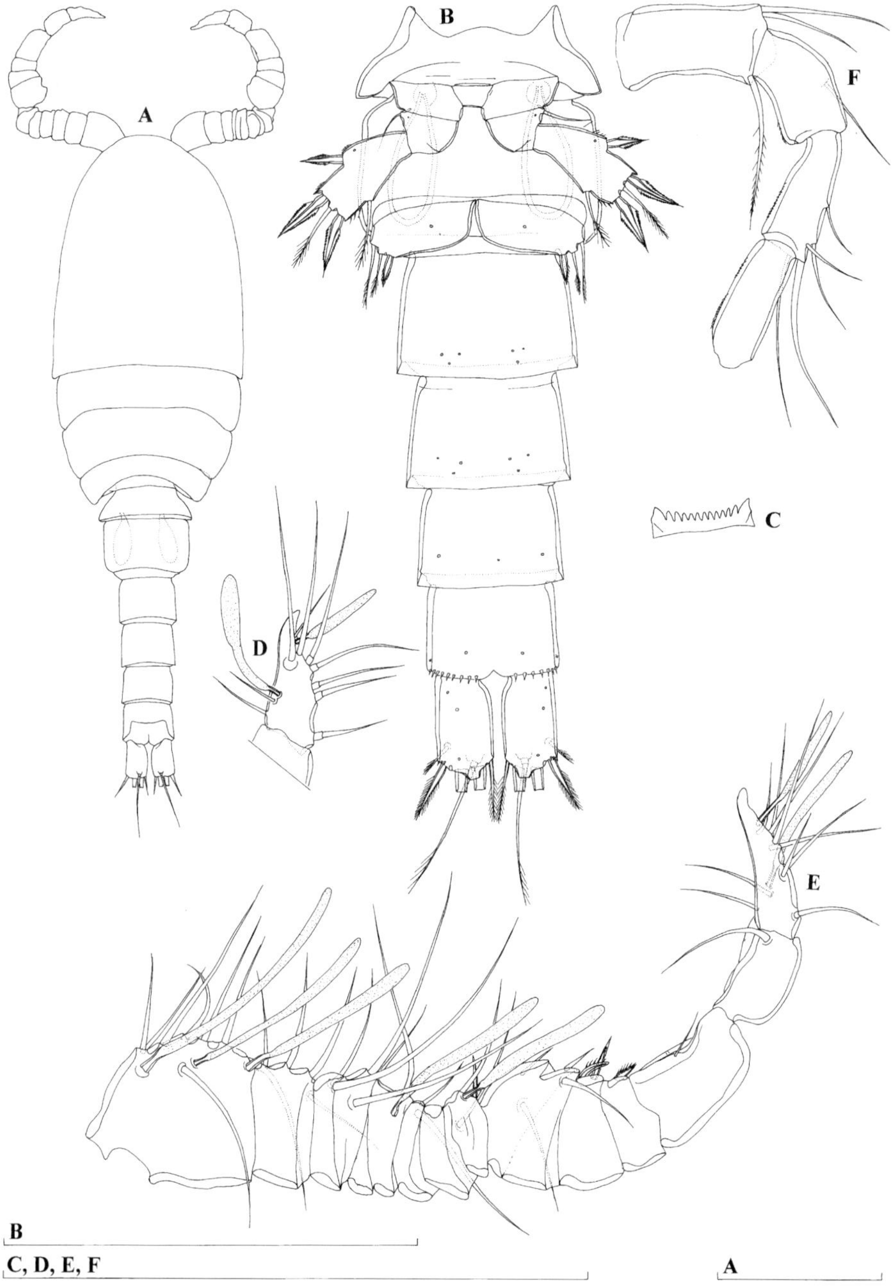

Fig. 66. *Neocyclops tropicus* n. sp., male from Hervey Bay (0.379 mm). A, habitus, dorsal view; B, urosome, ventral view; C, cutting edge of labrum; D, terminal segment of antennula; E, opposite antennula; F, antenna. Scales = 0.1 mm.

and nauplius eye similar to those of female. Hyaline fringes of all somites smooth.

Genital somite (fig. 66B) with two completely formed, elongated spermatophores, 1.7 times as wide as long (dorsal view), ornamented with four pores dorsally and two pores laterally; third and fourth urosomites ornamented with six pores ventrally, two dorsally, and two laterally on each side; fifth urosomite with three ventral, two dorsal, and one lateral pore on each side. Anal somite similar to that of female.

Caudal rami (fig. 66B) about 1.8 times as long as wide, parallel, with space between them less than half of a ramus' width. Armature and ornamentation similar as in female, except additional ventral pore present and medial cuticular pore clearly visible .

Antennula (fig. 66E) slightly shorter than cephalothorax, 14-segmented, geniculate, with geniculation between twelfth and thirteenth segments. Aesthetasc formula: 3.0.0.0.1.0.0.1.0.0.0.0.0.2. All aesthetascs slender and not very long; that on eighth segment longest. Setal formula: 8.4.4.1.1.2.0.3.4.2.1.1.1.10. All setae on tenth and eleventh segments, as well as one seta on eight and one on ninth segment, very short and spiniform. Only one seta on eleventh and one seta on tenth segment pinnate; all other setae smooth.

Antenna (fig. 66F) with exopodal seta on coxobasis present, otherwise similar to that of female.

Labrum (fig. 66C) similar to that of female, but cutting edge with 13 strong and sharp teeth.

Mandibula, maxillula, maxilla, maxilliped, and swimming legs all similar to those in female.

Fifth leg (fig. 66B) with one-segmented exopod (without any trace of subdivision), but with additional seta on its inner margin; intercoxal sclerite narrower and apical exopodal seta much shorter than in female, about as long as outermost apical spine; both inner and outer exopodal margins ornamented with several spinules. Basis similar to that of female, as is the ratio of the exopodal spines. Exopod about twice as long as its greatest width.

Sixth leg (fig. 66B) a distinct, large plate, armed with one short and pinnate spine and two pinnate setae; ornamented with one cuticular pore. Outermost seta 1.4 times as long as middle seta and twice as long as innermost spine.

Variability. — Body length of females ranging from 0.476 mm to 0.592 mm (0.529 mm average; n = 14), while in males it ranges from 0.379 mm to

0.413 mm (0.392 mm average; n = 3). The cuticular pore pattern is variable and sometimes even asymmetrical on the same specimen (figs. 63B, 64A, 65A). The cuticular pore pattern of the genital double somite is very variable both ventrally (figs. 64A, D, 65I) and dorsally (figs. 63B, 65A), although its posterior part seems to be always with eight characteristic pores. The shape of the seminal receptacle can also vary significantly (figs. 64A, D, 65B, I) and in females the habitus can be either more, or less robust (figs. 63A, 64C, 65H). The seta representing the exopod on the antenna can be present or absent in females (figs. 63D, 64F, 65D, K), and even the number of setae on the distal inner corner of the coxobasis can be reduced (fig. 65K). The caudal rami can have a more, or less constricted basal part (figs. 64A, E, 65J). One female from Exmouth has a number of abnormalities, which include the left caudal ramus (fig. 65A), seminal receptacle (fig. 65B), and fourth swimming legs (fig. 65E, F), but all other details, including the right caudal ramus, antennula (fig. 65C), and fifth leg (fig. 65G) are similar to those of the holotype female. The fifth legs vary only slightly in the shape of the exopod (figs. 64B, G, 65G).

Etymology. — The name of this species is the Latin adjective "tropicus", meaning "tropical", and refers to the distribution of this species in Australia.

Distribution. — The species was recorded very often in the tropical part of Australia, from Exmouth in Western Australia to Hervey Bay in Queensland. In the Pilbara region of Western Australia it occurs sympatrically with *N. pilbarensis* n. sp. on all three beaches the latter species was recorded from, and it was also found sympatrically with *N. australiensis* n. sp. at the following four localities: Exmouth in Western Australia; Dundee Beach in the Northern Territory; Yeppoon in Queensland; and Hervey Bay in Queensland. The latter species has a very wide Australian distribution, as mentioned before.

Discussion. — As mentioned in the discussion section of the previous species, *N. tropicus* n. sp. seems to be most closely related to *N. trajani* n. sp. The two species have an allopatric distribution, which further supports their close relationship, and could be reliably distinguished primarily by the shape, armature, and ornamentation of the caudal rami (see above). The only other *Neocyclops* species that is relatively similar to *N. tropicus*, is *N. herbsti* Petkovski, 1986, which was established as a new species for the specimens from the Red Sea mentioned by Herbst (1964) as *N. salinarum* (Gurney). However, *N. herbsti* has a much shorter dorsal seta on the caudal rami and also a four-segmented male fifth leg. Additionally, *N. herbsti* does not have any transformed setae on the fourth leg endopod, but this state is also recorded in some specimens of *N. tropicus* (see fig. 65E, F). The range of variability of *N. tropicus* is much wider than that of the more widely distributed *N. australiensis* n. sp. (or any other Australian species) and is not restricted to one or two characters (see the variability section). Most importantly, the cuticular pore pattern proved to be very variable, suggesting that we have to be very cautious when using this character in species delineation. Da Rocha (1995a) used this character to distinguish his new species from Belize from two closely related but more widely distributed congeners.

23. **Neocyclops pilbarensis** n. sp.
(figs. 67, 68)

Material examined. — Types: Holotype female dissected on 1 slide (WAM C37290); allotype male dissected on 1 slide (WAM C37291); paratypes, 3 females (1 on 1 slide in toto (TMAG G5790), other 2 in alcohol (WAM C37292)): Australia, Western Australia, Dampier, 4 August 2003, leg. T. Karanovic, 20°40.051′S 116°42.244′E.

Other material: Australia, Western Australia, Karratha, near jetty, 4 August 2003, leg. T. Karanovic, 20°43.298′S 116°53.458′E: 3 females in alcohol (WAM C37293).

Australia, Western Australia, Port Hedland, town beach, 4 August 2003, leg. T. Karanovic, 20°18.553′S 118°35.359′E: 4 females (1 on a slide in toto (TMAG G5791), others in alcohol (WAM C37294)).

Description. — Female (holotype). Body length, excluding caudal setae, 0.654 mm. Habitus (fig. 67A) robust, with prosome/urosome ratio of 1.3 and greatest width at posterior end of cephalothorax. Body length/width ratio about 2.9; cephalothorax about 2.3 times as wide as genital double somite. Posterior corners of cephalothorax not expanded laterally; other free pedigerous somites also without pronounced lateral expansions. Fifth pedigerous somite narrower than genital double somite. Colour of preserved specimen milky white. Nauplius eye not visible. Rostrum broadly rounded and furnished with two large sensilla.

Cephalothorax (fig. 67A) 1.2 times as long as its greatest width, representing 41% of total body length. Surface of cephalic shield with several large sensilla; no other ornamentation visible. Hyaline fringe of prosomites narrow and smooth. First pedigerous somite completely fused to cephalosome. Fifth pedigerous somite ornamented with only two large dorsal sensilla, and with a smooth fringe dorsally and ventrally. Sclerotized joint between fifth pedigerous somite and genital double somite well developed, almost as a pseudosomite (fig. 67B).

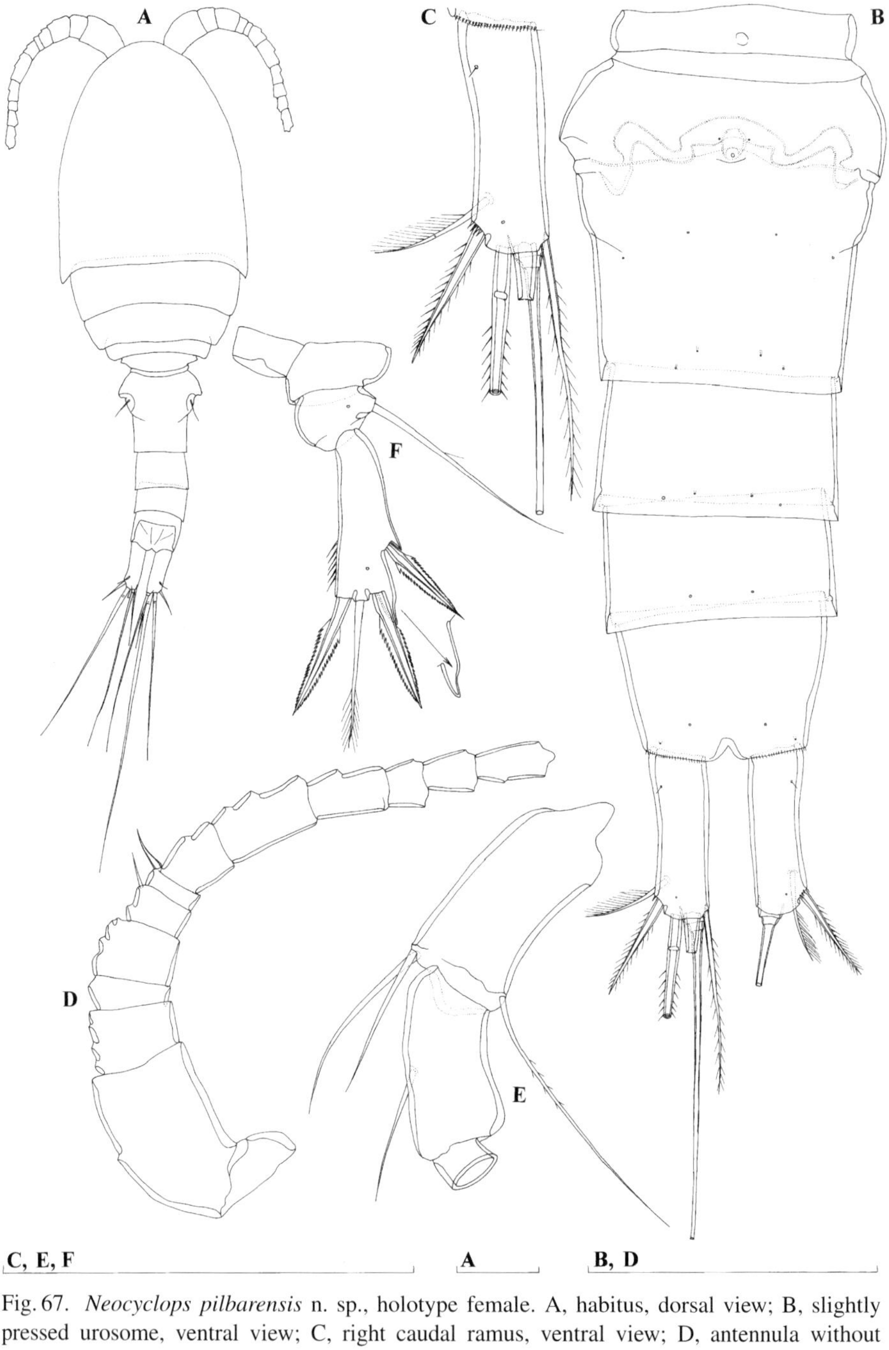

Fig. 67. *Neocyclops pilbarensis* n. sp., holotype female. A, habitus, dorsal view; B, slightly pressed urosome, ventral view; C, right caudal ramus, ventral view; D, antennula without armature; E, antenna; F, fifth leg. Scales = 0.1 mm.

Genital double somite (fig. 67B) with well developed lateral expansions at first third but without spiniform projections, about 0.9 times as long as wide, ornamented with a characteristic pattern of ten pores ventrally (two of these close to and anteriorly of copulatory pore) and 14 pores dorsally. Hyaline fringe of genital double somite and two succeeding somites narrow, not frilled, and completely smooth. No cuticular recesses, windows, or sensilla visible. Copulatory pore small, ovoid, situated just at 1/4 of the somite's length; copulatory duct wide, short, and rigidly sclerotized, directed dorsally and anteriorly, almost ovoid from ventral view. Seminal receptacle (fig. 67B) complex, narrow, curved, and tubular, but wholly situated anteriorly of copulatory pore; the central part connected to the copulatory duct is almost cylindrical, positioned close to ventral cuticular surface, and about four times as wide as long; constricted part on each side only slightly narrower than central part, and turning backwards (dorsally) and posteriorly; lateral arms not much expanded, about as wide as central part and connected to dorsolateral genital areas through another constriction. Ovipores situated somewhat dorsolaterally, covered by the reduced sixth legs, bearing two minute spines and one pinnate seta. Third urosomite ornamented with four ventral, one lateral, and six dorsal pores in posterior half, while fourth urosomite with only two ventral and two dorsal pores. Anal somite (fig. 67B) with smooth, broad, and slightly convex anal operculum, which represents 82% of the somite's width, and reaches to only 2/5 of the somite's length; ornamented with two large sensilla and two pores dorsally, one pore laterally on each side, four pores ventrally, and a transverse row of spinules on the posterior margin. Anal sinus widely open dorsally, ornamented with four diagonal (two on each side) rows of minute spinules.

Caudal rami (fig. 67B, C) cylindrical, about three times as long as wide (ventral view), parallel, with space between them more than one half of a ramus' width; they represent 8% of total body length. Distal margin ventrally smooth and convex, with medial cuticular pore hardly visible. Ornamentation consisting of several small spinules at base of outermost apical seta, one sensillum at anterior part ventrally, and one ventral, one lateral, and three dorsal pores. Dorsal seta very long, almost as long as outer principal apical seta and 3.5 times as long as ramus, inserted on a large protuberance that reaches slightly beyond posterior border of ramus, uniarticulate at base, and unipinnate distally. Lateral seta relatively long, although shortest on ramus, arising at 4/5, positioned dorsolaterally, uniplumose, and about 1.6 times as long as a ramus' width. Outermost apical seta stout and spiniform, bipinnate, about 0.7 times as long as ramus, and 0.6 times as long as innermost apical seta, which is also bipinnate, but more slender than outermost one. Principal apical setae with breaking planes and bipinnate; inner seta about 1.7 times as long as outer one and half as long as body.

Antennula (fig. 67D) slender, almost reaching to 2/3 of cephalothorax, 12-segmented, unornamented, with one slender aesthetasc on ninth, one on eleventh, and one on twelfth segment, and setal formula: 8.4.2.6.4.2.2.3.2.2.2.7. No setae with breaking planes and only one seta on tenth, one on eleventh, and three on twelfth segment articulating on basal part. One seta on sixth segment spiniform, all other setae slender and smooth. Length ratio of antennular segments, from proximal end and along caudal margins, 1 : 0.3 : 0.1 : 0.4 : 0.3 : 0.2 : 0.4 : 0.8 : 0.6 : 0.3 : 0.4 : 0.7.

Antenna (fig. 67E) four-segmented, comprising coxobasis and a three-segmented endopod; all endopodal segments of about the same length, coxobasis somewhat longer. Coxobasis about 2.2 times as long as wide, unornamented, and armed with two smooth setae at distal inner corner and the seta representing the exopod at distal outer corner; exopodal seta 1.5 times as long as longer seta at distal inner corner and 1.2 times as long as coxobasis. First endopodal segment armed with one smooth seta at middle, unornamented. Second endopodal segment about 2.4 times as long as wide, ornamented with longitudinal row of spinules along external margin, armed with five smooth setae (three lateral, two subapical; one subapical seta somewhat longer and much stronger than any other seta on that segment). Third endopodal segment 2.8 times as long as wide, armed with seven smooth apical setae (four of which strong, geniculate, and with breaking planes), and ornamented as previous segment.

Labrum trapezoidal, but with very broad cutting edge and distolateral corners almost on the same plane, ornamented with two diagonal rows of 13 long and slender spinules on ventral surface, and with finely serrated distolateral margin. Cutting edge slightly concave, with eight strong and sharp teeth (outermost ones strongest).

Mandibula, maxillula, maxilla, and maxilliped very similar to those of four preceding species.

All swimming legs with three-segmented exopod and endopod, although second and third endopodal segments of fourth leg not completely divided (fig. 68A); their armature formula as follows (legend: inner/outer spine or seta; inner/terminal/outer):

	Exopod			Endopod		
Segments	1	2	3	1	2	3
First leg	1/1	1/1	3/2/3	1/0	2/0	3/2/1
Second leg	1/1	1/1	4/2/3	1/0	2/0	3/2/1
Third leg	1/1	1/1	4/2/3	1/0	2/0	3/2/1
Fourth leg	1/1	1/1	4/2/2	1/0	2/0	2/2/1

Third exopodal segment spine formula 3.4.4.3 and third endopodal segment spine formula 2.3.3.3. All intercoxal sclerites without ornamentation and with rounded protrusions at both sides of distal margin; that of first leg most pronounced. Coxae ornamented with a few minute spinules along outer margin, and armed with a plumose seta at inner distal corner. Spine inserted at inner, protruded corner of basis of first leg reaching halfway second endopodal segment, stout and bipinnate. Basis of each leg with a row of hair-like spinules along inner margin; inner distal corners of second, third, and fourth leg pronounced as blunt spiniform processes; each basis armed with slender unipinnate seta on outer margin. Most exopodal and endopodal setae slender and plumose; outer apical seta on third exopodal segment of first leg pinnate along outer margin, plumose along inner; all setae on third endopodal and exopodal segments of fourth leg, as well as distal seta on second endopodal segment of the same leg, modified into very strong, spiniform armature elements, with a few hair-like spinules at their base and sharply serrated lateral margins (fig. 68A). All exopodal and endopodal spines stout and with sharply serrated lateral margins. All endopodal segments with hair-like spinules along outer margins; all exopodal segments with minute spinules along outer margins. Cuticular pores visible only on first leg exopod and endopod. Third endopodal segment of fourth swimming leg about 1.5 times as long as wide; inner apical spine longer than the segment and about 1.5 times as long as outer apical spine.

Fifth leg (fig. 67F) with intercoxal sclerite and three-segmented, comprising coxa, basis, and a one-segmented exopod. Intercoxal sclerite quadrangular, with somewhat convex distal margin, about 1.7 times as wide as long, and unornamented. Coxa unarmed and unornamented, almost 1.7 times as wide as long. Basis nearly triangular, about 1.3 times as wide as its greatest length, armed with a single unipinnate seta on short outer margin, and ornamented with a cuticular pore at base of seta. Exopod 2.6 times as long as its greatest width, narrower at basal part, with strongly pronounced distal outer corner, armed with three spines and a single seta, and ornamented with one pore between proximal spine and outer apical spine, and several long

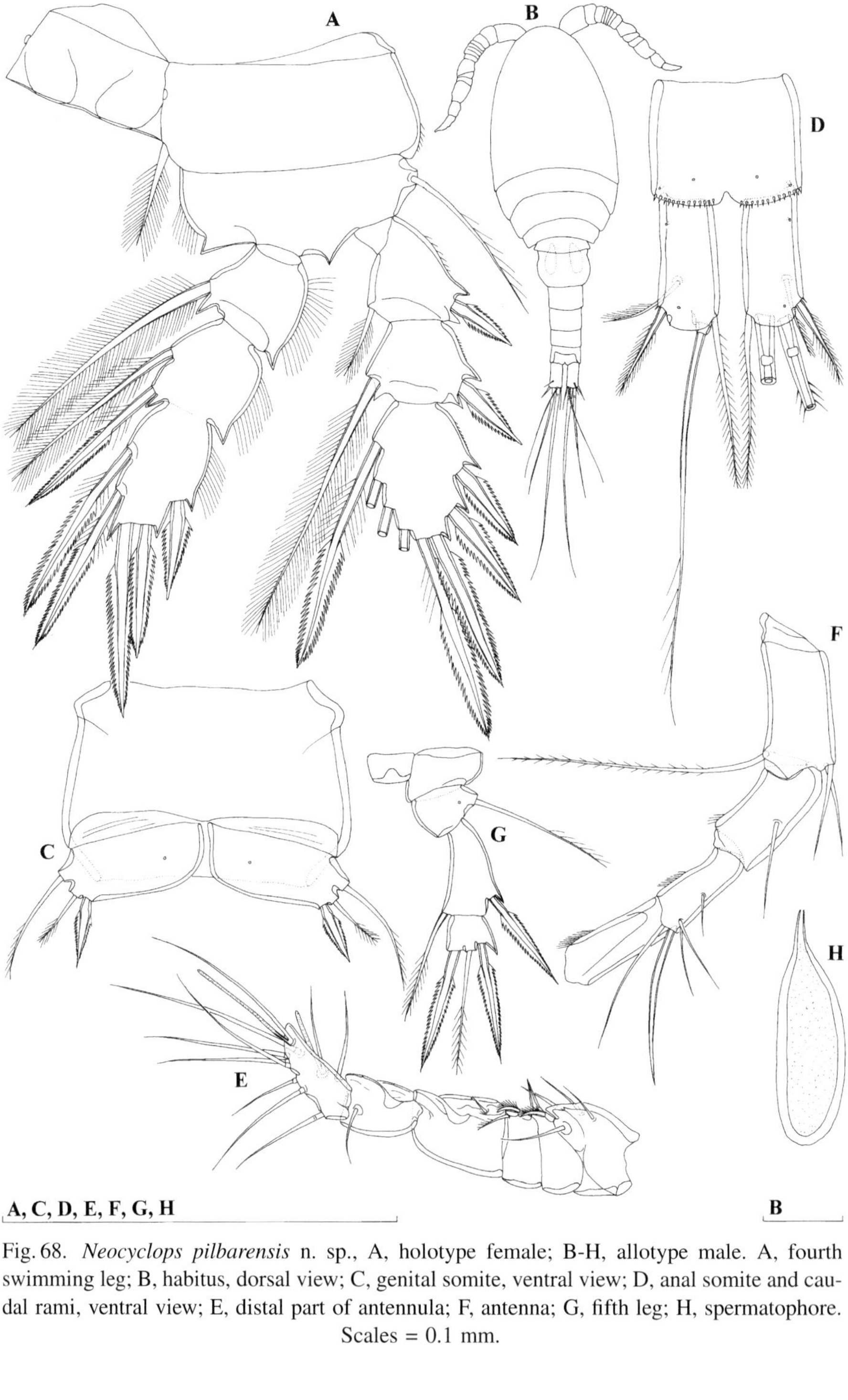

Fig. 68. *Neocyclops pilbarensis* n. sp., A, holotype female; B-H, allotype male. A, fourth swimming leg; B, habitus, dorsal view; C, genital somite, ventral view; D, anal somite and caudal rami, ventral view; E, distal part of antennula; F, antenna; G, fifth leg; H, spermatophore. Scales = 0.1 mm.

spinules along inner margin distally; seta plumose, positioned centrally and as the most apical element, only slightly longer than inner apical spine, about 1.2 times as long as outer apical spine, and 1.4 times as long as lateral spine; all spines stout and with sharply serrated lateral margins.

Sixth leg not clearly distinct, triangular plate, armed with two almost equal, smooth spines (inner one fused to plate) and one pinnate, much longer seta.

Male (allotype). Body length, excluding caudal setae, 0.446 mm. Habitus (fig. 68B) less robust than in female, with prosome/urosome ratio of 1.4, and greatest width near posterior end of cephalothorax. Body length/width ratio about 2.8; cephalothorax 2.3 times as wide as genital somite.

Cephalothorax (fig. 68B) somewhat longer than its greatest width, with posterior corners not pronounced laterally. Ornamentation of prosomites, colour, and nauplius eye similar to those of female. Hyaline fringes of all somites smooth.

Genital somite 1.5 times as wide as long (dorsal view), ornamented with four pores dorsally and two pores ventrally; third and fourth urosomites ornamented with two pores ventrally, two dorsally, and two laterally on each side; fifth urosomite with two dorsal pores and one lateral pore on each side. Anal somite similar to that of female, although somewhat narrower and with smaller anal operculum.

Caudal rami (fig. 68D) twice as long as wide, parallel, with space between them about one half ramus' width. Armature and ornamentation similar as in female, except anterior ventral pore instead of sensillum.

Antennula (fig. 68E) slightly shorter than cephalothorax, 14-segmented, geniculate, with geniculation between twelfth and thirteenth segments. Aesthetasc formula: 3.0.0.0.1.0.0.1.0.0.0.0.0.2. All aesthetascs slender and not very long. Setal formula: 8.4.4.0.1.4.1.2.4.2.2.1.1.10. All setae on tenth, eleventh, and twelfth segment very short and spiniform. Only setae on eleventh and one seta on tenth segment pinnate; all other setae smooth.

Antenna (fig. 68F) with exopodal seta on coxobasis much longer than in female and with a few spinules along outer margin of first endopodal segment, otherwise similar to that of female.

Labrum, mandibula, maxillula, maxilla, maxilliped, and swimming legs similar to those of female.

Fifth leg (fig. 68G) with two-segmented (completely divided) exopod and with additional seta on inner margin of first exopodal segment; intercoxal sclerite narrower and apical seta on second exopodal segment shorter than in female; pores on exopod absent, as are spinules, and outer distal corner not so well pronounced as in female. First exopodal segment 2.8 times as long as second, and seta on first exopodal segment only slightly shorter than seta on second.

Sixth leg (fig. 68C) a distinct, large plate, armed with one spine with sharply serrated lateral margins and two pinnate setae; ornamented with one cuticular pore. Outermost seta twice as long as middle seta and innermost spine.

Spermatophore (fig. 68H) 3.3 times as long as wide, bottle-shaped.

Variability. — Body length of females ranging from 0.596 mm to 0.662 mm (0.632 mm average; n = 10), while only one male was collected and studied. No other type of variability was observed.

Etymology. — The specific name comes from the Pilbara region in Western Australia, and was made with the Latin suffix "-ensis". Consequently, the name is an adjective for place.

Distribution. — This species is restricted to the Pilbara region in Western Australia, where it was always found sympatrically with the more widely distributed *N. tropicus* n. sp. (see above).

Discussion. — *Neocyclops pilbarensis* n. sp. differs from all other species of *Neocyclops* by a large projection on the outer distal corner of the fifth leg exopod, as well as by an unusually large setophore of the dorsal seta on the caudal ramus. Also, the dorsal caudal seta is unusually long, which is the character that *N. pilbarensis* shares with only two other species: *N. sharkbayensis* n. sp. and *N. geltrudeae* Pesce & Galassi, 1993. However, the three species are not very closely related, and could be distinguished by many characters. *Neocyclops pilbarensis* is also the only species of the genus *Neocyclops* with a lateral sensillum on the caudal ramus, which is most probably a plesiomorphic character state. From the other Australian representatives of the genus, *N. pilbarensis* could also be easily distinguished by its long innermost apical setae on the caudal rami.

24. **Neocyclops petkovskii** De Laurentiis, Pesce & Halse, 1997 (figs. 69-71)

Synonymy. —

Neocyclops (*Neocyclops*) *petkovskii* De Laurentiis, Pesce & Halse, 1997: 15, figs. 1-18.

Material examined. — Australia, South Australia, Streaky Bay, 25 December 2002, leg. T. Karanovic, 32°47′S 134°13′E: 2 males + 7 females (3 ovigerous) + 2 copepodids (1 female (WAM C37295) dissected on 2 slides; 1 female (WAM C37296) dissected on 1 slide; 1 male (WAM C37297) dissected on 1 slide; 1 male + 1 female on 1 slide (TMAG G5792) in toto; others (WAM C37298) in alcohol).

Redescription. — Female. Body length, excluding caudal setae, 0.765 mm. Habitus (fig. 69A) typical cyclopoid shape, very robust, with prosome/urosome ratio of 1.5 and greatest width at posterior end of cephalothorax. Body length/width ratio about 2.6; cephalothorax about 2.5 times as wide as genital double somite. Posterior corners of cephalothorax only slightly expanded laterally; other free pedigerous somites without pronounced lateral expansions. Fifth pedigerous somite as wide as genital double somite. Colour of preserved specimen milky white. Nauplius eye not visible. Rostrum broadly rounded and furnished with two large sensilla.

Cephalothorax (fig. 69A) almost triangular, somewhat shorter than its greatest width, representing 35% of total body length. Surface of cephalic shield with several large sensilla; no other ornamentation visible. Hyaline fringe of prosomites narrow and smooth. First pedigerous somite completely fused to cephalosome. Fifth pedigerous somite ornamented with only two large dorsal sensilla, and with a smooth fringe dorsally and ventrally. No sclerotized joint between fifth pedigerous somite and genital double somite visible.

Genital double somite (fig. 69B) with lateral expansions bearing not very sharp spiniform projections at first third, about 0.9 times as long as wide, ornamented with a characteristic pattern of eight pores ventrally (two of these in line with the copulatory pore), and 14 pores dorsally. Hyaline fringe of genital double somite and two succeeding somites narrow, not frilled, and completely smooth. No cuticular recesses or sensilla visible. Copulatory pore small, ovoid, situated just anterior to 1/3 of the somite's length, with two slender spermatophores attached; copulatory duct short and wide, rigidly sclerotized, directed dorsally and anteriorly, almost ovoid. Seminal receptacle (fig. 69B) complex, curved and tubular, but wholly situated anteriorly of copulatory pore; central part connected to copulatory duct almost cylindrical, positioned close to ventral cuticular surface, and 3.4 times as wide as long; constricted part on each side about 0.4 times as wide as central part and turning backwards (dorsally); lateral arms expanded, wider than central part, and connected to dorsolateral genital areas through another constriction. Ovipores situated somewhat dorsolaterally, covered by the reduced sixth legs, and with

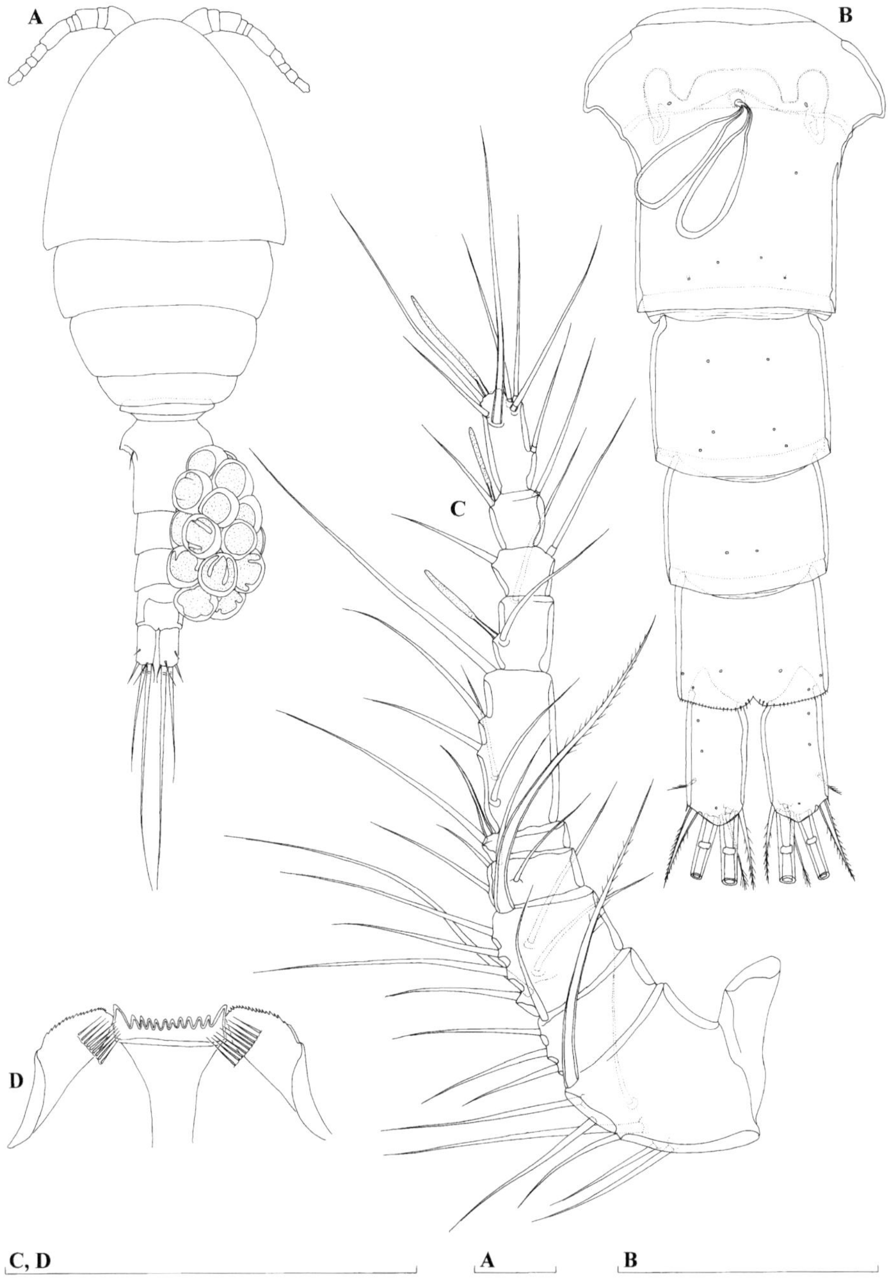

Fig. 69. *Neocyclops petkovskii* De Laurentiis, Pesce & Halse, 1997, ovigerous female (0.765 mm). A, habitus, dorsal view; B, abdomen, ventral view; C, antennula; D, labrum. Scales = 0.1 mm.

egg sacs attached, each with 21 eggs and some eggs already developed; egg sacs almost reaching posterior margin of anal somite. Third urosomite (fig. 69B) ornamented with six ventral and six dorsal pores (four in posterior half), while fourth urosomite with only two ventral and two dorsal pores. Anal somite with smooth, broad, and slightly convex anal operculum, which represents 65% of the somite's width, and reaches to only 1/3 of the somite's length; ornamented with two large sensilla and two pores dorsally, one pore laterally on each side, four pores ventrally, and a transverse row of spinules on posterior margin. Anal sinus widely open dorsally, ornamented with four diagonal (two on each side) rows of minute spinules.

Caudal rami (fig. 70A) cylindrical, 2.4 times as long as wide (ventral view), parallel, with space between them less than half a ramus' width; they represent 6% of total body length. Distal margin ventrally smooth and convex, with medial cuticular pore clearly visible. Ornamentation consisting of several small spinules at base of outermost apical seta and three ventral, one lateral, and three dorsal pores. Dorsal seta short, about 0.7 times as long as ramus, inserted on a small protuberance reaching border of ramus, uniarticulate at base, and bipinnate distally. Lateral seta the smallest, arising at 3/4, positioned dorsolaterally, uniplumose, and about 0.6 times as long as a ramus' width. Outermost apical seta stout and spiniform, bipinnate, about 0.6 times as long as ramus, and as long as innermost apical seta, which is also bipinnate, but more slender than outermost one. Principal apical setae with breaking planes and bipinnate; inner seta about twice as long as outer one and 0.35 times as long as body.

Antennula (fig. 69A) not very slender, reaching to half of cephalothorax, 10-segmented, unornamented, with one slender aesthetasc on seventh, one on ninth, and one on tenth segment, and setal formula: 8.4.8.4.2.5.2.2.2.7. No setae with breaking planes and only one seta on eighth, one on ninth, and four on tenth segment articulating on basal part. One seta on fifth segment spiniform but smooth, while subapical seta on first segment and subapical seta on third segment very strong and bipinnate; all other setae slender and smooth. Length ratio of antennular segments, from proximal end and along caudal margins, 1 : 0.7 : 1 : 0.6 : 0.4 : 2 : 1 : 0.7 : 0.8 : 1.4.

Antenna (fig. 70B) four-segmented, comprising coxobasis and a three-segmented endopod; all endopodal segments of about the same length, coxobasis somewhat longer. Coxobasis about 1.9 times as long as wide, unornamented, and armed with two smooth setae at distal inner corner, and seta representing the exopod at distal outer corner; exopodal seta bipinnate, reaching beyond posterior margin of appendage, 2.3 times as long as longer seta at distal inner corner, and about twice as long as coxobasis. First endopodal segment armed with one smooth seta at middle, and ornamented with several spinules along external margin. Second endopodal segment about 1.6 times as long as wide, ornamented with longitudinal row of spinules along external margin, and armed with five smooth setae (three lateral, two subapical; one subapical seta much stronger than any other seta on that segment). Third endopodal segment 2.4 times as long as wide, armed with seven smooth apical setae (four of which strong, geniculate, and with breaking planes), and ornamented as previous segment.

Labrum (fig. 69D) trapezoidal, with very broad cutting edge and distolateral corners almost on the same plane, ornamented with two diagonal rows of nine long and slender spinules on ventral surface, and with finely serrated distolateral margin. Cutting edge slightly concave, with 12 strong and sharp teeth (outermost ones strongest). No other ornamentation visible.

Mandibula, maxillula, maxilla, and maxilliped very similar to those of previous five species, and especially to those of the first of these species, i.e., *N. australiensis* n. sp.

All swimming legs with three-segmented exopod and endopod, although second and third endopodal segments of fourth leg not completely divided; their armature formula as follows (legend: inner/outer spine or seta; inner/terminal/outer):

	Exopod			Endopod		
Segments	1	2	3	1	2	3
First leg	1/1	1/1	3/2/3	1/0	2/0	3/2/1
Second leg	1/1	1/1	4/2/3	1/0	2/0	3/2/1
Third leg	1/1	1/1	4/2/3	1/0	2/0	3/2/1
Fourth leg	1/1	1/1	4/2/2	1/0	2/0	2/2/1

Third exopodal segment spine formula 3.4.4.3 and third endopodal segment spine formula 2.3.3.3. All intercoxal sclerites without ornamentation and with rounded protrusions at both sides of distal margin; that of first leg most pronounced. Coxae also unornamented, and armed with a plumose seta at inner distal corner. Spine inserted at inner, protruded corner of basis of first leg almost reaching distal margin of second endopodal segment, stout and bipinnate (fig. 70C). Basis of each leg ornamented with a row of hair-like spinules along inner margin and with a few minute spinules along distal margin close to endopod; inner distal corners of second, third, and fourth leg pronounced as blunt spiniform processes; each basis armed with a slender seta on outer margin, which is smooth in the fourth leg and pinnate in the others. All exopodal and endopodal setae slender and plumose; outer apical seta on third exopodal segment of first leg pinnate along outer margin, plumose along inner; no modified setae on third endopodal and exopodal segments of fourth leg (fig. 70D, E). All exopodal and endopodal spines stout and bipinnate. All endopodal segments with hair-like spinules along outer margins, which are also present on inner margin of second exopodal segment of all legs and on first exopodal segment of first leg. Cuticular pores visible only on first leg exopod and endopod. Third endopodal segment of fourth swimming leg about 1.2 times as long as wide; inner apical spine about 1.3 times as long as outer one (fig. 70E).

Fifth leg (fig. 70F) with intercoxal sclerite and three-segmented, comprising coxa, basis, and a one-segmented exopod. Intercoxal sclerite trapezoidal, with slightly convex distal margin, about 1.3 times as wide as long, and unornamented. Coxa unarmed and unornamented, twice as wide as long. Basis 1.3 times as wide as its greatest length, armed with a single smooth seta on shorter outer margin, and unornamented. Exopod 2.2 times as long as its greatest width, much narrower at basal part, armed with three spines and a single seta, and ornamented with one pore between proximal and outer apical spine; seta plumose, positioned centrally, and as the most apical element, 1.4 times as long as inner apical spine, about 1.3 times as long as outer apical spine, and 1.4 times as long as lateral spine; all spines bipinnate, but innermost apical spine much stouter than the other two and with spinules only at its distal fourth.

Sixth leg not clearly distinct, triangular plate, armed with two almost equal, smooth spines (inner one fused to plate) and one pinnate, much longer seta.

Male. Body length, excluding caudal setae, 0.526 mm. Habitus (fig. 71A) less robust than in female, with prosome/urosome ratio of 1.5 and greatest width near posterior end of cephalothorax. Body length/width ratio about 3.1; cephalothorax 2.6 times as wide as genital somite.

Cephalothorax (fig. 71A) slightly longer than its greatest width, with posterior corners not pronounced laterally. Ornamentation of prosomites, colour, and nauplius eye similar to those of female. Hyaline fringes of all somites smooth.

Genital somite (fig. 71B) 1.3 times as wide as long (dorsal view), ornamented with three pores dorsally, two pores ventrally, and two pores laterally; third urosomite ornamented with two dorsal and two ventral pores; fourth and

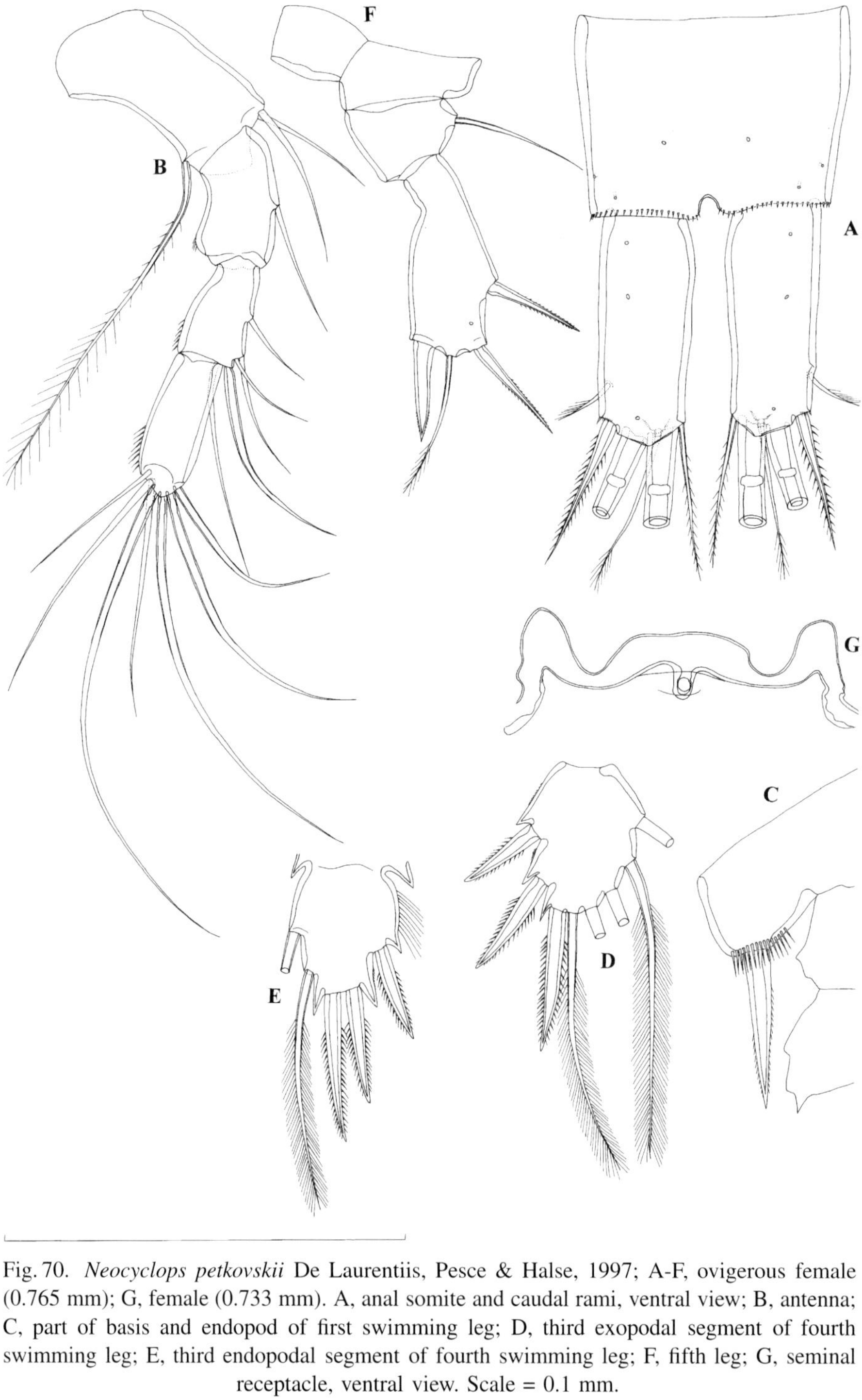

Fig. 70. *Neocyclops petkovskii* De Laurentiis, Pesce & Halse, 1997; A-F, ovigerous female (0.765 mm); G, female (0.733 mm). A, anal somite and caudal rami, ventral view; B, antenna; C, part of basis and endopod of first swimming leg; D, third exopodal segment of fourth swimming leg; E, third endopodal segment of fourth swimming leg; F, fifth leg; G, seminal receptacle, ventral view. Scale = 0.1 mm.

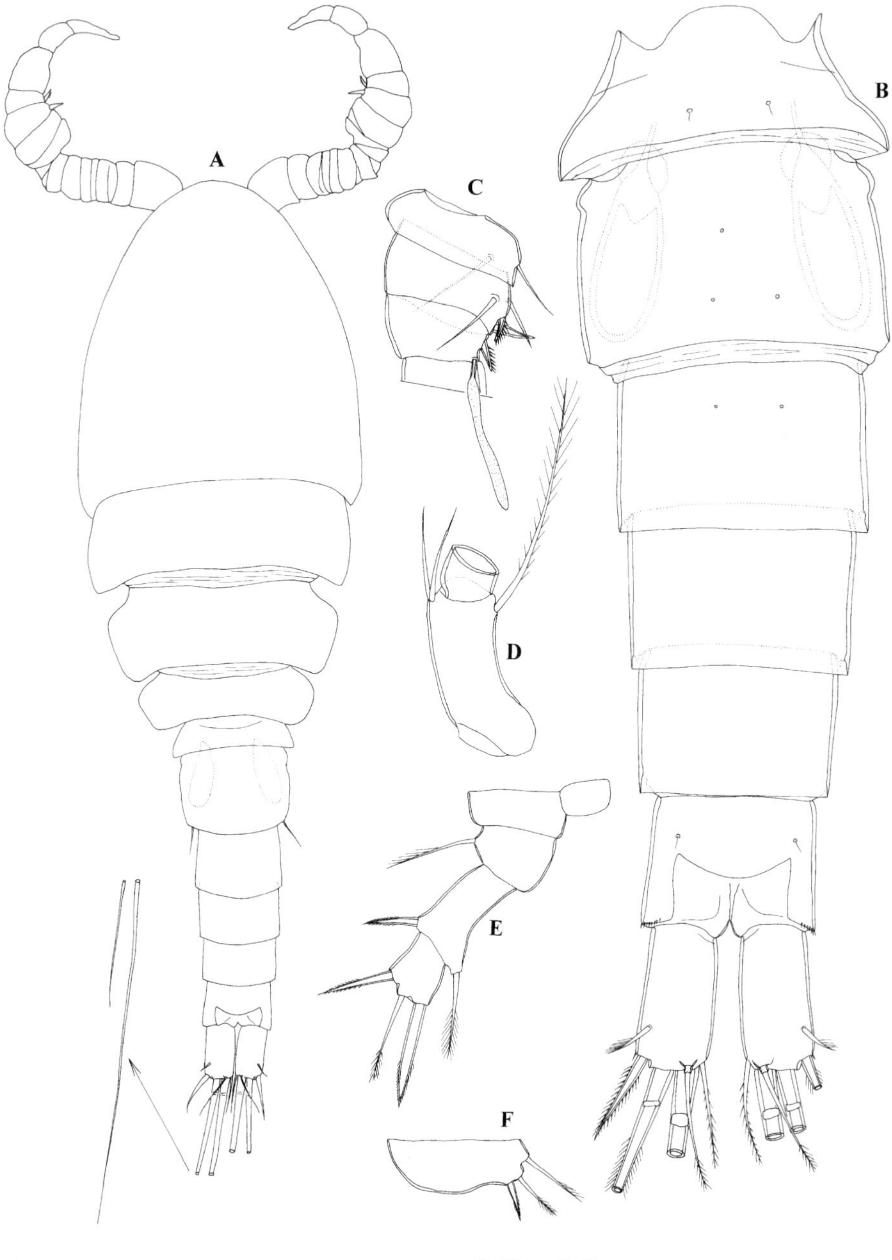

Fig. 71. *Neocyclops petkovskii* De Laurentiis, Pesce & Halse, 1997, male (0.526 mm). A, habitus, dorsal view; B, urosome, dorsal view; C, ninth, tenth, and eleventh segments of antennula; D, basis of antenna; E, fifth leg; F, sixth leg. Scales = 0.1 mm.

fifth urosomites ornamented with only two ventral pores each. Anal somite similar to that of female, but two dorsal pores missing.

Caudal rami (fig. 71B) about twice as long as wide, parallel, with space between them less than half a ramus' width. Armature and ornamentation similar as in female, except dorsal pores absent.

Antennula (fig. 71C) of length almost equal to that of cephalothorax, 14-segmented, geniculate, with geniculation between twelfth and thirteenth segments. Aesthetasc formula: 3.0.0.0.1.0.0.1.0.0.1.0.0.2. All aesthetascs slender and not very long. Setal formula: 8.4.4.0.1.4.1.2.2.4.1.1.1.10. Two setae on tenth, one on eleventh, and one on twelfth segment very short and spiniform. Only seta on eleventh and one seta on tenth segment pinnate; all other setae smooth.

Antenna (fig. 71D) with exopodal seta on coxobasis somewhat shorter, otherwise similar to that of female.

Labrum, mandibula, maxillula, maxilla, maxilliped, and swimming legs similar as in female.

Fifth leg (fig. 71E) with two-segmented (and completely divided) exopod and with additional seta on inner margin of first exopodal segment; intercoxal sclerite narrower and apical seta on second exopodal segment shorter than in female; pore on exopod absent. First exopodal segment 1.7 times as long as second one. Innermost apical spine much stouter than the other two, just like in female.

Sixth leg (fig. 71F) a distinct, large plate, armed with one pinnate spine and two pinnate setae, and unornamented. Outermost seta 1.3 times as long as middle seta and 1.7 times as long as innermost spine.

Variability. — Body length of females ranging from 0.733 mm to 0.765 mm (0.747 mm average; n = 7), while in males it ranges from 0.492 mm to 0.526 mm (0.509 mm average; n = 2). One female has a somewhat narrower seminal receptacle (fig. 70G), but no other types of variability were observed. All specimens have a very characteristic, stout innermost apical spine on the fifth leg, and all females have 10-segmented antennulae.

Distribution. — This species was described from the Western Australian saline Lake McLeod and my record from South Australia (some 2,200 km SE) widens its known distribution significantly.

Discussion. — De Laurentiis et al. (1997) described *N. petkovskii* from Lake McLeod in Western Australia, which is a 120 km long, saline claypan situated about 20 km from the coast and 40 km north of the city of Carnarvon. They also mentioned in the introduction section of their paper that "the species was also collected at Shark Bay", but they did not mention any material from there in their "Material examined" section. I have found three other species of *Neocyclops* in Shark Bay (*N. australiensis* n. sp.; *N. sharkbayensis* n. sp.; and *N. trajani* n. sp.), but no *N. petkovskii*, so it is possible that their casual note was actually referring to one of these species instead. *Neocyclops petkovskii* could be a warm-climate relict in South Australia, trapped by the Great Australian Bight and thus prevented from migrating north along the coast during the last ice-age, but invading saline continental lakes instead and continuing its northbound migration this way. Many other warm-climate relicts are known from South Australian bays and gulfs, including the most southern mangrove forest in the world. This could also be one of the models for invasion of the Australian continental waters (especially the subterranean ones), as many species of marine origin have been reported from arid Western Australia (Karanovic, 2004a, 2006).

The only significant difference between my material and that reported by De Laurentiis et al. (1997), is the segmentation of the male fifth leg. My specimens have a completely divided exopod, while those from Lake McLeod have a one-segmented exopod. Everything else, including a peculiarly stout inner apical spine on the fifth leg, as well as a 10-segmented female antennula, is the same in both populations. This, together with the shape of the male fifth leg in *N. australiensis* n. sp. (see above), made me abandon the subgeneric subdivision of the genus *Neocyclops* as established by Petkovski (1986). Consequently, I consider his subgenus *Protoneocyclops* Petkovski, 1986 to be just a junior synonym of *Neocyclops* Gurney, 1927. Maybe it should be mentioned here that also Petkovski's (1986) designation of *N. medius* Herbst, 1955 as the type species of the nominotypical subgenus was erroneous, as the type species of the nominotypical subgenus must be the same as the generic one (i.e., *N. salinarum* (Gurney, 1927)).

Petkovski (1986) synonymized the material from Cuba, reported by Pleşa (1981) as *Neocyclops* sp., with the widely distributed *N. stocki* Pesce, 1985, but reproduced only three out of 12 of Pleşa's (1981) drawings and also gave an erroneous locality for one of them (Pleşa's (1981) fig. 5 = Petkovski's (1986) fig. 13). However, closer examination of Pleşa's (1981) drawings shows that he was dealing with two separate species, which occurred sympatrically at Playa Juragua. While Pleşa's (1981) figs. 6, 8, and 11 could be identified as *N. stocki* Pesce, 1985, his figs. 5, 9, 10, and 12 represent an undescribed new species which I name here **Neocyclops plesai** n. sp. As this new species occurs both at Playa Juragua and Playa Yateritas, I choose Playa Yateritas as the type locality and the specimen illustrated on Pleşa's (1981) figs. 5, 9, and 12 as the holotype of *N. plesai*. Unfortunately, Pleşa's figs. 2, 3, 4, and 7 can not be attributed with certainty to either *N. stocki* or *N. plesai*. These two species differ by the shape of the female fifth leg (with inner exopodal margin convex at posterior part in *N. plesai* versus straight in *N. stocki*), as well as by the relative size of the innermost apical seta on the caudal ramus (which is shorter than the outermost one in *N. plesai* and much longer in *N. stocki*).

A careful examination of published literature revealed two more undescribed *Neocyclops* species, which I am going to name here. Monchenko (1974) reported a *Neocyclops* species from the Black Sea as *Eurycyclops remanei vicinus* (Herbst, 1955). However, his material differs greatly from *N. vicinus* (Herbst, 1955), which was also noted by Lotufo & Da Rocha (1993b), and I name it here **Neocyclops monchenkoi** n. sp. The new species has much longer caudal rami than *N. vicinus* (caudal rami index 3.3, instead of 2.4) and a longer caudal seta (1.3 times as long as ramus, instead of 0.9 times as in the latter species). The type locality for *N. monchenkoi* n. sp. is Tendivsjaya Kosa in the Black Sea and the holotype is the female illustrated in Monchenko's (1974) fig. 16.

Dussart (1974) described a new species after two females from Ethiopia as *Neocyclops affinis*. This name is a homonym of *N. affinis* (Pleşa, 1961), not a synonym as Dussart & Defaye (1985) erroneously presented it. The specimens, however, belong to an undescribed *Neocyclops* species, which I here name **N. dussarti** nomen novum. The type locality for *N. dussarti* is Rogue à Assab and the holotype is the female illustrated in Dussart's (1974) fig. 2a. *Neocyclops dussarti* has a much shorter caudal ramus than *N. affinis*, as well as a proportionally shorter dorsal caudal seta. Pesce (1985) reported *Neocyclops affinis* Dussart, 1974 from the Bahamas, but his drawings show that he was actually dealing with *N. affinis* (Pleşa, 1961), not with *N. dussarti*.

I also consider here **N. mediterraneus** (Kiefer, 1960), **N. remanei** (Herbst, 1952) and **N. vicinus** (Herbst, 1955) as valid species, not subspecies as they were sometimes regarded. In the preparation of the key to the species of the genus *Neocyclops* I also noted that Da Rocha (1995a) reported the caudal rami index of his new species (*N. ferrarii* Da Rocha, 1995) to be 3, while the measurements of his drawings show it to be rather 2.4. Also, in the key below, *N. ferrarii* Da Rocha, 1995 and *N. mediterraneus* (Kiefer, 1960) are only provisionally separated, and further studies (and a redescription of the female of the latter) are necessary to confirm the specific status of the former species.

KEY TO SPECIES OF THE GENUS *NEOCYCLOPS*

1. Antennula composed of more than 8 segments 2
– Antennula 8-segmented .... *N. salinarum* (Gurney, 1927)
2. Antennula composed of more than 10 segments 3
– Antennula 10-segmented .... *N. petkovskii* De Laurentiis, Pesce & Halse, 1997
3. Antennula 12-segmented 5
– Antennula 11-segmented 4
4. Dorsal seta on caudal ramus more than twice as long as ramus .... *N. geltrudeae* Pesce & Galassi, 1993
– Same seta as long as ramus .... *N. improvisus* Pleşa, 1973
5. Dorsal seta on caudal ramus more than twice as long as ramus 6
– Same seta less than twice as long as ramus 7
6. Second endopodal segment of all swimming legs with a single inner seta .... *N. sharkbayensis* n. sp.
– Same segment with two inner setae .... *N. pilbarensis* n. sp.
7. Dorsal seta on caudal ramus less than twice as long as ramus 8
– Same seta more than twice as long as ramus .... *N. tropicus* n. sp.
8. Caudal rami more than 1.7 times as long as wide 9
– Caudal rami 1.3 times as long as wide .... *N. trajani* n. sp.
9. Third endopodal segment of third leg with four setae and three spines 10
– Same segment with three setae and three spines .... *N. affinis* (Pleşa, 1961)
10. Third exopodal segment of fourth leg with five setae and three spines 11
– Same segment with four setae and three spines .... *N. herbsti* Petkovski, 1986
11. Third exopodal segment of first leg with five setae and three spines 12
– Same segment with four setae and three spines .... *N. wellsi* Petkovski, 1986
12. Antenna without exopod 13
– Antenna with exopodal seta 16
13. Caudal rami 1.8 times as long as wide 14
– Caudal rami 2.4 or 2.5 times as long as wide 15
14. Innermost seta on caudal ramus shorter than outermost .... *N. dussarti* nomen novum [syn. *N. affinis* Dussart, 1974: 96, fig. 2a]
– Innermost seta longer than outermost .... *N. medius* Herbst, 1955
15. Innermost seta on caudal ramus shorter than outermost .... *N. australiensis* n. sp.
– Innermost seta longer than outermost .... *N. parvus* (Sewell, 1949)
16. Dorsal seta on caudal ramus less than twice as long as ramus 17
– Same seta more than twice as long as ramus .... *N. plesai* n. sp. [syn. *Neocyclops* sp. Pleşa, 1981: 22, figs. 5, 9, 10, 12]
17. Dorsal seta on caudal ramus as long as or shorter than ramus 19
– Same seta longer than ramus (from 1.3 to 1.5 times) 18
18. Innermost apical seta on caudal ramus more than twice as long as outermost .... *N. stocki* Pesce, 1985

– Same seta shorter than outermost .... *N. monchenkoi* n. sp. [syn. *Eurycyclops remanei vicinus* (Herbst, 1955) – Monchenko, 1974: 30, fig. 16]
19. Dorsal caudal seta from 0.8 to 1.0 times as long as ramus 20
– Same seta 0.4 times as long as ramus .... *N. papuensis* Fiers, 1986
20. Innermost apical seta on caudal ramus as long as or shorter than outermost 22
– Same seta 1.4 to 1.5 times as long as outermost 21
21. Caudal rami 2.4 times as long as wide; inner seta on first exopodal segment of male fifth leg plumose .... *N. ferrarii* Da Rocha, 1995
– Caudal rami 2.8 times as long as wide, inner seta on first exopodal segment of male fifth leg spiniform .... *N. mediterraneus* (Kiefer, 1960)
22. Caudal rami between 2.4 and 2.7 times as long as wide 23
– Caudal rami 3.5 times as long as wide .... *N. remanei* (Herbst, 1952)
23. Inner spine on female fifth leg 1.1 times as long as outer.......................... .... *N. vicinus* (Herbst, 1955)
– Same spine 1.7 times as long as outer .... *N. magnus* (Sewell, 1949)

Genus **Halicyclops** Norman, 1903
Subgenus **Halicyclops** Norman, 1903

25. **Halicyclops (H.) thermophilus** Kiefer, 1929
(figs. 72-74)

Synonymy. —

Halicyclops thermophilus Kiefer, 1929a: 46, figs. 1-3; Kiefer, 1929b: 25; Heberer & Kiefer, 1932: 236, figs. 9-13, 13a; Kiefer, 1936: 283, figs. 37-43; Lindberg, 1949: 7: Lindberg, 1952: 57, fig. 1; Lindberg, 1957: 142; Wilson, 1958: 178; Wooldridge, 1977: 370; Reid, 1985: 30, fig. 50; Dussart & Defaye, 1985: 17.

Halicyclops thermophilus thermophilus – Yeatman, 1983: 59, fig. 1.

Halicyclops thermophilus spinifer Kiefer, 1935: 13, fig. 7; Kiefer, 1936: 285, figs. 49-52; Lindberg, 1941: 2, figs. 2-3; Sewell, 1949: 44, fig. 8; Yeatman, 1983: 59, figs. 2a-2f; Dussart & Defaye, 1985: 17.

Halicyclops spinifer – Lindberg, 1949: 7; Lindberg, 1957: 142; Wilson, 1958: 178; Pesce et al., 1996: 83, figs. 12-24; Mirabdullayev & Getz, 1996: 43, figs. 1-7; Karanovic, 2004a: 58.

Halicyclops (*Halicyclops*) *spinifer* – Karanovic, 2006: 20.

Halicyclops propinquus – Sewell, 1934: 87.

Halicyclops japonicus Itô, 1956: 473, fig. 3; Dussart & Defaye, 1985: 20; Ishida, 2002: 41, fig. 1a-e.

Halicyclops latus Shen & Tai, 1964: 369, figs. 6-10; Shen et al., 1979: 312, fig. 175; Dussart & Defaye, 1985: 20.

Halicyclops dedeckeri Brownell, 1983: 62, figs. 1-19; Dussart & Defaye, 1985: 21.

Halicyclops antiguaensis Herbst, 1983: 262, figs. 1-7; Dussart & Defaye, 1985: 21.

Material examined. — Australia, Western Australia, Karratha, near jetty, 4 August 2003, leg. T. Karanovic, 20°43.298′S 116°53.458′E: 2 males + 4 females (1 male (WAM C37299) and 1 female (WAM C37300) dissected on 1 slide each, 1 female on 1 slide in toto (TMAG G5793), others in alcohol (WAM C37301)).

Redescription. — Female. Habitus robust (fig. 72A) from dorsal view, with prosome/urosome ratio of 1.83 and greatest width near posterior end of cephalothorax. Body length/width ratio about 2.6; cephalothorax about 2.6 times as wide as genital double somite. Free pedigerous somites without pronounced lateral expansions. Colour of preserved specimen yellowish. Nauplius eye not visible. Rostrum well developed, membranous, broadly rounded, and furnished with two large sensilla.

Cephalothorax almost as long as its greatest width, representing 38% of total body length. Surface of cephalic shield with a few large sensilla; no other ornamentation visible. Hyaline fringe of prosomites narrow and smooth. Fifth pedigerous somite with four large dorsal sensilla, and with a smooth fringe ventrally and a finely serrated one dorsally (figs. 72B, 73A).

Genital double somite (figs. 72B, 73A) 0.94 times as long as wide, ornamented dorsally with two sensilla and two pores at middle, and two pores near posterior margin; two large spiniform processes laterally with their tips pointed posteriorly and reaching 3/4 of the double somite's length; two cuticular recesses ventrolaterally on posterior half of genital double somite. Hyaline fringe of genital double somite and two succeeding somites not frilled, but finely serrated. Copulatory pore small; copulatory duct almost as long as wide, rigidly sclerotized. Seminal receptacle very small, situated anteriorly of pore and duct, leaf-shaped (fig. 72B), with two narrow lateral arms exiting its lateroposterior corners. Ovipores situated dorsolaterally, at base of spiniform processes, covered by the reduced sixth legs, each bearing two minute spines and one unipinnate seta (fig. 73A). Third urosomite with a pair of ventral pores near posterior margin. Fourth urosomite without surface ornamentation; its fringe extended dorsally into sharply serrated pseudoperculum (fig. 73B), almost reaching posterior margin of anal somite. Anal somite with smooth, broad, and almost straight anal operculum (covered by pseudoperculum), which represents 46% of the somite's width; ornamented with two large sensilla and two pores dorsally, two pores ventrally, and a transverse row of long spinules on posterior margin both ventrally and laterally (fig. 72B). Anal sinus widely open, ornamented with several diagonal rows of minute spinules.

Caudal rami (fig. 72B) cylindrical, 1.5 times as long as wide, somewhat divergent, with space between them about one half of a ramus' width. Distal margin ventrally without cuticular tube pore. One slightly arched row of large spinules present ventrolaterally near posterior margin; two pores ventrally near middle and one spinule at base on innermost apical seta; no spinules

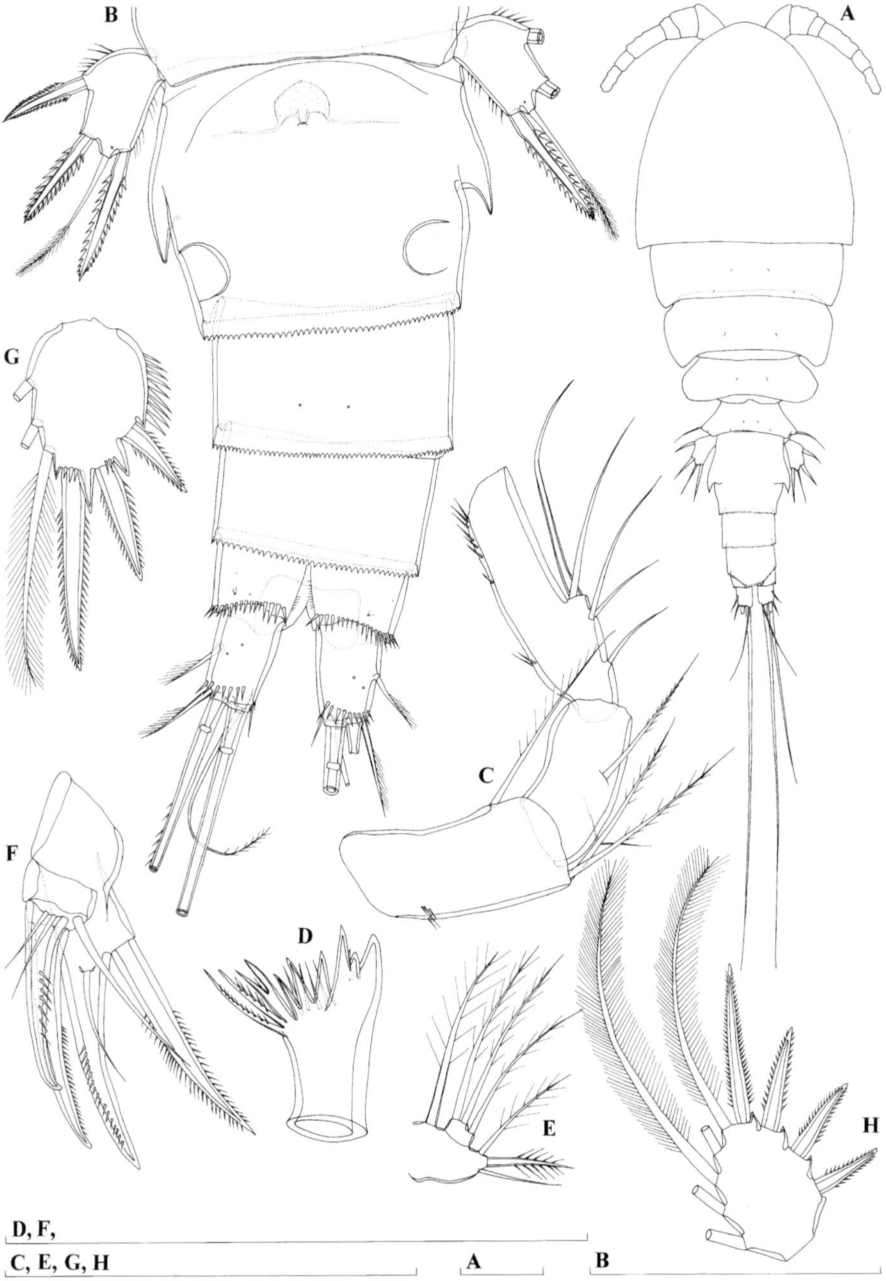

Fig. 72. *Halicyclops* (*H.*) *thermophilus* Kiefer, 1929, female from Karratha (0.690 mm). A, habitus, dorsal view; B, slightly pressed urosome, ventral view; C, antenna; D, cutting edge of mandibula; E, palp of maxillula; F, distal part of maxilla; G, third endopodal segment of second swimming leg; H, third exopodal segment of second swimming leg. Scales = 0.1 mm.

Fig. 73. *Halicyclops* (*H.*) *thermophilus* Kiefer, 1929, female from Karratha (0.690 mm). A, fifth pedigerous somite and genital double somite, dorsal view; B, pseudoperculum, dorsal view; C, antennula; D, labrum; E, maxilliped; F, basis and endopod of first swimming leg. Scales = 0.1 mm.

at base of dorsal seta. Dorsal seta long, about 2.2 times as long as ramus, inserted on protuberance overreaching border of ramus, uniarticulate at base and plumose distally. Lateral seta arising at middle, positioned only slightly dorsolaterally, uniplumose, and much shorter than ramus. Outermost apical seta spiniform, slightly shorter than ramus, and unipinnate distally along outer margin. Innermost apical seta minute, slender, and smooth. Principal apical setae with breaking plane, strong and bipinnate; inner seta twice as long as outer one, and 0.62 times as long as body.

Antennula (fig. 73C) short, hardly reaching middle of cephalothorax, six-segmented, with slender aesthetascs on sixth segment (fused basally to one seta) and setal formula: 8.12.5.5.4.9. One stout seta at distal outer corner of fourth segment, one on fifth, and five on sixth segment articulating on basal part; no setae with breaking plane and almost all setae slender (only one seta on third segment spiniform and very short). About 50% of all setae smooth, others plumose or pinnate in their distal parts. Transverse, short row of sharp spinules on first segment as only ornamentation. Length ratio of antennular segments, from proximal end and along caudal margins, 1 : 2 : 1.3 : 4.7 : 2.2 : 4.

Antenna (fig. 72C) three-segmented, comprising coxobasis and a two-segmented endopod. Coxobasis ornamented with short row of spinules at proximal inner (anterior) part; armed with two bipinnate setae at distal inner corner and with unipinnate seta representing exopod; exopodal seta about as long as coxobasis, while other two setae only slightly shorter. First endopodal segment armed with one bipinnate seta at middle, unornamented. Second endopodal segment about 1.6 times as long as first, ornamented with two longitudinal rows of spinules along caudal surface, and armed with five lateral and seven apical setae; one lateral seta spiniform and longer than the others, but shorter than the segment.

Labrum (fig. 73D) trapezoidal, with smooth ventral surface and with finely serrated distolateral margin. Cutting edge slightly concave, with 12 blunt teeth; second tooth on each side much stronger and longer than the others, which are subequal. No additional ornamentation visible.

Mandibula (fig. 72D) with palp reduced to a small protuberance, which bears two setae; distal seta pinnate distally, more than twice as long as proximal smooth seta but not reaching gnathobasal teeth. Cutting edge of coxal gnathobase clearly divided into three regions: inner group of four complex smooth teeth, innermost tooth strongest, outermost sharpest and longest; middle group of three simple smooth teeth and four spinules; outer group of one bifurcate strong tooth and two unipinnate setae (outermost seta longest).

Maxillula (fig. 72E) composed of well developed praecoxa and a two-segmented palp. Arthrite of praecoxa with four apical, strong, short, and smooth spines. Praecoxa armed with seven armature elements along inner margin, the longest one plumose. Palp with completely divided and very short endopod (bearing two apical and one lateral (or subapical) pinnate setae); armed laterally with one exopodal pinnate seta, and apically with two setae (one smooth, the other plumose) and one robust, strongly bipinnate spine.

Maxilla (fig. 72F) four-segmented but praecoxa fused to coxa on posterior surface. Proximal endite of praecoxa armed with two setae (one plumose, the other pinnate distally); distal endite very small, unarmed. Proximal endite of coxa with one short pinnate seta; distal endite highly mobile and bearing one proximal, completely fused, and very stout seta, and one distal, smaller seta. Seta fused to endite ornamented with three elongated spinules. Basis expanded into a spinulose claw and armed with two setae; strong seta somewhat longer than the claw. Endopod armed with two spinulose and three smooth setae. Longer endopodal spinulose seta about as long as the strong seta on basis; shorter endopodal spinulose seta about as long as the longest endopodal smooth seta. Both endopodal spinulose setae, as well as claw and strong seta on basis, prehensile and with all spinules of about the same length.

Maxilliped (fig. 73E) two-segmented, armed with three setae on protopod and five setae on endopod, and ornamented with a bunch of four very long spinules near distal margin of protopod. Proximal protopodal seta unipinnate and shortest of the three; middle seta strong and smooth and about 1.6 times as long as proximal one; distal protopodal seta about as long as proximal one and ornamented with three long spinules. The two lateral and innermost apical endopodal setae very stout and with several very long spinules. Middle apical seta the longest, smooth, and more than twice as long as the outermost, smooth, apical (actually subapical) seta. Protopod about 2.5 times as long as endopod.

All swimming legs with three-segmented exopod and endopod (figs. 72G, H, 73F, 74A). Armature formula (legend: inner/outer spine or seta; inner/terminal/outer):

	Exopod			Endopod		
Segments	1	2	3	1	2	3
First leg	1/1	1/1	3/2/3	1/0	1/0	3/2/1
Second leg	1/1	1/1	4/2/3	1/0	2/0	3/2/1
Third leg	1/1	1/1	4/2/3	1/0	2/0	3/2/1
Fourth leg	1/1	1/1	4/2/2	1/0	2/0	2/2/1

Third exopodal segment spine formula: 3.4.4.3. Intercoxal sclerite of first swimming leg with two bunches of delicate hairs; all other intercoxal sclerites without visible ornamentation. Coxae ornamented with a row of strong spinules along outer margin, and armed with a pinnate seta at inner distal corner; pinnate seta shortest on fourth leg. Spine inserted at inner, protruded corner of basis of first leg reaching beyond posterior margin of second endopodal segment (fig. 73F). Basis of each leg ornamented with posterior row of long spinules and armed with an outer slender and unipinnate seta. All exopodal and endopodal setae slender and plumose, except for two modified setae on endopod of fourth leg. Third endopodal segment of fourth swimming leg about 1.4 times as long as wide; inner apical spine about 1.5 times as long as outer one; two inner setae spiniform and strongly bipinnate (fig. 74A).

Fifth leg (fig. 72B) with basis and endopod completely fused to somite; outer basal seta inserted on elongate basal protuberance (about twice as long as wide) and unipinnate distally. Exopod about 1.6 times as long as wide, with spinules both along inner and outer margins; no spinules observed at base of outer spines; one cuticular pore at base of apical seta. Armature consisting of three strong spines and one apical seta; inner apical spine 1.2 times as long as the segment, about 1.3 times as long as outer apical spine and 1.5 times as long as outer proximal spine. Exopodal apical seta plumose at distal end, about 1.3 times as long as the segment and somewhat shorter than basal seta.

Sixth leg (fig. 73A) a distinct, triangular plate, armed with two almost equal, smooth spines fused to plate, and one unipinnate, much longer seta.

Male. Habitus (fig. 74B) less robust than that of female, with prosome/urosome ratio of 1.8 and greatest width at posterior end of cephalothorax. Body length/width ratio about 2.6; cephalothorax about 2.8 times as wide as genital somite.

Cephalothorax (fig. 74B) about as long as its greatest width. Ornamentation of prosomites, colour, and nauplius eye similar to those of female. Fifth pedigerous somite ornamented with a pair of pores ventrally, at bases of fifth legs. Hyaline fringe of fifth pedigerous somite smooth; that of genital somite sharply serrated; those of other urosomites (except anal one) sharply serrated both dorsally and ventrally. Dorsal hyaline fringe of preanal somite produced posteriorly in the form of a pseudoperculum, and more strongly serrated. Third and fourth urosomites with pair of ventral cuticular pores

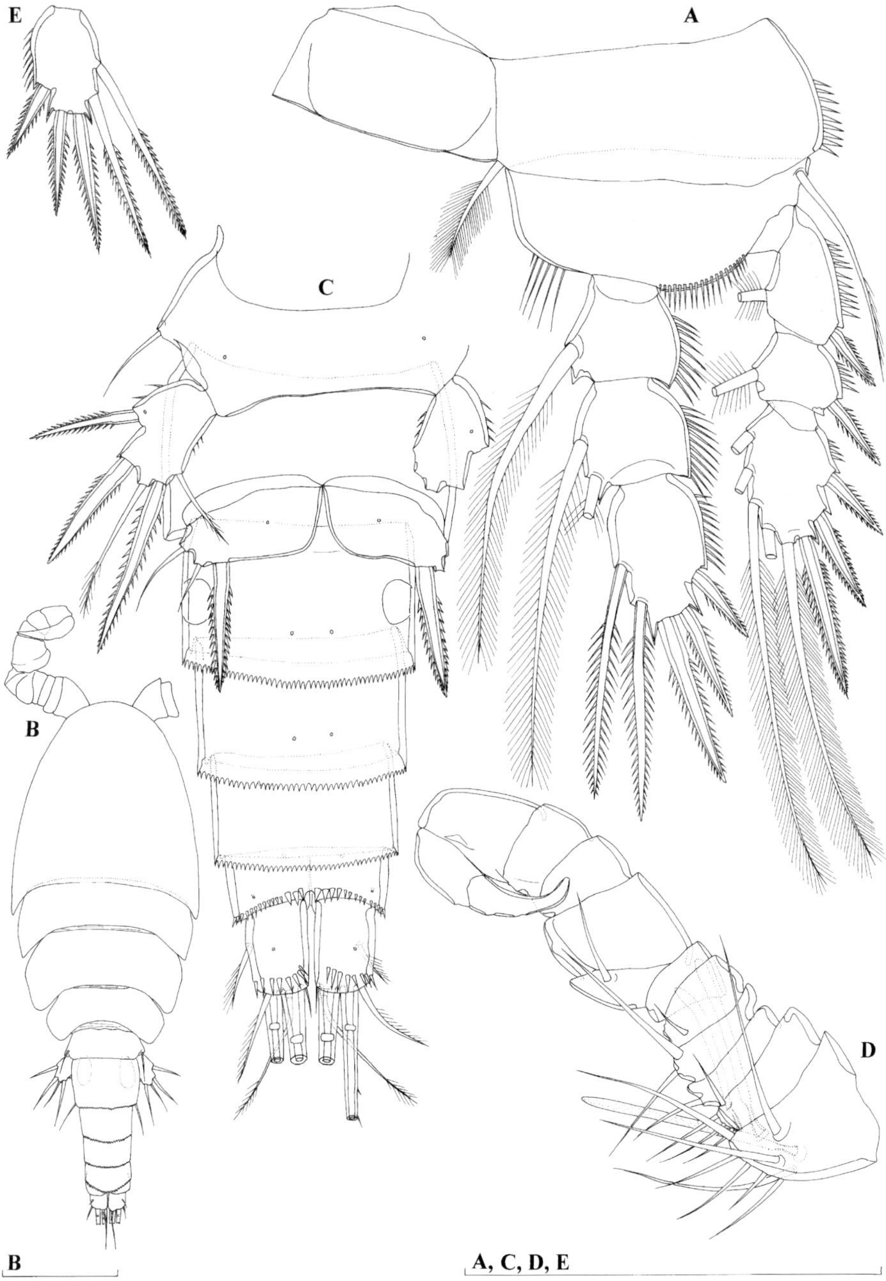

Fig. 74. *Halicyclops* (*H.*) *thermophilus* Kiefer, 1929, A, female from Karratha (0.690 mm); B-E, male from Karratha (0.425 mm). A, fourth swimming leg; B, habitus, dorsal view; C, urosome, ventral view; D, antennula with incomplete armature; E, third endopodal segment of fourth swimming leg. Scales = 0.1 mm.

each; third urosomite additionally ornamented with a pair of ventrolateral cuticular recesses (fig. 74C).

Anal somite (fig. 74C) with smooth, somewhat concave, and very broad anal operculum, ornamented with two sensilla dorsally, a pair of pores ventrally, and a transverse row of spinules on posterior margin. Anal sinus widely open, without surface ornamentation.

Caudal rami (fig. 74C) about 1.5 times as long as wide, almost parallel, with little space between them. Armature and ornamentation similar as in female, except only one ventral pore present.

Antennula (fig. 74D) shorter than cephalothorax, 13-segmented, digeniculate, with geniculations between eleventh and twelfth segments, and between sixth and seventh segments. First segment with three slender aesthetascs and eight smooth setae. Armature very tightly packed and difficult for positive identification.

Antenna, labrum, mandibula, maxillula, maxilla, maxilliped, and swimming legs all similar to those of female. Third endopodal segment of fourth leg (fig. 74E) somewhat more slender than in female.

Fifth leg (fig. 74C) exopod about 1.6 times as long as wide, ornamented with two large spinules on outer margin and several smaller spinules along inner margin. The armature consists of three spines and two setae. Inner apical spine almost 1.6 times as long as the segment, about 1.4 times as long as outer apical spine, and 1.6 times as long as outer proximal spine. Exopodal inner seta short, about 0.6 times as long as apical seta and 0.7 times as long as the segment; both setae bipinnate at distal end. Basal seta smooth and as long as exopodal inner one.

Sixth leg (fig. 74C) a distinct, large plate, armed with one serrated spine and two smooth setae; ornamented only with a cuticular pore on anterior surface. Innermost spine about 2.4 times as long as middle seta and 1.6 times as long as outer seta.

Variability. — Body length of females ranging from 0.581 mm to 0.690 mm (0.65 mm average; n = 4), while in males it ranges from 0.419 mm to 0.425 mm (0.422 mm average; n = 2). No other type of variability was observed.

Distribution. — *Halicyclops* (*H.*) *thermophilus* Kiefer, 1929 seems to be widely distributed in the tropical and subtropical regions of the world, and was reported so far from Java, India, Iran, Madagascar, Japan, China, South Africa, Antigua, Tonga, Western Samoa, Fiji, Brazil, Western Australia, and

Central Asia. Because the species often inhabits river estuaries and other brackish environments, it is possible that the naturally wide range of the species is even further enlarged by dispersal in ships' ballast waters. This may explain isolated populations in some parts of the world, and its restricted distribution in northwestern Western Australia.

Discussion. — The genus *Halicyclops* Norman, 1903 today encompasses more than 70 valid species (Karanovic, 2006), inhabiting estuaries, coastal lagoons, anchialine caves, and interstitial waters of beaches around the world, from approximately 60°N to 45°S (Da Rocha et al., 2000). Also, it has been found in Australia in continental saline and freshwater habitats (both in subterranean and surface waters) as far as 700 km from the nearest seashore (Karanovic, 2004a, 2006). Recently, I have subdivided the genus and established one new subgenus (*Rochacyclops* Karanovic, 2006) to accommodate two Australian representatives, along with three other non-Australian species (Karanovic, 2006). In Australia, the first record of the genus *Halicyclops* was that of Kiefer (1967), who described *H. ambiguus* as a new species from four near-coastal saline lakes in Victoria and South Australia. Pesce et al. (1996) reported *H. spinifer* Kiefer, 1935, a widely distributed species, for the first time from Australia from a cave and a well on the Cape Range peninsula (Western Australia). In the same paper they also described one new species from an anchialine cave on the same peninsula as *H. longifurcatus*. De Laurentiis et al. (1999) described *H. rochai* as a new species from four bores and one well from the Pilbara region in Western Australia. Two years later, the same authors (De Laurentiis et al., 2001) described another new species from three bores from the Murchison region in Western Australia as *H. eberhardi*. In a survey of the copepod fauna of the Western Australian Murchison region I (Karanovic, 2004a) recorded *H. eberhardi* from an additional 11 localities, reported *H. ambiguus* for the first time from Western Australia, and described two new species: *H. kieferi* and *H. pescei*. Finally, in the survey of the copepod fauna of the Western Australian Pilbara region (Karanovic, 2006) the subgenus *Rochacyclops* was established, *H.* (*R.*) *rochai* was recorded and redescribed from an additional 20 localities, *H.* (*R.*) *calm* was described as a new species, and a key to all eight Australian *Halicyclops* species was provided. Australia seemed to be a very promising place for *Halicyclops* diversity and I expected many species from the marine interstitial, as this genus is one of the major cyclopoid components in this ecosystem around the world. Thus, it was very surprising to find only two species here, both with a relatively limited distribution.

The story of *H.* (*H.*) *thermophilus* Kiefer, 1929 is one that went from a poor start to a worse finish, as can be seen from the synonymy in this monograph. Menzel (1925) reported a species of *Halicyclops* from a hot saline spring on Java (37-45°C), which Kiefer (1929a) formally described as *H. thermophilus*, giving a very short diagnosis and only three drawings: the female caudal rami and the male and female fifth legs. He did not mention the shape of the genital double somite, and in a different publication (Kiefer, 1929b) he admitted not observing it at all. Three years later, Heberer & Kiefer (1932) redescribed *H. thermophilus* from the type locality, although it is not clear whether they used the type or newly collected material. They provided a more comprehensive description and drawings of the antennula, parts of the fourth swimming leg and genital double somite, along with drawings already published in the original description. However, the drawing of the genital double somite was rudimentary, made from memory after an observation of live specimens, not an accurate representation made with drawing aids attached to the microscope. This was clearly stated both in the text and in the figure legend.

The real confusion, that will follow *H.* (*H.*) *thermophilus* to this day, started when Kiefer (1935) described a new subspecies, *H. thermophilus spinifer*, from Port Canning near Calcutta (India) in only two short sentences, and giving only a drawing of the genital double somite. This subspecies was justified on the basis of somewhat longer spiniform projections on the genital double somite. Next year, in his revision of the genus, Kiefer (1936) gave a few more details of both subspecies, including the drawings of the caudal rami, the female fifth leg, and the third endopodal segment of the fourth leg for *H. t. spinifer*, and the male sixth leg and a proper drawing of the genital double somite for *H. t. thermophilus*, repeating again all previously published drawings (some now for the third time). However, as Kiefer (1936) gave the sketch and the proper drawing of the genital double somite next to each other, the difference in proportions between the two becomes obvious. Both subspecies were illustrated with identical caudal rami and pseudoperculum, as well as with a very similar third endopodal segment of the fourth swimming leg (although the inner setae of *H. t. thermophilus* seem to be more spiniform than those of *H. t. spinifer*). The main difference between the two subspecies seemed to be the relative length of the lateral protrusions of the genital double somite, and the relative length of the inner apical spine of the female fifth leg (both shorter in the nominotypical subspecies). Lindberg (1941) redescribed *H. t. spinifer* from material collected on the western side of the Indian subcontinent (from Bandra, near Bombay: now Mumbai), as well as from Iran. He also noticed some variability in the length of the caudal rami and the lateral protrusions on the genital double somite, but found all specimens to have plumose inner setae on the third endopodal segment of the fourth leg. Sewell (1949) found and redescribed *H. t. spinifer* from salt lakes in Calcutta (very close to the type locality), but his specimens had spiniform setae on the fourth leg and relatively short lateral projections on the genital double somite (previously thought to be characteristics of the nominotypical subspecies), along with a long inner apical spine on the female fifth leg. This alone was probably enough to synonymize the two subspecies, but Sewell (1949) did not do so and, strangely enough, his publication remained unknown to all subsequent workers. Lindberg (1949) considered *spinifer* and *thermophilus* as valid species, as did most subsequent taxonomists. Lindberg (1952) reported, as *H. thermophilus*, specimens from Madagascar, which also combined the characters of both forms (large lateral spines on the genital double somite, short inner apical spine on the fifth leg, and spiniform setae on the third endopodal segment of the fourth leg) and remarked: "Du reste je dois dire que les différences entre *H. thermophilus* et *H. spinifer* semblent bien peu accusées et qu'on ne doit pas écarter la possibilité que l'étude d'un matériel plus complet pourrait montrer qu'il faut les considérer comme représentant une seule et même espèce". However, he did not formally synonymize them, possibly because of Kiefer's great authority. Not surprisingly, the "*thermophilus*"-group (as it was often called) soon began to fill with "new" species, which could be distinguished from *H. thermophilus*, *H. spinifer*, and other members usually only by a single character. *Halicyclops venezuelaensis*, described from Venezuela by Lindberg (1954), had very short caudal rami and very small lateral protrusions on the genital double somite and, in my opinion, is the only other valid species in this group (see also Da Rocha, 1995b for a redescription of this species). *Halicyclops japonicus*, described by Itô (1956) from Japan, had the proximal inner seta on the third endopodal segment of the fourth swimming leg plumose and the distal seta spiniform (see also Ishida, 2002), while in all other aspects it could not be distinguished from *H. spinifer*. Shen & Tai (1964) described *H. latus* as a new species from China, with a relatively short inner apical spine on the female fifth leg (see also Shen et al., 1979), although this was not much different from the specimens reported by Lindberg (1952) from Madagascar as *H. thermophilus*. Brownell (1983) described *H. dedeckeri* as a new species from South Africa, which had the proximal inner seta on the third endopodal segment of the fourth leg spiniform and the distal seta plumose (a reverse situation from that found in *H. japonicus*). Herbst (1983) described *H. antiguaensis* from Antigua (West Indies) as a new species, although he struggled to find any clear characters that would distinguish it from other species, and especially from *H. latus*. The shape of the inner setae on the third endopodal segment of the fourth leg (plumose or spiniform) could not be accepted as the only distinguishing character in the genus *Halicyclops*, as this was also shown to be variable in *H. venezuelaensis* by Da Rocha (1995b). Thus, I here consider *H. antiguaensis*, *H. dedeckeri*, *H. japonicus*, *H. latus*, and *H. spinifer* all as junior synonyms of *H.* (*H.*) *thermophilus*.

Wilson (1958) synonymized Lindberg's (1941) *H. thermophilus spinifer* with *H. thermophilus*, but she did not synonymize other findings, and her action may not have been intentional, as noted by Brownell (1983). Yeatman (1983) redescribed both *H. thermophilus* and *H. spinifer* from South Pacific islands, but insisted on their subspecific status "because there are only minor differences in structure", although he found them sympatrically on Tonga. Despite this and his pretty schematic drawings, his report extended the known range of the species significantly. Reid (1985) reported *H. thermophilus* from Brazil, based on an unpublished Ph.D. Thesis of Da Rocha. Pesce et al. (1996) recorded *H. spinifer* for the first time from Australia and redescribed it from the material collected on the Cape Range peninsula, but also reported this species from Brazil based on personal information provided by Da Rocha. Whether Da Rocha's information given to Reid (1985) and Pesce et al. (1996) was based on the same specimens, or if he found both forms in Brazil is unknown to me, but since they are synonyms now it is also irrelevant. I repeated Pesce's et al. (1996) finding in my key to the species (Karanovic, 2006) also under the name *H.* (*H.*) *spinifer*. Finally, Mirabdullayev & Getz (1996) redescribed *H. spinifer* from a lake in Central Asia.

The specimens of *H.* (*H.*) *thermophilus* presented here from Karratha in the Western Australian Pilbara region are identical with those reported by Pesce et al. (1996) from the Cape Range peninsula, which further strengthens my hypothesis (Karanovic, 2006: 230) about a close relationship of these two Western Australian regions. Because *H. thermophilus* often inhabits river estuaries it is possible that its current, very wide distribution is partly a result of passive dispersal in ships' ballast waters. This would explain its relatively restricted distribution in Australia, and also a lack of variability in some populations.

26. **Halicyclops (H.) lutum** n. sp. (figs. 75-77)

Material examined. — Types only: Holotype ovigerous female dissected on 2 slides (WAM C37302); allotype male dissected on 1 slide (WAM C37303); paratypes, 1 female + 2 copepodids (female on 1 slide in toto (TMAG G5794), copepodids in alcohol (WAM C37304)): Australia, Western Australia, Derby, near jetty, 5 August 2003, leg. T. Karanovic, 19°17.459′S 123°36.549′E.

Description. — Ovigerous female (holotype). Body length, excluding caudal setae, 0.561 mm. Habitus (fig. 75A) very robust, with prosome/urosome ratio of 1.7 and greatest width near posterior end of cephalothorax. Body length/width ratio about 2.4; cephalothorax about 2.6 times as wide as genital double somite. Free pedigerous somites without pronounced lateral expansions. Colour of preserved specimen yellowish. Nauplius eye not visible. Rostrum well developed, membranous, broadly rounded, and furnished with two large sensilla.

Cephalothorax (fig. 75A) almost triangular, slightly shorter than its greatest width; representing 37% of total body length. Surface of cephalic shield with a few large sensilla; no other ornamentation visible. Hyaline fringe of prosomites narrow and smooth. Fifth pedigerous somite with four large dorsal sensilla and with a smooth fringe dorsally and ventrally (fig. 75B).

Genital double somite (fig. 75B) 1.1 times as long as wide, ornamented with two large sensilla dorsally in middle, with two pores ventrally and two cuticular recesses ventrolaterally at posterior end. Hyaline fringe of genital double somite and two succeeding somites not frilled, but very sharply serrated both dorsally and ventrally. No spiniform processes laterally. Copulatory pore small; copulatory duct narrow and short, not rigidly sclerotized. Seminal receptacle reduced into two narrow lateral arms that form a large letter "T" with copulatory duct (fig. 75B). Ovipores situated dorsolaterally, covered by the reduced sixth legs, bearing two minute spines and one smooth seta. Third urosomite with a pair of ventral pores posteriorly; fourth urosomite without ornamentation. Dorsal hyaline fringe of fourth somite not produced posteriorly in form of pseudoperculum. Anal somite with smooth, broad, and slightly convex anal operculum, which represents 45% of the somite's width; ornamented with two large sensilla dorsally, two pores ventrally, and a transverse row of spinules on posterior margin. Anal sinus widely open, ornamented with several rows of minute spinules (fig. 75B).

Caudal rami (fig. 75B) almost 1.7 times as long as wide, parallel, with space between them about one half of a ramus' width. Distal margin ven-

Fig. 75. *Halicyclops* (*H.*) *lutum* n. sp., holotype ovigerous female. A, habitus, dorsal view; B, urosome, ventral view; C, antenna; D, third endopodal segment of second swimming leg; E, third exopodal segment of second swimming leg; F, third exopodal segment of third swimming leg. Scales = 0.1 mm.

trally convex, but without cuticular tube pore. Ornamentation consisting of one ventral pore at 1/3 of ramus' length and several small spinules at base of outermost apical seta. Dorsal seta uniarticulate, smooth, and relatively long, almost twice as long as ramus, inserted close to posterior end of ramus, although not on a protuberance. Lateral seta arising at 2/5, positioned dorsolaterally, uniplumose, and shorter than ramus. Outermost apical seta strong and spinulose, about 1.3 times as long as ramus, bipinnate distally. Innermost apical seta very short and slender, about 0.4 times as long as ramus and completely smooth. Principal apical setae bipinnate and with breaking planes; inner seta about 2.3 times as long as outer one and 0.6 times as long as body (fig. 75A).

Antennula (fig. 76A) hardly reaching middle of cephalothorax, six-segmented (although articulation between first and second, and second and third segments not complete), with slender aesthetasc on fifth segment and two aesthetascs on sixth segment, and setal formula: 8.11.6.5.3.9. Apical aesthetasc fused basally to one apical seta. Only two apical setae on sixth segment articulating on basal part; no setae with breaking planes; one seta on distal outer corner of third segment spiniform and short. Majority of setae strong and pinnate or plumose, only 14 setae (33%) smooth. Ornamentation consisting of several spinules at base of first segment. Length ratio of antennular segments, from proximal end and along caudal margins (note first and second segment fused along caudal margin), 1 : 1 : 2.7 : 1.2 : 2.2.

Antenna (fig. 75C) three-segmented, comprising coxobasis and a two-segmented endopod. Coxobasis cylindrical, about 2.5 times as long as wide, unornamented, and armed with two bipinnate setae at distal inner corner; seta representing exopod absent. First endopodal segment somewhat shorter than coxobasis, armed with one bipinnate seta at middle, unornamented. Second endopodal segment only slightly longer than first, ornamented with several arched rows of spinules and armed with five lateral and six apical smooth setae; one lateral and four apical setae geniculate.

Labrum trapezoidal, but not clearly observed.

Mandibula with palp reduced to a small protuberance, which bears two setae; distal seta pinnate distally, much longer than proximal smooth seta, but not reaching gnathobasal teeth. Cutting edge of coxal gnathobase similar to that of previous species, but mounted awkwardly on slide.

Maxillula composed of well developed praecoxa and a two-segmented palp (fig. 77B). Arthrite of praecoxa with four apical, strong, short, and smooth spines. Praecoxa armed with seven armature elements along inner margin, the

Fig. 76. *Halicyclops* (*H.*) *lutum* n. sp., holotype ovigerous female. A, antennula; B, maxilliped; C, first swimming leg; D, third endopodal segment of third swimming leg; E, fourth swimming leg. Scales = 0.1 mm.

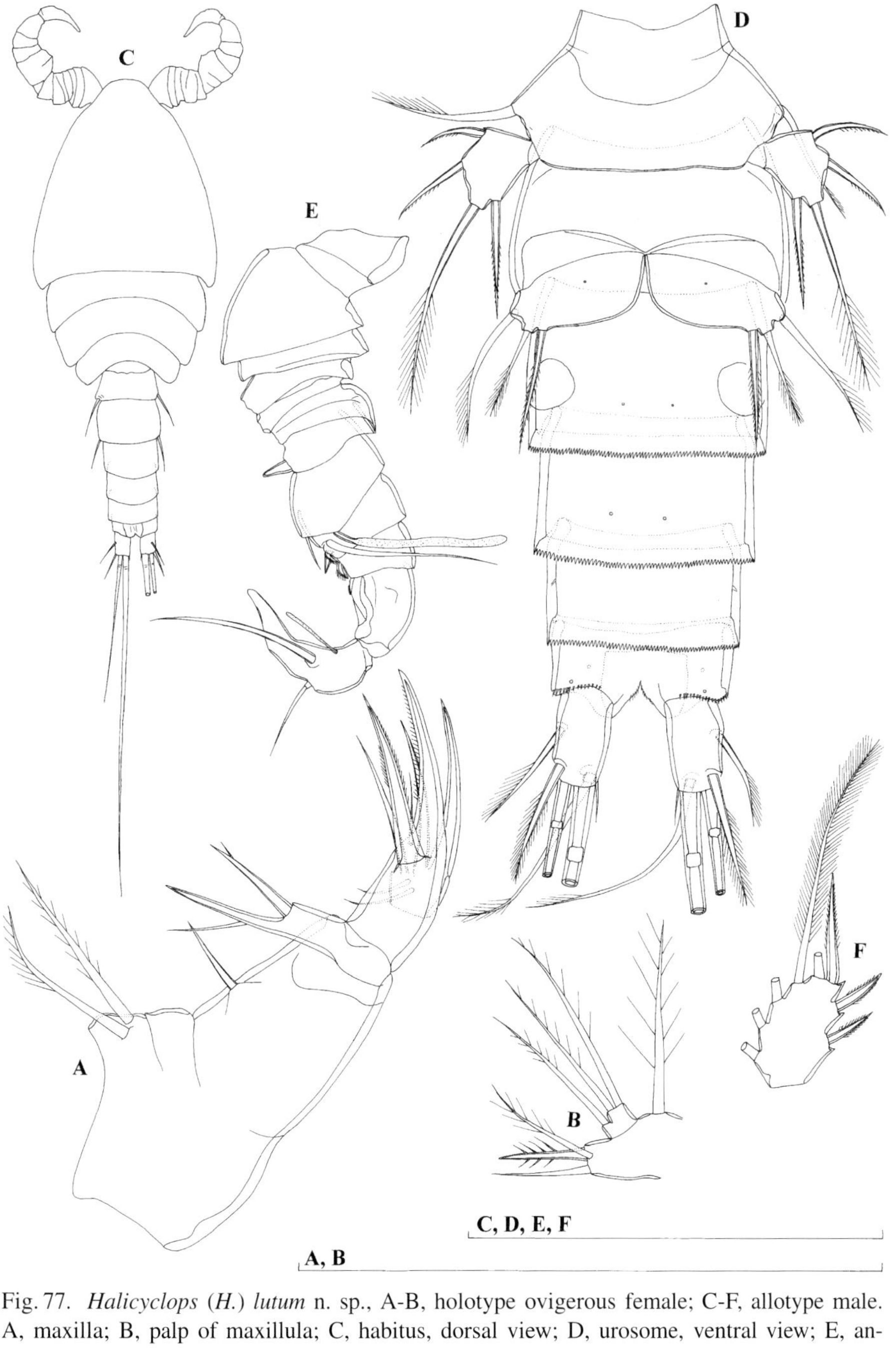

Fig. 77. *Halicyclops* (*H.*) *lutum* n. sp., A-B, holotype ovigerous female; C-F, allotype male. A, maxilla; B, palp of maxillula; C, habitus, dorsal view; D, urosome, ventral view; E, antennula with incomplete armature; F, third exopodal segment of third swimming leg. Scales = 0.1 mm.

longest one plumose. Palp with completely divided endopod, which bears two apical and one lateral pinnate setae, and armed laterally with one exopodal pinnate seta and apically with two slender (one smooth, the other pinnate) setae and one robust, strongly bipinnate spine.

Maxilla (fig. 77A) four-segmented but praecoxa fused to coxa on posterior surface; also mounted awkwardly on slide. Proximal endite of praecoxa armed with two pinnate setae; distal endite very small, unarmed. Proximal endite of coxa with one short and smooth seta; distal endite highly mobile and bearing one proximal, completely fused, and very stout spine, and one distal, slightly shorter stout seta, not fused to endite. Spine fused to endite ornamented with a single elongated spinule. Basis expanded into a spinulose claw and armed with two setae; the strong seta bipinnate and slightly shorter than claw, other seta smooth and small. Endopod armed with two strong and three slender setae, all of which smooth. Longer endopodal strong seta about as long as the strong seta on basis; shorter endopodal strong seta about twice as long as the longest endopodal slender seta. Basal claw with all spinules along concave margin of about the same size.

Maxilliped (fig. 76B) two-segmented, armed with three setae on protopod and five setae on endopod, but without any ornamentation. All three protopodal setae strong and geniculate, ornamented with several long spinules along convex margin; middle seta twice as long as proximal one and about three times as long as distal protopodal seta. Two lateral and the innermost apical endopodal setae very stout and with several very long spinules. Middle apical seta the longest, smooth, and about 1.3 times as long as the outermost, also smooth, apical seta. Protopod more than four times as long as endopod. Maxilliped also awkwardly mounted on slide, whence its somewhat unusual appearance when compared to the previous species.

All swimming legs with three-segmented exopod and endopod (figs. 75D, E, F, 76C, D, E). Armature formula of swimming legs as follows (legend: inner/outer spine or seta; inner/terminal/outer):

	Exopod			Endopod		
Segments	1	2	3	1	2	3
First leg	1/1	1/1	3/2/3	1/0	1/0	3/2/1
Second leg	1/1	1/1	4/2/3	1/0	2/0	3/2/1
Third leg	1/1	1/1	4/2/2	1/0	2/0	3/2/1
Fourth leg	1/1	1/1	4/2/2	1/0	2/0	2/2/1

Third exopodal segment spine formula: 4.4.3.3. Intercoxal sclerite of first leg with rounded protrusions at both sides, which are ornamented with an apical row of long spinules (fig. 76C); those of other swimming legs without ornamentation and with rounded protrusions much less pronounced. Coxae unornamented, except for coxa of first leg, which bears a row of spinules along posterior margin; armed with a plumose seta at inner distal corner; this seta very short on fourth leg (fig. 76E). Spine inserted at inner, protruded corner of basis of first leg very strong, with many strong spinules on both sides and almost reaching posterior margin of endopod (fig. 76C). Basis of each leg armed with outer smooth seta and ornamented with a row of spinules along distal margin; that of first leg with additional spinules at base of inner spine. All endopodal and exopodal setae slender and plumose, except for two modified setae on endopod of fourth leg. Third endopodal segment of fourth swimming leg (fig. 76E) about 1.5 times as long as wide; inner apical spine almost twice as long as outer one; two inner setae spiniform, bipinnate, and shorter than inner apical spine. All endopodal segments with hair-like spinules along outer margin and most exopodal segments with minute spinules along outer margin; their inner margins without ornamentation.

Fifth leg (fig. 75B) with basis and endopod completely fused to somite; outer basal seta inserted on short basal protuberance (which is about as long as wide) and unipinnate distally. Exopod about 1.7 times as long as wide, ornamented with several spinules at bases of outer spines. Armature consisting of three strong spines and one apical seta; inner apical (or subapical) spine as long as the segment, about 0.9 times as long as outer apical spine, and only slightly shorter than outer proximal spine. Exopodal apical seta plumose at distal end, about 1.7 times as long as the segment and 1.1 times as long as basal seta.

Sixth leg an indistinct, triangular plate, armed with two almost equal, smooth spines, which are fused to the plate, and one smooth, much longer seta.

Male (allotype). Body length, excluding caudal setae, 0.399 mm. Habitus (fig. 77C) less robust than in female, with prosome/urosome ratio of 1.6 and greatest width near posterior end of cephalothorax. Body length/width ratio about 2.5; cephalothorax about 2.9 times as wide as genital somite.

Cephalothorax (fig. 77C) only slightly longer than its greatest width. Ornamentation of prosomites, colour, and nauplius eye similar to those of female. Hyaline fringe of fifth pedigerous somite smooth both ventrally and dorsally;

that of genital somite sharply serrated dorsally; those of other urosomites (except anal somite) sharply serrated both dorsally and ventrally (fig. 77D).

Genital somite (fig. 77C) about 1.4 times as wide as long, ornamented with two dorsal sensilla; third urosomite (fig. 77D) with two cuticular pores ventrally and two cuticular recesses ventrolaterally; fourth urosomite with two dorsal sensilla and two ventral pores; fifth somite ornamented with only two lateral pores (one on each side). Anal somite with smooth, almost straight, and broad anal operculum, which represents 45% of the somite's width; ornamented with two sensilla dorsally and a transverse row of spinules on posterior margin, as well as with two ventral pores. Anal sinus widely open, ornamented with a few rows of minute spinules.

Caudal rami (fig. 77D) about 1.8 times as long as wide, almost parallel, with space between them about one ramus' width. Armature and ornamentation similar to those of female, although lateral seta much longer and no ventral cuticular pore visible.

Antennula (fig. 77E) of length almost equal to that of cephalothorax, 11-segmented, geniculate, with geniculation between eleventh and twelfth segments. Armature very tightly packed and difficult for positive identification. A few characteristic elements illustrated in fig. 77E.

Antenna, labrum, mandibula, maxillula, maxilla, maxilliped, and swimming legs (fig. 77F) all similar to those of female.

Fifth leg (fig. 77D) exopod about 1.5 times as long as wide, ornamented with two short rows of spinules on outer margin, and with several small spinules at base of spines. Armature same as in female, although inner apical spine proportionately longer. No additional inner setae present.

Sixth leg (fig. 77D) a distinct, large plate, armed with one spine and two pinnate setae; ornamented with a single cuticular pore. Outermost seta unipinnate, about 1.4 times as long as middle plumose seta, and 1.2 times as long as inner bipinnate spine.

Variability. — Body length of females ranging from 0.548 mm to 0.561 mm (0.555 mm average; n = 2). Only one male was collected and examined. No other type of variability or asymmetry was observed.

Etymology. — The species is named with the Latin noun "lutum" meaning "mud", and referring to the type of sediments this species was collected from. The name should be treated as a noun in apposition.

Distribution. — At present this species is known only from the type locality (Derby, Western Australia), although muddy habitats were not sampled extensively during this project and they are widespread across the northern part of the continent. I expect it to be found in the future in the littoral with muddy bottoms in the Kimberley region in Western Australia, as well as in most of the Northern Territory and most of northern Queensland.

Discussion. — *Halicyclops* (*H.*) *lutum* n. sp. could easily be distinguished from all previously described congeners by its spine formula (the number of spines on the third exopodal segment of the first to fourth swimming legs), which is 4.4.3.3. In most of the species of *Halicyclops* Norman, 1903, including the previously mentioned *H.* (*H.*) *thermophilus* Kiefer, 1929, this formula is 3.4.4.3, but many species have the formula variously reduced. For example, the formula 3.4.3.3 could be found in the following five species and subspecies: *H.* (*H.*) *brevispinosus meridionalis* Herbst, 1953; *H.* (*H.*) *eberhardi* De Laurentiis, Pesce & Humphreys, 2001; *H.* (*H.*) *laminifer* Herbst, 1982; *H.* (*H.*) *sinensis* Kiefer, 1928; and *H.* (*R.*) *tetracanthus* Da Rocha, 1996 (see Herbst, 1953; Monchenko, 1974; Shen et al., 1979; Herbst, 1982; Da Rocha, 1995b; De Laurentiis et al., 2001). However, a first swimming leg bearing four spines on the third exopodal segment, besides *H.* (*H.*) *lutum*, was so far reported only for *H.* (*H.*) *tenuispinus* Sewell, 1924, which was described from Chilka Lake in India by Sewell (1924). Bozic (1964) described *H.* (*H.*) *aquaesurgentis* as a new species from Réunion Island, stating that it has the same spine formula as another species described in the same paper (i.e., 3.4.4.3), but his drawing of the first leg clearly shows four spines on the third exopodal segment. Both *H.* (*H.*) *tenuispinus* and *H.* (*H.*) *aquaesurgentis* differ markedly from *H.* (*H.*) *lutum*, which further makes me label this character state as plesiomorphic.

Another character of *H.* (*H.*) *lutum* that is not commonly found in the genus *Halicyclops*, is a male fifth leg that is not sexually dimorphic. The following eight species have been described so far with this condition: *H.* (*H.*) *crassicornis* Herbst, 1955; *H.* (*H.*) *denticulatus* Kiefer, 1960; *H.* (*H.*) *exiguus* Kiefer, 1934; *H.* (*H.*) *gauldi* Pleşa, 1961; *H.* (*H.*) *korodiensis* Onabamiro, 1952; *H.* (*H.*) *maculatus* Da Rocha & Hakenkamp, 1993; *H.* (*H.*) *tageae* Lotufo & Da Rocha, 1993; and *H.* (*H.*) *ytororoma* Lotufo & Da Rocha, 1993 (see Kiefer, 1934, 1936, 1960b; Onabamiro, 1952; Herbst, 1955; Pleşa, 1961; Wells, 1967; Gabriel et al., 1987; Lotufo & Da Rocha, 1993a; and Da Rocha & Hakenkamp, 1993). As they do not have many characters in common, I believe this condition of the male fifth leg to be convergent in most of these species. It is very well possible that the additional inner seta on the male fifth leg (which is present in most of the species of *Halicyclops*) has been reduced independently a number of times. As mentioned above, the *Halicyclops* species with two inner setae on the male fifth leg have been allocated to the subgenus *Rochacyclops* Karanovic, 2006.

It is surprising that only two species of *Halicyclops* have been found in the Australian marine interstitial, and I could not think of another explanation than maybe the great diversity and abundance of the genus *Neocyclops* Gurney, 1927 here. Most of the described Australian *Halicyclops* species are inhabitants of continental waters. The genus now numbers nine species here, and below is the key to aid in their identification. It is basically the same key presented in Karanovic (2006), only emended for *H.* (*H.*) *lutum* and corrected for *H.* (*H.*) *thermophilus* instead of *H.* (*H.*) *spinifer* Kiefer, 1935.

KEY TO AUSTRALIAN SPECIES OF THE GENUS *HALICYCLOPS*

1. Innermost apical seta on caudal rami very short; cuticular recesses on urosome present; male fifth leg same as in female or with a single inner seta 3
– Innermost apical seta longer than caudal rami; cuticular recesses on urosome absent; male fifth leg with two inner setae 2
2. Swimming legs spine formula 3.4.4.3 .... *H.* (*R.*) *rochai* De Laurentiis, Pesce & Humphreys, 1999
– Spine formula 3(2).3.3.3 .... *H.* (*R.*) *calm* Karanovic, 2006
3. Genital double somite without lateral spiniform processes 4
– These processes present .... *H.* (*H.*) *thermophilus* Kiefer, 1929
4. Caudal rami less than twice as long as wide..................................... 5
– Rami more than four times as long as wide .... *H.* (*H.*) *longifurcatus* Pesce, De Laurentiis & Humphreys, 1996
5. Spine formula 3.4.4.3; male fifth leg with a single inner seta 6
– Spine formula 4.4.3.3; male fifth leg without inner seta .... *H.* (*H.*) *lutum* n. sp.
6. Cuticular tube on caudal rami present 7
– This tube absent 8
7. Integumental window on cephalothorax absent; caudal rami less than 1.5 times as long as wide .... *H.* (*H.*) *ambiguus* Kiefer, 1967
– Integumental window on cephalothorax present; caudal rami about 1.7 times as long as wide .... *H.* (*H.*) *pescei* Karanovic, 2004
8. Caudal rami 1.2 times as long as wide.. .... *H.* (*H.*) *eberhardi* De Laurentiis, Pesce & Humphreys, 2001
– Caudal rami twice as long as wide .... *H.* (*H.*) *kieferi* Karanovic, 2004

GENERAL DISCUSSION

Some three-quarters of the world's ice-free coastlines consist of sandy shores (Brown & McLachlan, 1990) and in Australia there are 9,416 mainland beaches: 3,411 in Western Australia; 1,650 in Queensland; 1,488 in the Northern Territory; 1,454 in South Australia; 721 in New South Wales; and 692 in Victoria (Short, 2006). In comparison to that, the 103 sampling sites made for this study, do not look like a very good coverage, and many interesting places and probably some species may have been missed. However, short-range endemism is not as frequent in the marine interstitial as it is in the freshwater subterranean habitats, and I believe most of the copepod diversity (certainly at the generic level) has now been sampled. For example, my very detailed and frequent sampling in Montenegro (unpublished Ph.D. Thesis) resulted in the discovery of only one new species of harpacticoid (Karanovic, 2000), after Petkovski's (1954, 1955a, b, 1957) studies of only several samples from the marine interstitial of this part of the Adriatic coast. Furthermore, most of the species described by Petkovski were later found in other parts of the Mediterranean Sea (see, for example, Apostolov & Marinov, 1988) and some have now even been recorded from Australia (Karanovic, 2006).

In order to sample as much of the copepod diversity from the marine interstitial of Australia as possible, all sampling sites were originally selected using maps, with the main criteria being that sites had to be accessible by road and approximately 100 km apart. It was hoped this would lead to samples from all types of sediments and in all conditions. Where a larger gap occurred (mostly due to inaccessibility) sampling sites were more frequent before and after the gap. Thus, I expect that future, more frequent sampling will serve to fine-tune the known distributions of some species, but will not result in any significant increase in the number of known species. The exception may be those parts of Australia that were very poorly sampled because of their inaccessibility (the Kimberley region in Western Australia, the majority of the Northern Territory coast, and northern Queensland), although these areas mostly support mangrove shores with muddy bottoms (Bridgewater, 1999), and the interstices in mud are generally too small to permit a proper metazoan interstitial fauna (Wells, 1986). Another promising place for further studies in Australia would be Tasmania, which may harbour some psychrophilic elements, but which was excluded from this study, as were all other continental islands. My estimate is that, at this stage, at least 70 to 80% of the copepod diversity has been sampled from this habitat in Australia.

As mentioned in the Introduction, only four harpacticoids described by Nicholls (1945a, b) were previously known from the Australian marine interstitial. Thus, this monograph provides the first records of poecilostomatoid and cyclopoid copepods from the Australian marine interstitial habitat. Poecilostomatoid copepods are not normally found in the marine interstitial, apart from a few species of the genus *Hemicyclops* Boeck, 1873, and they did not feature at all in Stygofauna Mundi, the synopsis of the global fauna inhabiting subterranean waters and the marine interstitial (Botosaneanu, 1986). In fact, two of the five poecilostomatoid species recorded in this study (*Onychocorycaeus pacificus* (Dahl, 1894) and *O. catus* (Dahl, 1894)) are actually planktonic elements, buried in the sand by the wave action. They were both collected dead, with all soft tissues completely disintegrated, and are only important as a potential food for other inhabitants of the marine interstitial. However, these planktonic species are rarely redescribed or examined in detail by taxonomists, and my specimens showed several abnormalities worth reporting. Three other poecilostomatoids seem to be true interstitial species. *Hemicyclops nichollsi* n. sp. was found, sometimes in large numbers, in southern Queensland and northern New South Wales (fig. 78) and seems to belong to a group of species with a western Pacific distribution. It is interesting to note that the part of southern Queensland and northern New South Wales, where *H. nichollsi* was recorded, was recognized as a well defined region by Giribet & Edgecombe (2006), who analysed the Gondwanan distribution of a centipede genus, as well as by Bridgewater (1999), who analysed the distribution of the mangroves in Australia. *Cemihyclops cedunensis* n. g. et sp. is the only representative of a newly established genus, so its zoogeographic affinities are hard to judge. However, this South Australian species (fig. 78) could well be a relict of an unsuccessful branch of the Clausidiidae Embleton, 1901, closely related to the previously mentioned genus *Hemicyclops*. *Polyankylis australis* n. sp. is only the second representative of the family Polyankyliidae Ho & Kim, 1997, the first one having been described from an intertidal pool in Korea. Both species were found at similar latitudes (at around 37°), which may suggest their preference for moderate to cold regions, but Korea and South Australia (fig. 78) may be some of the last refuges for this rare group of poecilostomatoid copepods.

Cyclopoids are much more successful colonizers of the marine interstitial than poecilostomatoids, although not as successful as harpacticoids. By

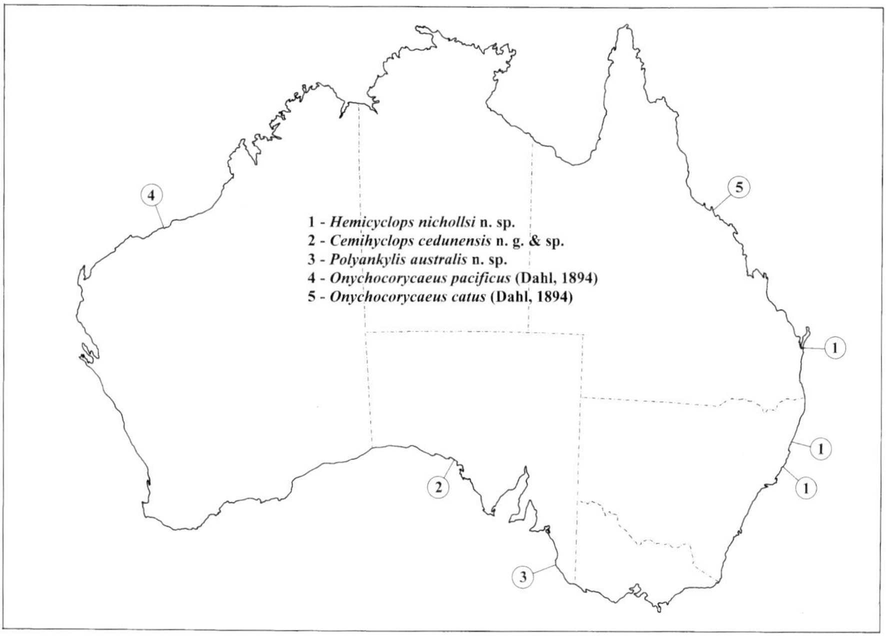

Fig. 78. Distribution of the poecilostomatoid copepods in the Australian marine interstitial.

1986, 54 cyclopoid and 338 harpacticoid species had been recorded globally (Herbst, 1986; Wells, 1986), cyclopoids thus representing less than 14%. In Australia the proportion seems to be comparable, as in Western Australia 10 cyclopoids are recorded and presented in this monograph, whereas (as mentioned in the Introduction) some 70 species of harpacticoids have been identified, which means that cyclopoids represent 12.5% of the total number. This is despite Australian beaches being predominately composed of fine to medium sand (0.1-1 mm), with a mean grain size of 0.4 mm, while only 4% of the beaches consist of coarse sand (Short, 2006), and with harpacticoids generally preadapted for smaller interstitial voids (Wells, 1986).

The only planktonic cyclopoid recorded in the Australian marine interstitial is *Oithona nana* Giesbrecht, 1893, which, like the two previously mentioned poecilostomatoids from the genus *Onychocorycaeus* Dahl, 1912, was probably also buried in the sand by the wave action. *Oithona nana* was found in South Australia (fig. 79) not far from where it was reported from the littoral by Nicholls (1944); the one damaged specimen I studied and redescribed proved useful for resolving the taxonomic status of this problematic species. As a result, another species (*O. attenuata* Faran, 1913) was established as its junior synonym (see the discussion section for *O. nana*). The other 20 species

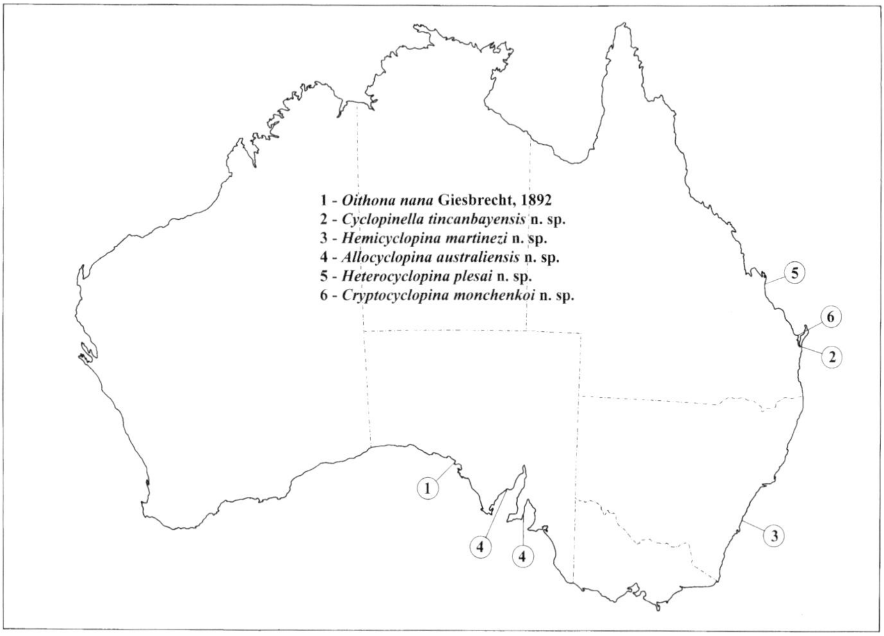

Fig. 79. Distribution of the families Oithonidae Dana, 1853 and Cyclopinidae G. O. Sars, 1913 (except the genus *Cyclopina* Claus, 1862) in the Australian marine interstitial.

of cyclopoids in this monograph seem to be either obligate dwellers in the marine interstitial, or use this ecosystem as a part of their habitat (i.e., seek and exploit resources here, either occasionally or permanently). The equivalent categories in freshwater subterranean habitats are called "stygobites" and "stygophiles" (Humphreys, 2000a), although the corresponding terms are not frequently used for the marine interstitial fauna. Precise terminology has always been problematic for these habitats (Giere, 1993). For example, Husman (1967) recognized no less than 27 biotopes (all with their different names) in the marine, brackish, and freshwater subterranean environment, but his complicated classification has never been widely accepted.

Only two cyclopoid species, other than *Oithona nana*, were known previously from other habitats in Australia, before being collected in the marine interstitial: *Neocyclops petkovskii* De Laurentiis, Pesce & Halse, 1997 and *Halicyclops* (*H.*) *thermophilus* Kiefer, 1929. The former species was recorded from the marine interstitial of South Australia (fig. 82) in this study, and from the Western Australian continental saline Lake McLeod by De Laurentiis et al. (1997). The latter species is widely distributed in tropical and subtropical regions of the world, but in Australia it was recorded only from the Pilbara

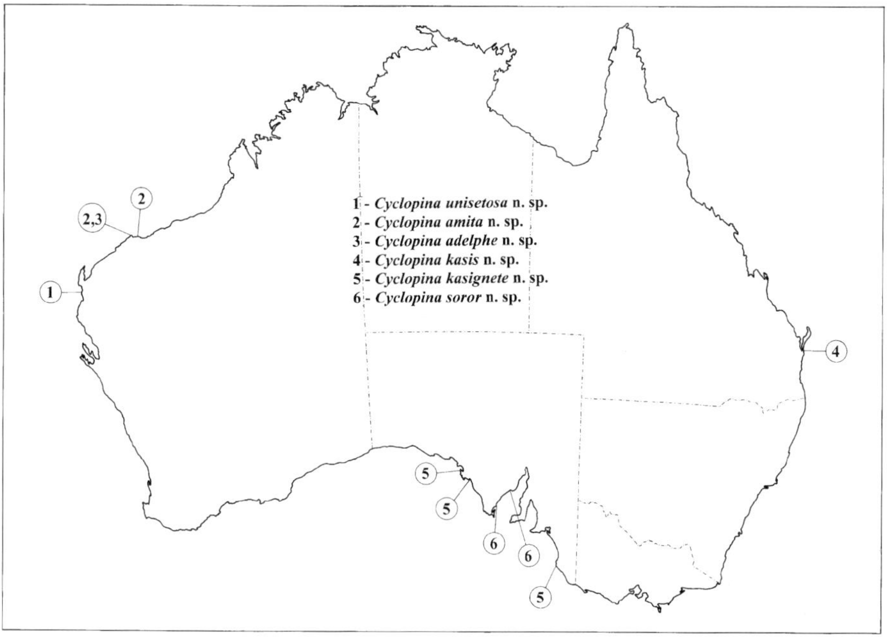

Fig. 80. Distribution of the genus *Cyclopina* Claus, 1862 in Australia.

region of Western Australia (fig. 81). I speculate that the distribution of *Halicyclops* (*H.*) *thermophilus* may be a consequence of passive dispersal in ship ballast waters (see the discussion section for this species), while the distribution of *Neocyclops petkovskii* may be used as a model for better understanding the colonization of freshwater habitats from the marine interstitial. The importance of the marine interstitial for colonizing the freshwater subterranean habitats has been frequently hypothesized (Giere, 1993; Humphreys, 2000b; Karanovic, 2004a), but, unfortunately, there is very little evidence for that in the copepods collected during this study. Only two cyclopoid genera reported here have some representatives in Australian continental waters: *Neocyclops* Gurney, 1927 and *Halicyclops* Norman, 1903. The only representative of the genus *Neocyclops* is the above mentioned *N. petkovskii*, which was found in continental saline lakes (although its precise habitat there is not quite clear). The genus *Halicyclops*, in addition to the two species from the marine interstitial, has seven species in continental waters, some of them in subterranean habitats. However, the marine interstitial species of this genus are not closely related to their freshwater congeners, so that any hypothesis for their origin from each other or from a common ancestor must be rejected. The Southern Hemisphere has traditionally been considered as having a fundamentally

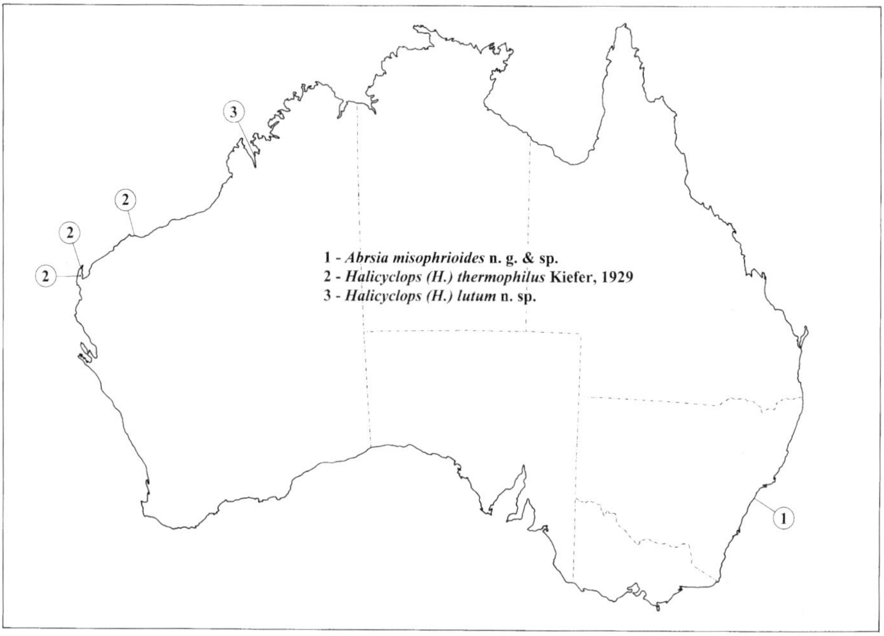

Fig. 81. Distribution of the genera *Abrsia* n. g. and *Halicyclops* Norman, 1903 in the Australian marine interstitial.

vicariant history (Sanmartín & Ronquist, 2004) and this is certainly true for many groups of Australian freshwater subterranean copepods (Karanovic, 2004a, 2006). However, dispersal seems to be a much more important mechanism in the marine interstitial, and ancestors of today's freshwater copepods of marine origin were probably replaced a long time ago by new colonizers.

Dispersal, facilitated by the West Wind Drift (see Waters & Roy, 2004), seems to be the most plausible explanation for the current distribution of the species of *Allocyclopina* Kiefer, 1954, which have been recorded from Madagascar, Réunion Island, and South Australia (fig. 79). Unfortunately, our knowledge about distribution patterns of copepods is hindered by a lack of data. For example, Jaume & Boxshall (1990) interpreted the lack of variability between three "disjunct" anchialine island populations of *Cyclopinella multispinosa* (Jaume & Boxshall, 1990) n. comb. as a consequence of their highly conservative morphology over evolutionary time (thus employing the vicariance model), which was very soon dismissed by Holmes & Gotto (2000), who found this species in the littoral as well. Thus, my hypothesis about a Tethyan origin of the genus *Cryptocyclopina* Monchenko, 1979 should be taken very cautiously (see the discussion section for *Cryptocyclo-*

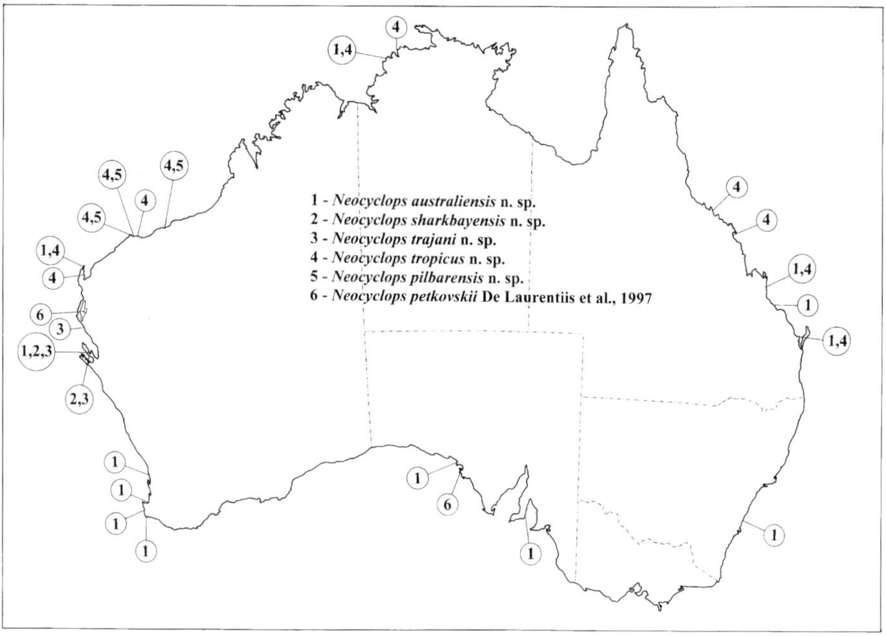

Fig. 82. Distribution of the genus *Neocyclops* Gurney, 1927 in Australia.

pina monchenkoi n. sp.). This genus is so far known only from two species, one from the interstitial of the Caspian and Black seas, and the other from the interstitial of Queensland (fig. 79).

The cyclopoid genera, *Cyclopinella* G. O. Sars, 1913, *Hemicyclopina* Herbst, 1953, *Cyclopina* Claus, 1862, *Heterocyclopina* Pleşa, 1968, and *Neocyclops* Gurney, 1927 are either cosmopolitan or very widely distributed and all, except *Hemicyclopina*, are known from a wide range of marine habitats. For example, the genus *Cyclopinella* has its four representatives in littoral mud, anchialine caves, deep hydrothermal sites, and the marine interstitial, and other cyclopoids exhibit even more extraordinary shifts in utilized habitats, as documented by Boxshall & Strong (2006). The affinities of the newly established and monospecific family Abrsiidae n. fam. naturally holds very few clues about its zoogeographic connections. It is even difficult to compare Australian cyclopoids from the marine interstitial to any other region in the world, as no comparable comprehensive study has been done so far. Of course, our knowledge of the marine interstitial fauna is biased towards that of beaches in Europe and the Mediterranean Sea, both in taxonomic and in ecological terms (Wells, 1986). In Stygofauna Mundi (Botasaneanu, 1986) Australia was divided into only two regions (Western Australia and

"Other areas of Australia and Tasmania"), while Europe consisted of no less than 34 different regions and subregions (actually "zones", "provinces", and "districts" in the original terminology).

Notwithstanding the difficulties in establishing global zoogeographic connections of the Australian marine interstitial cyclopoid fauna, a number of zoogeographic regions can clearly be recognized in Australia, based solely on subterranean copepods. The surprising differences between two neighbouring Western Australian regions (Murchison and Pilbara) on the generic level for copepods, and as also reported for other invertebrate groups, prompted me to develop a "pulsating desert hypothesis" as a possible explanation for the evolution of different faunas (Karanovic, 2006). The importance of looking at small-scale patterns when inferring Gondwanan biogeography was recently demonstrated by Giribet & Edgecombe (2006) for a centipede genus. They found that different parts of Australia have closer affinities to other Gondwanaland fragments than to each other. For example, they analysed northern and southern regions of both New South Wales and Queensland as distinct units, and subsequently recognized northern New South Wales and southern Queensland as a well-defined region. The discovery that Australian regions have different relationships to other Gondwanan areas was anticipated by Weston & Crisp (1994), who cautioned that assumptions of "monophyly" of large continental blocks as Australia were unwarranted. Bridgewater (1999) divided Australian mangroves into nine distinct biogeographic regions, which seems to be the most suitable model for the marine interstitial copepods as well. However, he did not recognize Shark Bay in Western Australia as a distinct region, probably because "on the western coastline of Australia mangol is relatively scarce, occurring in suitable habitats as a thin band between the sea and the highly arid interior" (Bridgewater, 1999). Also, he did not recognize eastern South Australia as a distinct region, as is indicated by the copepod data. Distributions of cyclopoid and poecilostomatoid copepods (presented here), as well as those of harpacticoid copepods (T. Karanovic, in prep.), strongly suggest that these two regions are distinct from other parts of Australia. Marine interstitial copepods strongly suggest the Pilbara region of Western Australia (together with Cape Range) is distinct from Shark Bay, despite the two areas neighbouring each other. A similar pattern is apparent in the freshwater subterranean habitats in the adjacent Hinterlands. The distribution patterns of the species of *Neocyclops* (fig. 82), the predominant cyclopoid genus in the Australian marine interstitial, provides support for separating the Pilbara and Shark Bay. While *N. australiensis* n. sp. has a circum-Australian distribution and *N. tropicus* n. sp. is distributed in tropical northern Australia (north from Cape Range in Western Australia), *N. pilbarensis* n. sp. is endemic to the Pilbara region (including Cape Range), while two species are endemic to Shark Bay: *N. sharkbayensis* n. sp. and *N. trajani* n. sp. I have discussed before (Karanovic, 2006) that the Pilbara region is inseparable from Cape Range, as far as subterranean and anchialine copepods are concerned, and the marine interstitial copepods confirm this conclusion.

The distribution of the *Cyclopina* species is also relevant (fig. 80). This genus has the same number of species in the Australian marine interstitial as *Neocyclops*, but the *Cyclopina* species are less frequent and always present in lower numbers. *Cyclopina* has no representatives in Shark Bay, while no less than three species are endemic to the Pilbara region: *C. unisetosa* n. sp., *C. amita* n. sp., and *C. adelphe* n. sp. Also, two *Cyclopina* species are endemic to South Australia (*C. kasignete* n. sp. and *C. soror* n. sp.), which, together with the previously mentioned endemic South Australian species (*Cemihyclops cedunensis*, *Polyankylis australis*, and *Allocyclopina australiensis*) strongly suggest this region to be separate from both southern Western Australia and from eastern Australia, in contrast to the respective proposals of Bridgewater (1999) and Botosaneanu (1986). Distributions of harpacticoid copepods from the marine interstitial (T. Karanovic, in prep.) also substantiate this view, but will not be discussed here.

Dispersal and vicariance are often considered competing hypotheses in historical biogeography (Sanmartín & Ronquist, 2004), but both processes appear to have shaped the recent distribution of Australian copepods. Vicariance was considered to be a more acceptable hypothesis for explaining zoogeographic connections of the freshwater subterranean representatives with disjunct distribution patterns (Karanovic, 2004a, 2006), while dispersal has been traditionally regarded as a better model for explaining recent distributions of marine and continental surface-water animals (Wilson, 1999; Reid, 2001; Waters & Roy, 2004; Karanovic, 2005b; Waters & Craw, 2006). However, dispersal cannot be completely rejected even for some subterranean copepods with disjunct distribution (see Karanovic & Ranga Reddy, 2004). It is interesting to note that the three most speciose cyclopoid genera (*Halicyclops*, *Cyclopina*, and *Neocyclops*) in the marine interstitial on a global scale (Herbst, 1986) are also the most speciose in the Australian marine interstitial. However, while the genus *Halicyclops* is the most speciose in the world, it is the least speciose of the three in Australia. On the other hand,

Halicyclops has many species in Australian continental waters (some as far as 700 km from the nearest seashore), while no *Halicyclops* species has been reported from inland waters of other continents. Of the two species presented in this monograph, one is postulated to be a recent arrival, and the other was collected from a muddy beach and may actually have been inhabiting crab burrows. The near absence of shore dwelling *Halicyclops* species in Australia may be associated with the great diversity and abundance of the genus *Neocyclops*, which may be competing for the same resources. *Neocyclops* may also be a relatively recent arrival (although not as a single event, because Australian species are not more closely related to each other than to other congeners). Most *Neocyclops* species are known from the Neotropical region and could have colonized, and still colonize, Australia using the West Wind Drift.

The macrofauna and the interstitial fauna of sandy beaches comprise distinct communities, with a few or no trophic links (Brown & McLachlan, 1990). The prime function of the interstitial system is the processing of organic materials flushed into the sand, which is mineralized by a food chain having heterotrophic bacteria at its base and predatory meiofauna (including copepods) at the apex. In the process, nutrients are recycled to the sea. The interstitial system should, therefore, be seen as a carbon sink (Brown & McLachlan, 1990), which has significant implications in the age when we are trying to fight carbon levels in the atmosphere. The interstitial system may be seen as a gigantic natural filter, cleansing and purifying surface waters: it mineralizes the organic materials it receives, and returns the nutrients to the sea. However, we know very little about the dynamics of this system and are just starting to investigate the species compositions of some of the more important animal groups. Copepods, as all top predators in an ecosystem, are crucially important for the health of the marine interstitial.

ACKNOWLEDGMENTS

This study was financially supported by the Australian Biological Resource Study (ABRS Grant No. 204-05). Thanks are extended to the Western Australian Museum for granting me the status of a Research Associate during the preparation of this book. I am very grateful to Dr. Stuart Halse, from the Department of Enviroment and Conservation in Perth, for his valuable comments on an early draft of this monograph. Dr. William F. Humphreys, from the Western Australian Museum in Perth, also checked part of the early draft. My wife and daughter accompanied me on all field trips and provided much needed support.

Figs. 83-114

Fig. 83. Western Australia, Nanga Bay. Type locality of *Neocyclops sharkbayensis* n. sp. and *N. trajani* n. sp.

Fig. 84. Western Australia, Shell Beach. Home of *Neocyclops australiensis* n. sp., *N. sharkbayensis* n. sp. and *N. trajani* n. sp.

Fig. 85. Western Australia, Shell Beach, detail of sampling hole.

Fig. 86. Western Australia, Coral Bay. Type and only locality of *Cyclopina unisetosa* n. sp.

Fig. 87. Western Australia, Exmouth. Home of *Neocyclops australiensis* n. sp. and *N. tropicus* n. sp.

Fig. 89. Western Australia, Pebble Beach, detail of sampling hole.

Fig. 108. New South Wales, Harrington. Home of *Hemicyclops nichollsi* n. sp.

Fig. 88. Western Australia, Pebble Beach. Home of *Neocyclops tropicus* n. sp.

Fig. 90. Western Australia, Dampier. Type locality of *Cyclopina amita* n. sp., *C. adelphe* n. sp., *Neocyclops tropicus* n. sp., and *N. pilbarensis* n. sp.

Fig. 91. Western Australia, Karratha. Home of *Cyclopina amita* n. sp., *Neocyclops tropicus* n. sp., *N. pilbarensis* n. sp., and *Halicyclops* (*H.*) *thermophilus* Kiefer, 1929.

Fig. 92. Western Australia, Point Samson. Home of *Neocyclops tropicus* n. sp. and a new family of Harpacticoida (in preparation).

Fig. 93. Western Australia, Point Samson, detail of sampling hole.

Fig. 94. Western Australia, Port Hedland. Home of *Onychocorycaeus pacificus* (Dahl, 1894), *Neocyclops tropicus* n. sp., and *N. pilbarensis* n. sp.

Fig. 95. Western Australia, Derby. The type and only locality of *Halicyclops* (*H.*) *lutum* n. sp.

Fig. 96. Northern Territory, Dundee Beach. Home of *Neocyclops australiensis* n. sp. and *N. tropicus* n. sp.

Fig. 97. Northern Territory, Dundee Beach, detail of sampling hole.

Fig. 98. Northern Territory, Darwin. Home of *Neocyclops tropicus* n. sp.

Fig. 99. Queensland, Midge Point. Another home of *Neocyclops tropicus* n. sp.

Fig. 100. Queensland, Midge Point, detail of sampling hole.

Fig. 101. Queensland, Yeppoon. Type and only locality of *Heterocyclopina plesai* n. sp. and home of *Neocyclops australiensis* n. sp. and *N. tropicus* n. sp.

Fig. 102. Queensland, Gladstone. Home of *Neocyclops australiensis* n. sp.

Fig. 103. Queensland, Hervey Bay. Type and only locality of *Cryptocyclopina monchenkoi* n. sp. and also home of *Neocyclops australiensis* n. sp. and *N. tropicus* n. sp.

Fig. 104. Queensland, Hervey Bay, detail of sampling hole.

Fig. 105. Queensland, Tin Can Bay. Type locality of *Hemicyclops nichollsi* n. sp., *Cyclopinella tincanbayensis* n. sp., and *Cyclopina kasis* n. sp.

Fig. 106. New South Wales, Moonee Beach. Home of *Hemicyclops nichollsi* n. sp.

Fig. 107. New South Wales, Moonee Beach, detail of sampling hole.
(For fig. 108, see p. 302.)

Fig. 109. New South Wales, Swansea. Type and only locality of *Abrsia misophrioides* n. g., n. sp., the sole representative of a new family of Cyclopoida.

Fig. 110. New South Wales, Swansea, detail of sampling hole.

Fig. 111. New South Wales, Wollongong. Type and only locality of *Hemicyclopina martinezi* n. sp. and also home of *Neocyclops australiensis* n. sp.

Fig. 112. South Australia, Robe. Type locality of *Polyankylis australis* n. sp. and *Cyclopina kasignete* n. sp.

Fig. 113. South Australia, Venus Bay. Home of *Cyclopina kasignete* n. sp.

Fig. 114. South Australia, Smoky Bay, with jetty. Home of *Oithona nana* Giesbrecht, 1893 and *Neocyclops australiensis* n. sp.

REFERENCES

AGUESSE, P.-C. & B.-H. DUSSART, 1957. Sur quelques Crustacés de Camargue et leur écologie. Vie et Milieu, **7**: 481-520.

APOSTOLOV, A. M. & T. M. MARINOV, 1988. Copepoda, Harpacticoida [Morski harpaktikoidi]. Fauna na Bulgaria, **18**: 1-383. (Bulgarska Akademia na Naukite, Sofia).

BODIN, P., 1997. Catalogue of the new marine harpacticoid copepods (1997 edition). Documents de Travail de l'Institut Royal des Sciences Naturelles de Belgique, **89**: 1-304. (Brussels).

BOESGAARD, T. M. & R. M. KRISTENSEN, 2001. Tardigrades from Australian marine caves with a redescription of *Actinarctus neretinus* (Arthrotardigrada). Zoologischer Anzeiger, **240**: 253-264.

BOSSANYI, J. & H. O. BULL, 1971. The marine fauna of the Cullercoats District, 5. Arthropoda 3c: Crustacea; Copepoda. Report of the Dove Marine Laboratory, **3**(17): 1-59.

BOTOSANEANU, L., 1986. General introduction. In: L. BOTOSANEANU (ed.), Stygofauna mundi, a faunistic, distributional, and ecological synthesis of the world fauna inhabiting subterranean waters (including the marine interstitial): 1-4. (E. J. Brill / Dr. W. Backhuys, Leiden).

BOXSHALL, G. A. & S. H. HALSEY, 2004. An introduction to copepod diversity **1-2**: 1-966. (The Ray Society, London).

BOXSHALL, G. A. & A. G. HUMES, 1987. A new species of *Hemicyclops* (Copepoda: Poecilostomatoida) associated with as echiuran worm in Hong Kong. Asian Marine Biology, **4**: 61-66.

BOXSHALL, G. A. & E. E. STRONG, 2006. An extraordinary shift in life habitat within a genus of cyclopoid copepods in Lake Tanganyika. Zoological Journal of the Linnean Society, London, **146**: 275-285.

BOZIC, B., 1964. Copépodes Harpacticoïdes et Cyclopoïdes de la Réunion, II. Plage St. Pierre. Bulletin du Muséum National d'Histoire Naturelle, Paris, **36**: 481-499.

BRIAN, A., 1938. Description d'une nouvelle espèce de Copépode Cyclopoïde du genre *Cyclopina*. Bulletin de la Société Zoologique de France, **63**: 13-18.

BRIDGEWATER, P., 1999. Mangroves. In: A. E. ORCHARD (ed.), Flora of Australia (2nd ed.), **1**, Introduction: 416-420. (Australian Biological Resources Study, Canberra).

BROWN, A. C. & A. MCLACHLAN, 1990. Ecology of sandy shores: 1-328. (Elsevier, Amsterdam).

BROWNELL, C. L., 1983. A new euryhaline copepod from Cape Town: *Halicyclops dedeckeri* n. sp. (Copepoda: Cyclopoida). South African Journal of Zoology, **18**: 62-66.

BRUCE, A. J. & P. J. F. DAVIE, 2006. A new anchialine shrimp of the genus *Procaris* from Christmas Island: the first occurrence of the Procarididae in the Indian Ocean (Crustacea: Decapoda: Caridea). Zootaxa, **1238**: 23-33.

CAMPOS-HERNÁNDEZ, A. & E. SUÁREZ-MORALES, 1993. Clave para la identificación de las especies de Corycaeidae (Copepoda; Poecilostomatoida) del Golfo de México y zonas adyacentes del Mar Caribe. Anales del Instituto de Ciencias del Mar y Limnología, Universidad Nacional Autónoma de México, **20**: 117-124.

CHAPPUIS, P. A., 1942. Eine neue Methode zur Untersuchung der Grundwasserfauna. Acta Scientifica Mathematisch-Naturwissenschaftlichen Universität Francisco-Josephinae, Koloszvar, **6**: 1-7.

CHAPPUIS, P. A. & M. SERBAN, 1953. Copépodes de la nappe phréatique de la plage d'Agigea près Constanza. Notes Biospéologiques, **8**: 91-102.

CHEN, Q., S. ZHANG & C. ZHU, 1974. On planktonic copepods of the Yellow Sea and the East China Sea. Studia Marina Sinica, **9**: 27-76.

COULL, B. C. & B. W. DUDLEY, 1985. Dynamics of meiobenthic copepod populations: a long-term study (1973-1983). Marine Ecology Progress Series, **24**: 219-229.

DAHL, M., 1912. Die Copepoden der Plankton-Expedition, I. Die Corycaeinen, mit Berücksichtigung aller bekannten Arten. Ergebnisse der Plankton-Expedition, **2**(G)(f)(1): 1-134. (Verlag von Lipsius & Tischer, Kiel and Leipzig).

DAMKAER, D. M., 1996. Copepod taxonomy: discovery vs. recognition. Proceedings of the Biological Society of Washington, **109**: 687-694.

DA ROCHA, C. E. F., 1995a. *Neocyclops* (*Protoneocyclops*) *ferrarii*, a new species of cyclopoid (Crustacea: Cyclopoida) from Belize, with remarks on the morphology of the genus *Neocyclops*. Contributions to Zoology, Amsterdam, **65**: 45-51.

——, 1995b. Copepods of the genus *Halicyclops* (Cyclopidae) from Belize. Hydrobiologia, **308**: 1-11.

DA ROCHA, C. E. F. & C. C. HAKENKAMP, 1993. New species of *Halicyclops* (Copepoda Cyclopidae) from the United States of America. Hydrobiologia, **259**: 145-156.

DA ROCHA, C. E. F. & T. M. ILIFFE, 1993. New cyclopoids (Copepoda) from anchialine caves in Bermuda. Sarsia, **78**: 43-56.

DE LAURENTIIS, P., G. L. PESCE & S. A. HALSE, 1997. Discovery of the first representative of the genus *Neocyclops* Gurney in Australia, and description of *Neocyclops* (*Neocyclops*) *petkovskii* n. sp. (Copepoda, Halicyclopinae). Bulletin Zoölogisch Museum Universiteit van Amsterdam, **16**: 15-19.

DE LAURENTIIS, P., G. L. PESCE & W. F. HUMPHREYS, 1999. Copepods from ground waters of Western Australia, IV. Cyclopoids from basin and craton aquifers (Crustacea: Copepoda: Cyclopidae). Records of the Western Australian Museum, **19**: 243-257.

——, —— & ——, 2001. Copepods from ground waters of Western Australia, VI. Cyclopidae (Crustacea: Copepoda) from the Yilgarn Region and the Swan Coastal Plain. In: W. F. HUMPHREYS & M. S. HARVEY (eds.), Subterranean biology in Australia 2000. Records of the Western Australian Museum, (Suppl.) **64**: 115-131.

DUSSART, B. H., 1974. Contribution à l'étude des Copépodes des eaux douces d'Éthiopie. Bulletin de l'Institut Français d'Afrique Noire, **36**: 92-116.

DUSSART, B.[H.] & D. DEFAYE, 1985. Répertoire mondial des Copépodes Cyclopoïdes: 1-236. (Centre National de la Recherche Scientifique, Bordeaux).

ESTERLY, C. O., 1905. The pelagic Copepoda of the San Diego region. University of California Publications in Zoology, **2**: 113-233.

FARRAN, G. P., 1911. Plankton from Christmas Island, Indian Ocean, I. On Copepoda of the family Corycaeidae. Proceedings of the Zoological Society of London, **1911**: 282-296.

——, 1913. Plankton from Christmas Island, Indian Ocean, II. On Copepoda of the genera *Oithona* and *Paroithona*. Proceedings of the Zoological Society of London, **1913**: 181-193.

——, 1936. Copepoda. Scientific Reports of the Great Barrier Reef Expedition 1928-1929, British Museum (Natural History), **5**: 73-142.

FIERS, F., 1986. New and interesting copepods (Crustacea, Copepoda) from brackish waters of Laing Island (northern Papua New Guinea). Bulletin de l'Institut Royal des Sciences Naturelles de Belgique, Brussels, (Biologie) **56**: 99-120.

FOSSHAGEN, A., 1970. *Thespesiopsyllus paradoxus* (Sars) (Copepoda, Cyclopoida) from western Norway. Sarsia, **42**: 33-40.

FRÜCHTL, F., 1923. Notizen über die Varibilität nordadriatischer Planktoncopepoden. Verhandlungen der Zoologisch-Botanischen Gesellschaft in Wien, **73**: 135-157.

——, 1024. Die Cladoceren- und Copepoden-Fauna des Aru-Archipels, mit Beiträgen zur Kenntnis der strukturellen Anomalien indo-pazifischer Plankton-copepoden. Arbeiten aus dem Zoologischen Institut der Universität Innsbruck, **1**: 1-18.

GABRIEL, C., S. MAAS, H. J. DUMONT & A. B. M. EGBORGE, 1987. *Halicyclops korodiensis* Onabamiro (Crustacea, Copepoda) in the estuary of the Warri River, Nigeria, West Africa. Hydrobiologia, **144**: 155-161.

GIERE, O., 1993. Meiobenthology, the microscopic fauna in aquatic sediments: 1-328. (Springer-Verlag, Berlin).

GIRIBET, G. & G. D. EDGECOMBE, 2006. The importance of looking at small-scale patterns when inferring Gondwanan biogeography: a case study of the centipede *Paralamyctes* (Chilopoda, Lithobiomorpha, Henicopidae). Biological Journal of the Linnean Society, London, **89**: 65-78.

GOMEZ, S. & P. MARTÍNEZ ARBIZU, 2004. First record of the genus *Cyclopina* (Copepoda: Cyclopoida), and fully illustrated redescription of *Cyclopina caissara* from northwestern Mexico. Anales del Instituto de Biología, Universidad Nacional Autónoma de México, (Zoología) **75**: 121-134.

GONZÁLES, J. G. & T. E. BOWMAN, 1965. Planktonic copepods from Bahia Fosforescente, Puerto Rico, and adjacent waters. Proceedings of the United States National Museum, **117**: 241-303.

GOODING, R. U., 1960. North and South American copepods of the genus *Hemicyclops* (Cyclopoida: Clausidiidae). Proceedings of the United States National Museum, Smithsonian Institution, Washington, D.C., **112**: 159-195.

GRANDORI, R., 1926. Nuove specie di Copepodi della Laguna Veneta. Bollettino dell'Istituto Zoologico dell R. Università di Roma, **3**: 38-70.

GURNEY, R., 1927a. IX Report on the Crustacea Copepoda of brine-pools at Kabret. Transactions of the Zoological Society of London, **22**: 173-177.

——, 1927b. Report on Crustacea, Copepoda (littoral and semiparasitic). Transactions of the Zoological Society of London, **22**: 451-577.

HAMOND, R., 1968. Some marine copepods (Misophrioida, Cyclopoida, and Notodelphyoida) from Norfolk, Great Britain. Crustaceana, (Suppl.) **1**: 37-60.

——, 1973. The marine and brackish-water copepods of Norfolk: Calanoida, Misophrioida, Cyclopoida, Monstrilloida, Notodelphyoida and incertae sedis. Cahiers de Biologie Marine, **14**: 335-360.

HARRIS, R. P., 1972a. Horizontal and vertical distribution of the interstitial harpacticoid copepods of a sandy beach. Journal of the Marine Biological Association of the United Kingdom, **52**: 375-388.

——, 1972b. Seasonal changes in population density and vertical distribution of harpacticoid copepods on an intertidal sand beach. Journal of the Marine Biological Association of the United Kingdom, **52**: 493-505.

——, 1972c. Reproductive activity of the interstitial copepods of a sandy beach. Journal of the Marine Biological Association of the United Kingdom, **52**: 507-524.

HEBERER, G. & F. KIEFER, 1932. Zur Kenntnis der Copepodenfauna der Sunda-Inseln. Archiv für Naturgeschichte Berlin, (Neue Folge) **1**: 225-274.

HERBST, H. V., 1952. Neue Cyclopoida Gnathostoma (Crustacea Copepoda) des Küstengrundwassers. Kieler Meeresforschungen, **9**: 94-111.

——, 1953. Weitere Cyclopoida Gnathostoma (Crustacea Copepoda) des Küstengrundwassers. Kieler Meeresforschungen, **9**: 257-270.

——, 1955. Cyclopoida Gnathostoma (Crustacea Copepoda) von der brasilianischen Atlantikküste. Kieler Meeresforschungen, **11**: 214-229.

——, 1962. Marine Cyclopoida Gnathostoma (Copepoda) von der Bretagne-Küste als Kommensalen von Polychaeten. Crustaceana, **4**: 191-206.

——, 1964. Cyclopoida Gnathostoma (Crustacea Copepoda) aus dem Litoral und Küstengrundwasser des Roten Meeres. Kieler Meeresforschungen, **20**: 155-169.

——, 1972. Drei neue marine Cyclopoida Gnathostoma (Crustacea: Copepoda) aus dem nordamerikanischen Küstenbereich. Gewässer und Abwässer, **68/69**: 107-124.

——, 1983. Ein neuer Copepode (Cyclopoida: Gnathostoma) von den Westindischen Inseln: *Halicyclops antiguaensis* n. sp. Bijdragen tot de Dierkunde, Amsterdam, **53**: 262-266.

——, 1986. Copepoda: Cyclopoida aus dem Meeres- und Brackwasser-Interstitial. In: L. BOTOSANEANU (ed.), Stygofauna mundi. A faunistic, distributional and ecological synthesis of the world fauna inhabiting subterranean waters (including the marine interstitial): 313-320. (E. J. Brill / Dr. W. Backhuys, Leiden).

HICKS, G. R. F. & B. C. COULL, 1983. The ecology of marine meiobenthic harpacticoid copepods. Annual Review of Oceanography and Marine Biology, **21**: 67-175.

HO, J.-S., M. DOJIRI, G. HENDLER & G. B. DEETS, 2003. A new species of Copepoda (Thaumatopsyllidae) symbiotic with a brittle star from California, U.S.A., and designation of a new order Thaumatopsylloida. Journal of Crustacean Biology, **23**: 582-594.

HO, J.-S. & I.-H. KIM, 1991. Two new species of the genus *Hemicyclops* (Copepoda, Poecilostomatoida, Clausidiidae) from crab burrows in the Yellow Sea. Korean Journal of Zoology, **34**: 289-299.

—— & ——, 1990. *Hemicyclops ctenidis*, a new poecilostomatoid copepod (Clausidiidae) associated with a polychaete in Korea. Korean Jounal of Zoology, **33**: 231-237.

—— & ——, 1997. A new family of poescilostomatoid copepods (Polyankyliidae) from a tide pool on mud flat in Korea. Korean Journal of Biological Sciences, **1**: 429-434.

—— & ——, 2004. A new genus of the Clausidiidae (Copepoda: Poecilostomatoida) associated with a polychaete from Korea, with discussion of the taxonomic status of *Hersiliodes* Canu, 1888. Proceedings of the Biological Society of Washington, **117**: 95-105.

HO, J.-S., S. OHTSUKA & N. NAKADACHI, 2006. A new family of poecilostomatoid copepods (Umazuracolidae) based on specimens parasitic on the black scraper (*Thamnaconus modestus*) in Japan. Zoological Science, **23**: 483-496.

HO, J.-S. & V. E. THATCHER, 1898. A new family of cyclopoid copepods (Ozmanidae) parasitic in the hemocoel of a snail from the Brazilian Amazon. Journal of Natural History, London, **23**: 903-911.

HO, J.-S. & W. J. WARDLE, 1992. *Pholadicola intestinalis*, new genus and species, a clausidiid copepod parasitic in a deep-burrowing clam from Texas. Bulletin of Marine Science, **51**: 37-44.

HOLMES, J. M. C. & R. V. GOTTO, 2000. A checklist of the Cyclopoida (Crustacea: Copepoda) of Ireland. Bulletin of the Irish Biogeographical Society, **24**: 2-42.

HOLTHUIS, L. B. 1954. C. S. Rafinesque as a carcinologist, an annotated compilation of the information on Crustacea contained in the works of that author. Zoologische Verhandelingen, Leiden, **25**: 1-43.

HOWART, F. G., 1980. The zoogeography of specialized cave animals: a bioclimatic model. Evolution, **34**: 394-406.

HULINGS, N. C. & J. S. GRAY, 1971. A manual for study of meiofauna. Smithsonian Contributions to Zoology, **78**: 1-83.

HUMES, A. G., 1965. New species of *Hemicyclops* (Copepoda, Cyclopoida) from Madagascar. Bulletin of the Museum of Comparative Zoology, Harvard, **134**: 159-260.

——, 1984. *Hemicyclops columnaris* sp. n. (Copepoda, Poecilostomatoida, Clausidiidae) associated with a coral in Panama (Pacific side). Zoologica Scripta, **13**: 33-39.

——, 1995. Three new species of *Hemicyclops* (Copepoda: Poecilostomatoida: Clausidiidae) from northwestern Madagascar. Bulletin du Muséum National d'Histoire Naturelle, Paris, **17**: 141-162.

——, 1999. Copepoda (Cyclopinidae and Misophriidae) from a deep-sea hydrothermal site in the northeastern Pacific. Journal of Natural History, London, **33**: 961-987.

HUMPHREYS, W. F., 2000a. Background and glossary. In: H. WILKENS, D. C. CULVER & W. F. HUMPHREYS (eds.), Ecosystems of the world, **30**, Subterranean ecosystems: 3-14. (Elsevier, Amsterdam).

——, 2000b. Relict faunas and their derivation. In: H. WILKENS, D. C. CULVER & W. F. HUMPHREYS (eds.), Ecosystems of the world, **30**, Subterranean ecosystems: 417-432. (Elsevier, Amsterdam).

——, 2001. Groundwater calcrete aquifers in the Australian arid zone: the context to an unfolding plethora of stygal biodiversity. In: W. F. HUMPHREYS & M. S. HARVEY (eds.), Subterranean biology in Australia 2000. Records of the Western Australian Museum, (Suppl.) **64**: 63-83.

——, 2006. Aquifers: the ultimate groundwater-dependent ecosystems. Australian Journal of Botany, **54**: 115-132.

HUMPREYS, W. F. & D. L. DANIELOPOL, 2006. *Danielopolina* (Ostracoda, Thaumatocyprididae) on Christmas Island, Indian Ocean, a sea mount island. Crustaceana, **78**: 1339-1352.

HUSMANN, S. 1967. Klassifizierung mariner, brackiger und limnischer Grundwasserbiotope. Helgoländer Wissenschaftliche Meeresuntersuchungen, **16**: 271-278.

HUYS, R., 1988. Gelyelloida, a new order of stygobiont copepods from European karstic systems. Hydrobiologia, **167/168**: 485-495.

HUYS, R. & G. A. BOXSHALL, 1990. The rediscovery of *Cyclopicina longifurcata* (Scott) (Copepoda: Cyclopinidae) in deep water in the North Atlantic, with a key to genera of the subfamily Cyclopininae. Sarsia, **75**: 17-32.

—— & ——, 1991. Copepod evolution: 1-468. (The Ray Society, London).

HUYS, R., J. LLEWELLYN-HUGHES, P. D. OLSON & K. NAGASAWA, 2006. Small subunit rDNA and Bayesian inference reveal *Pectenophilus ornatus* (Copepoda incertae sedis) as highly transformed Mytilicolidae, and support assignment of Chondracanthidae and Xarifiidae to Lichomolgoidea (Cyclopoida). Biological Journal of the Linnean Society, London, **87**: 403-425.

ISHIDA, T., 2002. Illustrated fauna of the freshwater cyclopoid copepods of Japan. Bulletin of the Biogeographical Society of Japan, **57**: 37-106.

ITÔ, T., 1956. Three new copepods from brackish-water lakes of Japan. Pacific Science, **10**: 468-473.

ITOH, H. & S. NISHIDA, 1997. Naupliar stages of *Hemicyclops japonicus* (Copepoda: Poecilostomatoida) reared in the laboratory. Journal of Crustacean Biology, **17**: 162-173.

—— & ——, 1998. A new species of *Hemicyclops* (Copepoda, Poecilostomatoida) from burrows of the ocypodid crab *Macrophthalmus japonicus* in an estuarine mud-flat in Tokyo Bay, Japan. Hydrobiologia, **379**: 85-92.

IVANENKO, V. N. & D. DEFAYE, 2004. A new genus and species of deep-sea cyclopoid (Crustacea, Copepoda, Cyclopinidae) from the Mid-Atlantic Ridge (Azores Triple Junction, Lucky Strike). Zoosystema, **26**: 49-64.

JAUME, D. & G. A. BOXSHALL, 1996a. Rare cyclopoid copepods (Crustacea) from Mediterranean littoral caves. Bulletin of the Natural History Museum London, (Zoology) **62**: 83-99.

—— & ——, 1996b. Two new genera of cyclopinid copepods (Crustacea) from anchialine caves on western Mediterranean and eastern Atlantic islands. Zoological Journal of the Linnean Society, London, **117**: 283-304.

—— & ——, 1997. Two new genera of cyclopinid copepods (Cyclopoida: Cyclopinidae) from anchialine caves of the Canary and Balearic Islands, with a key to genera of the family. Zoological Journal of the Linnean Society, London, **120**: 79-101.

KANG, Y. S., S.-H. HUH & S. S. LEE, 1990. Taxonomy and distribution of Corycaeidae (Copepoda: Cyclopoida) in the Korean waters in summer. The Journal of the Oceanological Society of Korea, **25**: 49-61.

KARANOVIC, T., 2000. *Arenopontia* (*Neoleptastacus*) *huysi* sp. nov. (Crustacea, Copepoda, Harpacticoida) from marine interstitial of Montenegro (S.E. Europe). Helgoland Marine Research, **54**: 33-38.

——, 2003. First representative of the genus *Allocyclops* Kiefer, 1932 (Crustacea, Copepoda, Cyclopoida) from Australian subterranean waters. Annales de Limnologie, **39**: 141-149.

——, 2004a. Subterranean Copepoda from arid Western Australia. Crustaceana Monographs, **3**: i-vi, 1-366. (Brill, Leiden).

——, 2004b. The genus *Metacyclops* Kiefer in Australia (Crustacea: Copepoda: Cyclopoida), with description of two new species. Records of the Western Australian Museum, **22**: 193-212.

——, 2005a. Two new subterranean Parastenocarididae (Crustacea, Copepoda, Harpacticoida) from Western Australia. Records of the Western Australian Museum, **22**: 353-374.

——, 2005b. Two new genera and three new species of subterranean cyclopoids (Crustacea, Copepoda) from New Zealand, with redescription of *Goniocyclops silvestris* Harding, 1958. Contributions to Zoology, Amsterdam, **74**: 223-254.

——, 2006. Subterranean copepods (Crustacea, Copepoda) from the Pilbara region in Western Australia. Records of the Western Australian Museum, (Suppl.) **70**: 1-239.

KARANOVIC, T. & G. L. PESCE, 2002. Copepods from ground waters of Western Australia, VII. *Nitokra humphreysi* sp. nov. (Crustacea: Copepoda: Harpacticoida). Hydrobiologia, **470**: 5-12.

KARANOVIC, T., G. L. PESCE & W. F. HUMPHREYS, 2001. Copepods from ground waters of Western Australia, V. *Phyllopodopsyllus wellsi* n. sp. (Crustacea: Copepoda: Harpacticoida), with a key to world species. Records of the Western Australian Museum, **20**: 333-344.

KARANOVIC, T. & Y. RANGA REDDY, 2004. First record of *Phyllognathopus bassoti* Rouch, 1972 from India, with remarks on the family Phyllognathopodidae Gurney, 1932 (Crustacea, Copepoda, Harpacticoida). Annales de Limnologie, **40**: 121-132.

KIEFER, F., 1929a. Neue Ruderfusskrebse von den Sunda-Inseln. Zoologischer Anzeiger, **84**: 46-49.

——, 1929b. Crustacea Copepoda, 2. Cyclopoida Gnathostoma. Das Tierreich, **53**: 1-102. (Walter de Gruyter & Co., Berlin and Leipzig).

——, 1934. Neue Ruderfusskrebse von der Insel Haiti. Zoologischer Anzeiger, **108**: 227-233.

——, 1935. Zur Kenntnis der Halicyclopen (Crustacea Copepoda). Zoologischer Anzeiger, **110**: 10-13.

——, 1936. Freilebende Süss- und Salzwassercopepoden von der Insel Haiti mit einer Revision der Gattung *Halicyclops* Norman. Archiv für Hydrobiologie, **30**: 263-317.

——, 1954. Neue Cyclopoida Gnathostoma (Crust. Cop.) aus Madagascar, I. Cyclopininae und Halicyclopinae. Zoologischer Anzeiger, **153**: 308-313.

——, 1956. Recherches sur la faune interstitielle des sediments marins et d'eau douce à Madagascar, VI. Cyclopides de Madagascar (Crustacea Copepoda). Mémoires de l'Institut Scientifique de Madagascar, **10**: 43-68.

——, 1960a. Neue Cyclopoida Gnathostoma (Crust. Cop.) von den Inseln Madagaskar und Reunion. Zoologischer Anzeiger, **165**: 226-232.

——, 1960b. Beiträge zur Copepodenkunde (XX). Zoologischer Anzeiger, **165**: 37-45.

——, 1967. Cyclopiden aus salzhaltigen Binnengewässern Australiens (Copepoda). Crustaceana, **12**: 292-302.

KIM, I.-H., 2000. Poecilostomatoid copepods from an intertidal mud flat in the Yellow Sea. Journal of Natural History, London, **34**: 367-432.

——, 2001. *Foliomolgus cucullus*, a new genus and species of Clausidiidae (Crustacea: Copepoda: Poecilostomatoida) associated with a polychaete in Korea. Proceedings of the Biological Society of Washington, **114**: 660-666.

KIM, I.-H. & J.-S. HO, 1992. Copepodid stages of *Hemicyclops ctenidis* Ho and Kim, 1990 (Clausidiidae), a poecilostomatoid copepod associated with a polychaete. Journal of Crustacean Biology, **12**: 631-646.

LANG, K., 1946. Einige für die schwedische Fauna neue marine Cyclopoida Gnathostoma nebst Bemerkungen über die Systematik der Unterfamilie Cyclopininae. Arkiv för Zoologi, **38**: 1-15.

LINDBERG, K. 1940. Cyclopoïdes (Crustacés Copépodes) nouveaux de l'Inde. Records of the Indian Museum, Calcutta, **45**: 47-55.

——, 1941. Cyclopides (Crustacés Copépodes) de l'Inde, V. Contribution à l'étude du genre *Halicyclops* Norman. Records of the Indian Museum, Calcutta, **63**: 1-7.

——, 1949. Contributions à l'étude des Cyclopides (Crustacés Copépodes). Kungl. Fysiografiska Sällskapets i Lund Förhandligar, **19**: 1-24.

——, 1950. Cyclopoïdes nouveaux ou peu connus (Crustacés Copépodes). Mémoires du Muséum National d'Historie Naturelle, Paris, (n. sér.) **29**: 259-297.

——, 1952. Cyclopides (Crustacés Copépodes) de Madagascar. Mémoires de l'Institut Scientifique de Madagascar, **7**: 53-67.

——, 1953. Cyclopides (Crustacés Copépodes) de l'Australie. Acta Universitatis Lundensis, **49**: 1-8.

——, 1954. Cyclopides (Crustacés Copépodes) de l'Amérique du Sud. Arkiv för Zoologi, **7**: 193-222.

——, 1957. Cyclopides (Crustacés Copépodes) de la Côte d'Ivoire. Bulletin de l'Institut Français d'Afrique Noire, **19**: 134-178.

——, 1961. Une *Cyclopinina* nouvelle du sable de la côte Atlantique du Portugal. Kungl. Fysiografiska Sällskapets i Lund Förhandlingar, **31**: 127-132.

LOTUFO, G. R., 1994. *Cyclopina* (Copepoda, Cyclopoida) from Brazilian sandy beaches. Zoologica Scripta, **23**: 147-159.

——, 1995. Psammic *Procyclopina* (Copepoda, Cyclopoida) from the coast of Brazil. Zoologica Scripta, **24**: 245-255.

LOTUFO, G. R. & C. E. F. DA ROCHA, 1991. Copepods from intertidal interstitial water of Salvador, Brazil, I. *Cuipora janaina* gen. n., sp. n. & *Cyclopina caiala* sp. n. (Cyclopoida: Cyclopinidae). Bijdragen tot de Dierkunde, Amsterdam, **61**: 107-118.

—— & ——, 1993a. Intertidal interstitial *Halicyclops* from Brazilian coast (Copepoda: Cyclopoida). Hydrobiologia, **264**: 175-184.

—— & ——, 1993b. *Neocyclops* Gurney from Brazilian sandy beaches (Copepoda: Cyclopoida). Bijdragen tot de Dierkunde, Amsterdam, **63**: 163-172.

MARCUS, A. 1971. Copépodes du mésopsammon de la zone médiolittorale de la mer Noire (littoral roumain). La description d'une nouvelle sous-espèce *Klieonychocamptus kliei ponticus* n. ssp. Travaux du Muséum d'Historie Naturelle "Grigore Antipa", Bucuresti, **11**: 71-82.

——, 1975. Observations écologiques sur les Copépodes trouvés dans le "champ de *Phyllophora*" de la mer Noire. Travaux du Muséum d'Historie Naturelle "Grigore Antipa", Bucuresti, **16**: 71-75.

MARTÍNEZ ARBIZU, P., 2000a. A new species of *Cyclopetta* from the Laptev Sea (Arctic Ocean) with the recognition of Cyclopettidae fam. nov., a new monophylum of free-living Cyclopoida (Copepoda). Bulletin de l'Institut Royal des Sciences Naturelles de Belgique, (Biologie) **70**: 91-101.

——, 2000b. Giselinidae fam. nov., a new monophyletic group of cyclopoid copepods (Copepoda, Crustacea) from the Atlantic deep sea. Helgoland Marine Research, **54**: 190-212.

——, 2001a. Psammocyclopinidae fam. n., a new monophyletic group of marine Cyclopoida (Copepoda, Crustacea) with the description of *Psammocyclopina georgei* sp. n. from the Magellan Region. Revista Brasileira de Zoologia, **18**: 1325-1339.

——, 2001b. Hemicyclopinidae n. fam., a new monophyletic group of marine cyclopinid Cyclopoida, with description of one new genus and two new species. Senckenbergiana Biologica, **81**: 37-54.

MCKINNON, A. D., 1994. *Australopsyllus fallax* gen. et sp. nov., the third known species of the family Thaumatopsyllidae (Copepoda: Cyclopoida). Sarsia, **79**: 27-32.

——, 2000. Two new species of *Oithona* (Copepoda: Cyclopoida) from mangrove waters of north Queensland, Australia. Plankton Biology and Ecology, **47**: 100-113.

MENZEL, R., 1925. Beiträge zur Kenntnis der Mikrofauna von Niederländisch-Ostindien, VI. Bewohner salzhaltiger Thermalgewässer. Treubia, **6**: 450-454.

MIELKE, W. 1976. Ökologie der Copepoda eines Sandstrandes der Nordseeinsel Sylt. Mikrofauna des Meeresbodens, **59**: 451-536.

MIRABDULLAYEV, I. M. & I. A. GETZ, 1996. *Halicyclops spinifer* (Kiefer, 1935) (Crustacea, Copepoda) - Noviy tropicheckii predstavitel fauni Centralnoi Azii. Doklady Akademii Nauk Respubliki Uzbekistan, **4**: 43-44.

MONCHENKO, V. I., 1974. Schelepnoroti Ciklopodibni Ciklopi (Cyclopidae). Fauna Ukraïni, **27**(3): 1-452.

——, 1976. Novii dlia fauny Sovietskogo Soiuza vid ciklopinidy *Cyclopina* cf. *steueri* Früchtl (Crustacea, Copepoda). Doklady Akademii Nauk Ukrainskoi SSR, (B) **1976**(9): 842-845.

——, 1979a. *Cryptocyclopina inopinata* gen. et sp. n. (Crustacea, Copepoda) iz intersticiali Kaspijskogo Moria. Zoologicheskii Zhurnal, **58**: 1470-1477.

——, 1979b. Vtoroe obnaruzhenie *Cyclopina esilis* Brian (Crustacea, Copepoda) i pereopisanie vida. Doklady Akademii Nauk Ukrainskoi SSR, (B) **1979**(5): 387-391.

——, 1980. *Cyclopina parapsammophila* - novij vid veslonogih rakov iz Chernogo Moria. Biologia Moria, **6**: 35-40.

——, 2006. The Caspian endemic genus *Cryptocyclopina* (Crustacea, Copepoda) in the Black Sea. Vestnik Zoologii, **40**: 371-374.

MOORE, B. P., 1964. Present-day cave beetle fauna of Australia: a pointer to past climatic change. Helictite, **3**: 3-9.

MORI, T., 1937. The pelagic Copepoda from neighbouring waters of Japan: 1-150. (Yokendo Co., Tokyo).

MULYADI, 2005. Two new species of *Hemicyclops* (Copepoda, Clausidiidae) and a new species of *Paramacrochiron* (Copepoda, Macrochironidae) from Indonesian waters. Crustaceana, **78**: 917-929.

NICHOLAS, W. L. & A. C. STEWART, 1995. New genera, species and new subfamily of Xyalidae (Nematoda: Monhysterida) from ocean beaches in Australia and Thailand. Transactions of the Royal Society of South Australia, **119**: 47-66.

NICHOLAS, W.[L.] & M. A. TODARO, 2005. Observations on Gastrotricha from a sandy beach in southeastern Australia with a description of *Halichaetonotus australis* sp. nov. (Gastrotricha: Chaetonotida). New Zealand Journal of Marine and Freshwater Research, **39**: 973-980.

—— & ——, 2006. Two new species of Tetranchyroderma (Gastrotricha, Macrodasyida) from a sandy beach in southeastern Australia. New Zealand Journal of Marine and Freshwater Research, **40**: 249-258.

NICHOLLS, A. G., 1935. Copepods from the interstitial fauna of a sandy beach. Journal of the Marine Biological Association of the United Kingdom, (New Series) **20**: 397-406.

——, 1939. Some new sand-dwelling copepods. Journal of the Marine Biological Association of the United Kingdom, (New Series) **23**: 327-341.

——, 1944. Littoral Copepoda from South Australia, II. Calanoida, Cyclopoida, Notodelphyoida, Monstrilloida and Caligoida. Records of the South Australian Museum, **8**: 1-62.

——, 1945a. Marine Copepoda from Western Australia, 4. Psammophilous harpacticoids. Journal of the Royal Society of Western Australia, **29**: 17-24.

——, 1945b. Marine Copepoda from Western Australia, 5. A new species of *Paramesochra*, with an account of a new harpacticoid family, the Remaneidae, and its affinities. Journal of the Royal Society of Western Australia, **29**: 91-105.

NISHIDA, S. 1985. Taxonomy and distribution of the family Oithonidae (Copepoda, Cyclopoida) in the Pacific and Indian Oceans. Bulletin of the Ocean Research Institute, University of Tokyo, **20**: 1-167.

NISHIDA, S., O. TANAKA & M. OMORI, 1977. Cyclopoid copepods of the family Oithonidae in Suruga Bay and adjacent waters. Bulletin of the Plankton Society of Japan, **24**: 119-158.

ONABAMIRO, S. D., Four new species of *Cyclops* sensu lat. (Crustacea: Copepoda) from Nigeria. Proceedings of the Zoological Society of London, **122**: 253-266.

PESCE, G. L., 1985. Cyclopoids (Crustacea, Copepoda) from West Indian groundwater habitats. Bijdragen tot de Dierkunde, Amsterdam, **55**: 295-323.

PESCE, G. L. & D. P. GALASSI, 1987. Discovery of the first representative of the genus *Neocyclops* Gurney (Copepoda, Halicyclopinae) in groundwater of Italy. Crustaceana, **52**: 209-212.

—— & ——, 1993. The genus *Neocyclops* Gurney in the West Indies: an update including the description of *Neocyclops* (*Protoneocyclops*) *geltrudeae* n. sp. (Crustacea, Copepoda, Cyclopidae). Bijdragen tot de Dierkunde, Amsterdam, **63:** 115-120.

PESCE, G. L. & P. DE LAURENTIIS, 1996. Copepods from ground waters of Western Australia, III. *Diacyclops humphreysi* n. sp. and comments on the *Diacyclops crassicaudis*-complex (Copepoda, Cyclopidae). Crustaceana, **69**: 524-531.

PESCE, G. L., P. DE LAURENTIIS & W. F. HUMPHREYS, 1996a. Copepods from ground waters of Western Australia, I. The genera *Metacyclops*, *Mesocyclops*, *Microcyclops* and *Apocyclops* (Crustacea: Copepoda: Cyclopidae). Records of the Western Australian Museum, **18**: 67-76.

——, —— & ——, 1996b. Copepods from ground waters of Australia, II. The genus *Halicyclops* (Crustacea: Copepoda: Cyclopidae). Records of the Western Australian Museum, **18**: 77-85.

PETKOVSKI, T. K., 1954. Harpacticiden des Grundwassers unserer Meeresküste. Acta Musei Macedonici Scientiarum Naturalium Skopje, **2**: 93-123.

——, 1955a. IV. Beitrag zur Kenntnis der Copepoden. Acta Musei Macedonici Scientiarum Naturalium Skopje, **3**: 71-104.

——, 1955b. Weitere Beiträge zur Kenntnis der Grundwasser-Copepoden der adriatischen Küste. Acta Musei Macedonici Scientiarum Naturalium Skopje, **3**: 209-225.

——, 1957. VI. Beitrag zur Kenntnis der Grundwasser-Copepoden unserer Meeresküste. Izdanija Institut de Pisciculture de la PR Macédonie Skopje, **2**: 1-16.

——, 1975. Revision von *Acanthocyclops*-Formen der *vernalis*-Gruppe aus Jugoslawien. Acta Musei Macedonici Scientiarum Naturalium Skopje, **14**: 93-142.

——, 1986. Zur Taxonomie des Genus *Neocyclops* Gurney, 1927 (Crustacea, Copepoda, Cyclopoida). Acta Musei Macedonici Scientiarum Naturalium Skopje, **18**: 27-46.

PETKOVSKI, T. K. & T. KARANOVIC, 1997. Two new copepod species (Crustacea: Copepoda) from the Ohrid Lake. Annales de Limnologie, **33**: 245-253.

PLEŞA, C. 1961. New cyclopoids (Crustacea, Copepoda) of the interstitial fauna from the beaches of Ghana. Journal of the West African Science Association, **7**: 1-13.

——, 1963. Étude sur la faune interstitielle littorale de la mer Noire, III. Résultats préliminaires des recherches sur la côte Roumaine, avec aperçu spécial sur les Cyclopoïdes Gnathostomes (Crustacea, Copepoda). Vie et Milieu, **14**: 775-813.

——, 1968. Un nouveau Cyclopoïde interstitiel de la mer de Chine: *Heterocyclopina vietnamensis* n. g., n. sp. (Crustacea, Copepoda). Vie et Milieu, **19**: 329-344.

——, 1973. Un nouveau Cyclopide interstitiel de la mer des Caraïbes: *Neocyclops improvisus* n. sp. (Crustacea, Copepoda). Résultats des Expéditions Biospéologiques Cubano-Roumaines à Cuba, **1**: 119-122.

——, 1981. Cyclopides (Crustacea, Copepoda) de Cuba. Résultats des Expeditions Biospéologiques Cubano-Roumaines à Cuba, **3**: 17-34.

POR, F. D., 1979. The Copepoda of Di Zahav Pool (Gulf of Elat, Red Sea). Crustaceana, **37**: 13-34.

PREKER, M., 2005. Meiofauna: life between sand grains. Queensland Naturalist, **43**: 33-38.

REID, J. W., 1985. Chave de identificação e lista de referências bibliográficas para as espéces continentais Sulamericanas de vida livre da ordem Cyclopoida (Crustacea, Copepoda). Boletim de Zoologia Universidade de São Paulo, **9**: 17-143.

——, 2001. A human challenge: discovering and understanding continental copepod habitats. Hydrobiologia, **453/454**: 201-226.

REMANE, A., 1933. Verteilung und Organisation der benthonischen Mikrofauna der Kieler Bucht. Wissenschaftliche Meeresuntersuchungen, Kiel, **21**: 163-221.

ROCHA, C. E. F. [DA], T. M. ILIFFE, J. W. REID & E. SUÁREZ-MORALES, 2000. *Prehendocyclops*, a new genus of the subfamily Halicyclopinae (Copepoda, Cyclopoida, Cyclopidae) from cenotes of the Yucatan Peninsula, Mexico. Sarsia, **85**: 119-140.

ROSENDORN, I., 1917. Copepoda 1, der Deutschen Tiefsee Expedition, die Gattung *Oithona*. Wissenschaftliche Ergebnisse der Deutschen Tiefsee-Expedition "Valdivia", **23**: 1-58.

ROUCH, R. & F. LESCHER-MOUTOUÉ, 1977. *Gelyella droguei* n. g., n. sp., curieux Harpacticide des eaux souterraines continentales de la nouvelle famille des Gelyellidae. Annales de Limnologie, **13**: 1-14.

SANMARTÍN, I. & F. RONQUIST, 2004. Southern Hemisphere biogeography inferred by event-based models: plant versus animal patterns. Systematic Biology, **53**: 216-243.

SARS, G. O., 1913. Copepoda Cyclopoida, Parts I & II: Oithonidae, Cyclopinidae, Cyclopidae (part). An Account of the Crustacea of Norway, with Short Descriptions and Figures of all the Species, **6**: 1-32.

SEWELL, R. B. S., 1924. Fauna of the Cilka Lake, Crustacea Copepoda. Memoirs of the Indian Museum, **5**: 771-851.

——, 1934. A study of the fauna of the Salt Lakes, Calcutta. Records of the Indian Museum, Calcutta, **36**: 45-121.

——, 1947. The free-swimming planktonic Copepoda, systematic account. Scientific Reports of the John Murray Expedition 1933-34, **8**(1): 1-303.

——, 1949. The littoral and semi-parasitic Cyclopoida, the Monstrilloida and Notodelphyoida. Scientific Reports of the John Murray Expedition 1933-34, **9**(2): 17-199.

SHEN, C.-J. & S.-O. BAI, 1956. The marine Copepoda from the spawning ground of *Pneumatophorus japonicus* (Houttuyn) off Chefoo, China. Acta Zoologica Sinica, **8**: 217-234.

SHEN, C.-J. & A.-Y. TAI, 1964. Descriptions of new species of freshwater Copepoda from Kwangtung Province, south China. Acta Zootaxonomica Sinica, **1**: 367-396.

SHEN, C.-J., A.-Y. TAI, C.-Z. ZHANG, Z.-Y. LI, D.-X. SONG, Y.-Z. SONG & G.-X. CHEN, 1979. Fauna Sinica, Crustacea, freshwater Copepoda: 1-450. (Science Press, Peking [Beijing]).

SHORT, A. D., 2006. Australian beach systems - nature and distribution. Journal of Coastal Research, **22**: 11-27.

SMIRNOV, S., 1929. Zur Systematik der Copepoden-Familie Cyclopinidae G. O. Sars. Zoologischer Anzeiger, **81**: 203-210.

SØRENSEN, M. V., A. JØRGENSEN & T. M. BOESGAARD, 2000. A new *Echinoderes* (Kinorhyncha: Cyclorhagida) from a submarine cave in New South Wales, Australia. Cahiers de Biologie Marine, **41**: 167-179.

STEUER, A., 1940. Über einige Copepoda Cyclopoida der mediterranen *Amphioxus*sande. Note dell'Istituto Italo-Germanico di Biologia Marina di Rovigno d'Istria, **2**: 1-27.

STEWART, A. C. & W. L. NICHOLAS, 1994. New species of Xyalidae (Nematoda: Monhysterida) from Australian ocean beaches. Invertebrate Taxonomy, **8**: 91-115.

STOCK, J. H., 1973. Notes on copepods associated with invertebrates, I, II. Bulletin Zoologisch Museum Universiteit van Amsterdam, **3**: 39-46.

STOCK, J. H. & J. C. VON VAUPEL KLEIN, 1996. Mounting media revisited: the suitability of Reyne's fluid for small crustaceans. Crustaceana, **69**: 794-798.

TANAKA, O., 1957. On Copepoda of the family Corycaeidae in Japanese waters. Journal of the Faculty of Agriculture Kyushu University, **11**: 77-97.

THOMPSON, I. C. & A. SCOTT, 1903. Report on the Copepoda collected by Professor Herdman at Ceylon in 1902. Ceylon Pearl Oyster Fisheries, Supplementary Reports, **7**: 227-307.

UMMERKUTTY, A. N. P., 1960. Studies on Indian copepods, 3. *Nearchinotodelphys indicus*, a new genus and species of archinotodelphyid copepod from Indian seas. Journal of the Marine Biological Association of India, **2**: 165-178.

——, 1961. Studies on Indian copepods, 5. On eleven new species of marine cyclopoid copepods from the south-east coast of India. Journal of the Marine Biological Association of India, **3**: 19-69.

VERVOORT, W., 1964. Free-living Copepoda from Ifaluk Atoll in the Caroline Islands. Bulletin of the United States National Museum, **236**: 1-431.

VERVOORT, W. & F. RAMÍREZ, 1966. *Hemicyclops thalassius* nov. spec. (Copepoda, Cyclopoida) from Mar del Plata, with revisionary notes on the family Clausidiidae. Zoologische Mededelingen, Leiden, **41**: 195-220.

VILELA, M. H., 1968. Copépodes da Campanha do N.R.P. "Faial", 1958-1959. Notas e Estudos do Instituto de Biologia Marítima, Lisboa, **35**: 1-46.

VINCX, M., 1996. Meiofauna in marine and freshwater sediments. In: G. S. HALL (ed.), Methods for the examination of organismal diversity in soils and sediments: 187-195. (Cab International, Wallingford / New York).

WATERS, J. M. & D. CRAW, 2006. Goodbye Gondwana? New Zealand biogeography, geology, and the problem of circularity. Systematic Biology, **55**: 351-356.

WATERS, J. M. & M. S. ROY, 2004. Out of Africa: the slow train to Australasia. Systematic Biology, **53**: 18-24.

WELLS, J. B. J., 1965. Copepoda (Crustacea) from the meiobenthos of some Scottish marine sub-littoral muds. Proceedings of the Royal Society of Edinburgh, (Biology) **69**: 1-33.

——, 1967. Littoral Copepoda (Crustacea) of Inhaca Island, Mozambique. Transactions of the Royal Society of Edinburgh, **67**: 189-358.

——, 1986. Copepoda: marine-interstitial Harpacticoida. In: L. BOTOSANEANU (ed.), Stygofauna mundi, a faunistic, distributional and ecological synthesis of the world fauna inhabiting subterranean waters (including the marine interstitial): 356-381. (E. J. Brill / Dr. W. Backhuys, Leiden).

WESTON, P. H. & M. D. CRISP, 1994. Cladistic biogeography of waraths (Proteaceae: Embothrieae) and their allies across the Pacific. Australian Systematic Botany, **7**: 225-249.

WILSON, C. B., 1935a. A new and important copepod habitat. Smithsonian Miscellaneous Collections, **94**: 1-13.

——, 1935b. Parasitic copepods from the Pacific coast. American Midland Naturalist, **16**: 776-797.

——, 1950. Copepods gathered by the United States Fisheries Steamer "Albatross" from 1887 to 1909, chiefly in the Pacific Ocean. Bulletin of the United States National Museum, **100**: 141-461.

WILSON, G. D. F., 1999. Some of the deep-sea fauna is ancient. Crustaceana, **72**: 1019-1029.

WILSON, M. S., 1958. The copepod genus *Halicyclops* in North America, with description of a new species from Lake Pontchartrain, Louisiana, and the Texas coast. Tulane Studies in Zoology, **6**: 176-189.

WOOLDRIDGE, T., 1977. A new species of *Halicyclops* (Copepoda, Cyclopoida) from estuaries in Transkei, Southern Africa. Annals of the South African Museum, **73**: 361-371.

YEATMAN, H. C., 1982. Description of the male *Hemicyclops adhaerens*. Journal of the Elisha Mitchell Scientific Society, **98**: 27-31.

——, 1983. Copepods from microhabitats in Fiji, western Samoa, and Tonga. Micronesica, **19**: 57-90.

TAXONOMIC INDEX

Published in this series:

CRM 001 - *Stephan G. Bullard*	Larvae of anomuran and brachyuran crabs of North Carolina
CRM 002 - *Spyros Sfenthourakis et al.*	The biology of terrestrial isopods, V
CRM 003 - *Tomislav Karanovic*	Subterranean Copepoda from arid Western Australia
CRM 004 - *Katsushi Sakai*	Callianassoidea of the world (Decapoda, Thalassinidea)
CRM 005 - *Kim Larsen*	Deep-sea Tanaidacea from the Gulf of Mexico
CRM 006 - *Katsushi Sakai*	Upogebiidae of the world (Decapoda, Thalassinidea)
CRM 007 - *Ivana Karanovic*	Candoninae (Ostracoda) from the Pilbara region in Western Australia
CRM 008 - *Frank D. Ferrari & Hans-Uwe Dahms*	Post-embryonic development of the Copepoda
CRM 009 - *Tomislav Karanovic*	Marine interstitial Poecilostomatoida and Cyclopoida (Copepoda) of Australia

These books can be ordered from:

Brill Academic Publishers
c/o Extenza-Turpin
Pegasus Drive
Stratton Business Park
Biggleswade, Bedfordshire SG18 8QB
United Kingdom
Tel.: +44 (0)1767 604 954
Fax: +44 (0)1767 601 640
E-mail: brill@turpin-distribution.com